D1786595

# Selected Titles in This Series

750 **Masaki Izumi and Hideki Kosaki,** Kac algebras arising from composition of subfactors: General theory and classification, 2002

749 **Nanhua Xi,** The based ring of two-sided cells of affine Weyl groups of type $\widetilde{A}_{n-1}$, 2002

748 **Jürgen Ritter and Alfred Weiss,** The lifted root number conjecture and Iwasawa theory, 2002

747 **Armand Borel, Robert Friedman, and John W. Morgan,** Almost commuting elements in compact Lie groups, 2002

746 **Peter Niemann,** Some generalized Kac-Moody algebras with known root multiplicities, 2002

745 **Mikhail A. Lifshits and Werner Linde,** Approximation and entropy numbers of Volterra operators with application to Brownian motion, 2002

744 **Roger Chalkley,** Basic global relative invariants for homogeneous linear differential equations, 2002

743 **Heng Sun,** Spectral decomposition of a covering of $GL(r)$: the Borel case, 2002

742 **J. E. Gilbert, Y. S. Han, J. A. Hogan, J. D. Lakey, D. Weiland, and G. Weiss,** Smooth molecular functions and singular integral operators, 2002

741 **Francisco Santos,** Triangulations of oriented matroids, 2002

740 **Rick Durrett,** Mutual invadability implies coexistence in spatial models, 2002

739 **Georgios K. Alexopoulos,** Sub-Laplacians with drift on Lie groups of polynomial volume growth, 2002

738 **Yasuro Gon,** Generalized Whittaker functions on $SU(2,2)$ with respect to the Siegel parabolic subgroup, 2002

737 **Arjen Doelman, Robert A. Gardner, and Tasso J. Kaper,** A stability index analysis of 1-D patterns of the Gray-Scott model, 2002

736 **Wojciech Chachólski and Jérôme Scherer,** Homotopy theory of diagrams, 2002

735 **Martina Brück, Xi Du, Joonsang Park, and Chuu-Lian Terng,** The submanifold geometries associated to Grassmannian systems, 2002

734 **Michel Van den Bergh,** Blowing up of non-commutative smooth surfaces, 2001

733 **Milé Krajčevski,** Tilings of the plane, hyperbolic groups and small cancellation conditions, 2001

732 **Jan O. Kleppe, Juan C. Migliore, Rosa Miró-Roig, Uwe Nagel, and Chris Peterson,** Gorenstein liaison, complete intersection liaison invariants and unobstructedness, 2001

731 **Jesús Bastero, Mario Milman, and Francisco J. Ruiz,** On the connection between weighted norm inequalities, commutators and real interpolation, 2001

730 **Suhyoung Choi,** The decomposition and classification of radiant affine 3-manifolds, 2001

729 **Michael Grosser, Eva Farkas, Michael Kunzinger, and Roland Steinbauer,** On the foundations of nonlinear generalized functions I and II, 2001

728 **Laura Smithies,** Equivariant analytic localization of group representations, 2001

727 **Anthony D. Blaom,** A geometric setting for Hamiltonian perturbation theory, 2001

726 **Victor L. Shapiro,** Singular quasilinearity and higher eigenvalues, 2001

725 **Jean-Pierre Rosay and Edgar Lee Stout,** Strong boundary values, analytic functionals, and nonlinear Paley-Wiener theory, 2001

724 **Lisa Carbone,** Non-uniform lattices on uniform trees, 2001

723 **Deborah M. King and John B. Strantzen,** Maximum entropy of cycles of even period, 2001

722 **Hernán Cendra, Jerrold E. Marsden, and Tudor S. Ratiu,** Lagrangian reduction by stages, 2001

721 **Ingrid C. Bauer,** Surfaces with $K^2 = 7$ and $p_g = 4$, 2001

*(Continued in the back of this publication)*

# Kac Algebras Arising from Composition of Subfactors: General Theory and Classification

MEMOIRS
of the
American Mathematical Society

Number 750

# Kac Algebras Arising from Composition of Subfactors: General Theory and Classification

Masaki Izumi
Hideki Kosaki

July 2002 • Volume 158 • Number 750 (first of 4 numbers) • ISSN 0065-9266

**American Mathematical Society**
Providence, Rhode Island

2000 *Mathematics Subject Classification.*
Primary 16L37; Secondary 46L89, 16W30.

**Library of Congress Cataloging-in-Publication Data**

Izumi, Masaki, 1965–
Kac algebras arising from composition of subfactors : general theory and classification / Masaki Izumi, Hideki Kasaki.
p. cm. — (Memoirs of the American Mathematical Society, ISSN 0065-9266 ; no. 750)
Includes bibliographical references and index.
ISBN 0-8218-2935-1 (alk. paper)
1. Kac-Moody algebras. I. Kosaki, Hideki. II. Title. III. Series.

QA3 .A57 no. 750
[QA252.3]
510 s—dc21
[512′.55] 2002018393

## Memoirs of the American Mathematical Society

This journal is devoted entirely to research in pure and applied mathematics.

**Subscription information.** The 2002 subscription begins with volume 155 and consists of six mailings, each containing one or more numbers. Subscription prices for 2002 are $524 list, $419 institutional member. A late charge of 10% of the subscription price will be imposed on orders received from nonmembers after January 1 of the subscription year. Subscribers outside the United States and India must pay a postage surcharge of $31; subscribers in India must pay a postage surcharge of $43. Expedited delivery to destinations in North America $35; elsewhere $130. Each number may be ordered separately; *please specify number* when ordering an individual number. For prices and titles of recently released numbers, see the New Publications sections of the *Notices of the American Mathematical Society.*

**Back number information.** For back issues see the *AMS Catalog of Publications.*

Subscriptions and orders should be addressed to the American Mathematical Society, P. O. Box 845904, Boston, MA 02284-5904. *All orders must be accompanied by payment.* Other correspondence should be addressed to Box 6248, Providence, RI 02940-6248.

**Copying and reprinting.** Individual readers of this publication, and nonprofit libraries acting for them, are permitted to make fair use of the material, such as to copy a chapter for use in teaching or research. Permission is granted to quote brief passages from this publication in reviews, provided the customary acknowledgment of the source is given.

Republication, systematic copying, or multiple reproduction of any material in this publication is permitted only under license from the American Mathematical Society. Requests for such permission should be addressed to the Acquisitions Department, American Mathematical Society, P. O. Box 6248, Providence, Rhode Island 02940-6248. Requests can also be made by e-mail to `reprint-permission@ams.org`.

*Memoirs of the American Mathematical Society* is published bimonthly (each volume consisting usually of more than one number) by the American Mathematical Society at 201 Charles Street, Providence, RI 02904-2294. Periodicals postage paid at Providence, RI. Postmaster: Send address changes to Memoirs, American Mathematical Society, P. O. Box 6248, Providence, RI 02940-6248.

© 2002 by the American Mathematical Society. All rights reserved.
This publication is indexed in *Science Citation Index*®, *SciSearch*®, *Research Alert*®, *CompuMath Citation Index*®, *Current Contents*®/*Physical, Chemical & Earth Sciences.*
Printed in the United States of America.

∞ The paper used in this book is acid-free and falls within the guidelines established to ensure permanence and durability.
Visit the AMS home page at URL: `http://www.ams.org/`

10 9 8 7 6 5 4 3 2 1 07 06 05 04 03 02

# Contents

**Abstract.** We deal with a map $\alpha$ from a finite group $G$ into the automorphism group $Aut(\mathcal{L})$ of a factor $\mathcal{L}$ satisfying (i) $G = N \rtimes H$ is a semi-direct product, (ii) the induced map $g \in G \to [\alpha_g] \in Out(\mathcal{L}) = Aut(\mathcal{L})/Int(\mathcal{L})$ is an injective homomorphism, and (iii) the restrictions $\alpha|_N, \alpha|_H$ are genuine actions of the subgroups on the factor $\mathcal{L}$. The pair $\mathcal{M} = \mathcal{L} \rtimes_\alpha H \supseteq \mathcal{N} = \mathcal{L}^{\alpha|_N}$ (of the crossed product $\mathcal{L} \rtimes_\alpha H$ and the fixed-point algebra $\mathcal{L}^{\alpha|_N}$) gives us an irreducible inclusion of factors with Jones index $\#G$. The inclusion $\mathcal{M} \supseteq \mathcal{N}$ is of depth 2 and hence known to correspond to a Kac algebra of dimension $\#G$.

A Kac algebra arising in this way is investigated in detail, and in fact the relevant multiplicative unitary (satisfying the pentagon equation) is described. We introduce and analyze a certain cohomology group (denoted by $H^2((N, H), \mathbf{T})$) providing complete information on the Kac algebra structure, and we construct an abundance of non-trivial examples by making use of various cocycles. The operator algebraic meaning of this cohomology group is clarified, and some related topics are also discussed.

Sector technique enables us to establish structure results for Kac algebras with certain prescribed underlying algebra structure. They guarantee that "most" Kac algebras of low dimension (say less than 60) actually arise from inclusions of the form $\mathcal{L} \rtimes_\alpha H \supseteq \mathcal{L}^{\alpha|_N}$, and consequently their classification can be carried out by determining $H^2((N, H), \mathbf{T})$. Among other things we indeed classify Kac algebras of dimension 16 and 24, which (together with previously known results) gives rise to the complete classification of Kac algebras of dimension up to 31. Partly to simplify classification procedure and hopefully for its own sake, we also study "group extensions" of general (finite-dimensional) Kac algebras with some discussions on related topics.

2000 *Mathematics Subject Classification.* 46L37; Secondary 46L89, 16W30

*Key words and phrases.* bimodule, depth 2 inclusion of factors, Goldman-type theorem, intermediate subfactor, intertwiners, intrinsic group, isometry, Jones index, Kac algebra, outer action, multiplicative unitary, pentagon equation, properly infinite factor, 2-cocycle, sector

CHAPTER 1

# Introduction

The Jones index theory ([**36**]) deals with inclusions $\mathcal{N} \subseteq \mathcal{M}$ of factors with finite index, and in related subfactor analysis the "position" of a subfactor $\mathcal{N}$ in the ambient factor $\mathcal{M}$ is of fundamental importance. Detailed account as well as the state of the art of the theory can be found in the recent textbook [**16**] by Evans and Kawahigashi. It is known that bimodules and sectors (see [**5, 25, 46, 47**] and also the references of [**16**] for more detailed information) play important roles in this subject matter. On the other hand, Hopf algebras or more precisely Kac algebras appear in many places of the theory of operator algebras as the right framework to handle duality phenomenon (see [**13, 57**] for example). Roughly speaking, a Kac algebra means a $*$-Hopf algebra $\mathcal{A}$ (with the coproduct $\Gamma : \mathcal{A} \longrightarrow \mathcal{A} \otimes \mathcal{A}$) equipped with an antipode $\kappa$ (which is an involutive anti-automorphism of $\mathcal{A}$) compatible with the $*$-operation. A finite group $G$ give us the most typical (but somewhat simple-minded) examples of (finite-dimensional) Kac algebras: the algebra $\ell^\infty(G)$ of $\mathbf{C}$-valued functions and the group ring $\mathbf{C}(G)$. Here, $\Gamma, \kappa$ (for the former) and $\hat{\Gamma}, \hat{\kappa}$ (for the latter) are given by

$$(\Gamma(f))(g,h) = f(gh), \quad (\kappa(f))(g) = f(g^{-1}),$$
$$\hat{\Gamma}(\lambda_g) = \lambda_g \otimes \lambda_g, \quad \hat{\kappa}(\lambda_g) = \lambda_{g^{-1}}.$$

(Here, $\mathcal{A} \otimes \mathcal{A}$ is identified with $\ell^\infty(G \times G)$, the functions of two varibles, for $\mathcal{A} = \ell^\infty(G)$.) The former is referred to as a commutative Kac algebra while the latter is known as a cocommutative Kac algebra, and these two types altogether are called trivial Kac algebras. At an early stage of the index theory it was observed by Ocneanu that an irreducible inclusion $\mathcal{N} \subseteq \mathcal{M}$ is of depth 2 if and only if $\mathcal{M}$ is the crossed product of $\mathcal{N}$ relative to a coaction of a Kac algebra (of dimension equal to the Jones index $[\mathcal{M} : \mathcal{N}]$). Proofs were supplied by several authors ([**9, 27, 29, 48, 62, 70**]), and bimodules and sectors of course played important roles. This characterization in particular means that the relevant Kac algebra structure (together with a coaction) can be recovered from the position of $\mathcal{N}$ inside of $\mathcal{M}$, and makes it possible to use subfactor analysis for study of finite-dimensional Kac algebras.

Assume that an outer action of a finite group $G$ on a factor $\mathcal{L}$ is given. When $G$ is a semi-direct product $G = N \rtimes H$ (or more generally a product group $G = N \cdot H$), by considering the crossed product $\mathcal{M} = \mathcal{L} \rtimes H$ and the fixed-point algebra $\mathcal{N} = \mathcal{L}^N$ we get the two-step inclusion

$$\mathcal{N} = \mathcal{L}^N \subseteq \mathcal{L} \subseteq \mathcal{M} = \mathcal{L} \rtimes H.$$

Received by the editor September 1, 2000

Then, $\mathcal{N} \subseteq \mathcal{M}$ is an irreducible inclusion of factors of depth 2, and hence a Kac algebra is available from the inclusion data. It is well-known that this way of obtaining depth 2 inclusions corresponds to the bicrossed product construction in Hopf algebra theory ([**49, 71**], and see also [**60**] for more operator algebraic approach). The depth 2 condition here can be easily seen based on the Frobenius reciprocity for bimodules or sectors. This proof indicates that the same reasoning works even when an outer action is replaced by an injective homomorphism (i.e., $G$-kernel) from $G$ into the quotient group $Out(\mathcal{L}) = Aut(\mathcal{L})/Int(\mathcal{L})$. However, we should note that a $G$-kernel in question must be a genuine action when restricted to the subgroups $N$ and $H$ because $\mathcal{M} = \mathcal{L} \rtimes H$ and $\mathcal{N} = \mathcal{L}^N$ have to be defined.

The discussion so far means that a certain injective $G$-kernel gives rise to "deformation" of a bicrossed product Kac algebra. It is known ([**35**]) that $G$-kernels are described in terms of 3-cocycles on $G$. However, since our $G$-kernels are required to be actions when restricted to the subgroups, we are forced to deal with very special 3-cocycles. Such a special 3-cocycle is described in term of a pair of certain 2-cocycles (see Chapter 2 and Chapter 3), but let us emphasize that one has to deal with a different notion of coboundaries to study resulting Kac algebras (see Lemma 2.6,(iii) and the example in §1 of Chapter 7). These cocycles were also studied in [**21, 22**], and the resulting cohomology group (denoted by $H^2((N,H),\mathbf{T})$) can be also identified as the one naturally attached to the related multiplicative unitary (dealt in Appendix to [**1**]) as shown in Chapter 3. We will also clarify the operator algebraic meaning of the cohomology group $H^2((N,H),\mathbf{T})$. Indeed it is the complete invariant for "cocycle conjugacy classes" of "actions of the pair $(N,H)$" when $\mathcal{L}$ is the hyperfinite $II_1$-factor (see Chapter 2 for the precise meaning). As was worked out in the Baaj-Skandalis article [**1**], every information on a Kac algebra can be deciphered once a relevant multiplicative unitary (satisfying the pentagon equation) is known. By making use of sector technique we will explicitly write down the multiplicative unitary (arising from the setting explained so far) in Chapter 4, and the structure of the resulting Kac algebra (coproduct, antipode, intrinsic group, and so on) will be clarified in Chapter 5 and Chapter 6. For example, the underlying algebra of our Kac algebra turns out to be a twisted crossed product $\ell^\infty(N) \rtimes H$ ([**76**]) while that of the dual Kac algebra is the direct sum of group rings of the normal subgroup $N$ twisted by certain cocycles.

It turns out that the above cohomology and corresponding deformation themselves have been known to specialists as a special case of the Hopf algebra extension of a commutative Hopf algebra by a cocommutative kernel (see [**37**] and also [**18, 65**] for example). However, we would like to emphasize that subfactor analysis viewpoint especially the bimodule or sector theory of factors provides us extra information. This viewpoint indeed enables us to obtain very useful structure results for certain Kac algebras, which will be carried out in Chapter 9. More precisely, a Kac algebra possessing a certain prescribed underlying algebra is shown to be captured via our construction, i.e., a Kac algebra arising from a semi-direct product $N \rtimes H$ possibly with a cocycle. (The idea behind these structure results will be explained at the beginning of Chapter 9.) Our method of constructing Kac algebras seems to be quite effective, and in fact many relevant structures are easily deformed by the presence of various cocycles as is demonstrated in Chapter 7.

In [**38**] Kac-Paljutkin discovered an 8-dimensional non-trivial Kac algebra (with the underlying algebra $\mathbf{C} \oplus \mathbf{C} \oplus \mathbf{C} \oplus \mathbf{C} \oplus M_2(\mathbf{C})$). It is the only non-trivial Kac algebra of dimension 8, and indeed the smallest non-trivial one. It will be referred to as the Kac-Paljutkin algebra, and will appear in several places in the present article (for instance in Chapter 7, §2). In Chapter 8 we will point out a close relationship between the Kac-Paljutkin algebra and one of (index 4) inclusions of (the hyperfinite type $II_1$) factors with the Coxeter-Dynkin graph $D_6^{(1)}$.

Results covered in the first half (Chapters 2-9) of the present article were announced in [**32**]. The last half (Chapters 10-14) is mainly devoted to classification of Kac algebras of low dimension. Structure results in Chapter 9 actually guarantee that "most" Kac algebras of low dimension (say dimension less than 60) arise from a two-step inclusion $\mathcal{L} \rtimes H \supseteq \mathcal{L} \supseteq \mathcal{L}^N$ with a suitable cocycle. This is shown in Chapter 10, and heavily depends on the prime factorization of the dimension of a Kac algebra in question (i.e., the index of the corresponding depth 2 inclusion). Therefore, these Kac algebras can be classified by making use of our cohomology group $H^2((N,H),\mathbf{T})$ (or more precisely its quotient $H^2((N,H),\mathbf{T})/\sim$). In this way Kac algebras of dimension $pq$, $pqr$, or $p^2q$ (less than 60) are classified. The product group $D_{14} \cdot \mathbf{Z}_3$ (where the dihedral group $D_{14}$ is normal in $D_{14} \cdot \mathbf{Z}_3$) gives us a non-trivial Kac algebra of dimension $42 = 2 \cdot 3 \cdot 7$. Among other things we will show that it (and its dual) is indeed the only non-trivial Kac algebra of dimension $pq$ or $pqr$ ($< 60$).

Classification of Kac algebras of dimension 16 and 24 requires more involved arguments, and will be worked out in Chapters 11 and 14. The symmetric group $\mathfrak{S}_4$ (with the subgroup $\mathfrak{S}_3$ and the cyclic permutations $\mathbf{Z}_4 = \langle(1,2,3,4)\rangle$) forms a typical product group, giving rise to a non-trivial Kac algebra of dimension 24. All the other Kac algebras of dimension either 16 or 24 will be shown to arise from a suitable semi-direct product $N \rtimes H$. All Kac algebras (more generally semi-simple Hopf algebras over an algebraically closed field of characteristic 0) of dimension $p$ are trivial (i.e., $\mathbf{C}(\mathbf{Z}_p)$) thanks to Zhu's article [**77**] while those of dimension either $p^2$ or $p^3$ have been classified by Masuoka ([**53, 54**]). Consequently, classification of Kac algebras of dimension up to 31 is completed in the present article.

In Chapter 12 we will deal with a two-step inclusion

$$\mathcal{N} \subseteq \mathcal{L} \subseteq \mathcal{M} = \mathcal{L} \rtimes H,$$

where $\mathcal{N} \subseteq \mathcal{L}$ is a general inclusion of depth 2 (i.e., $\mathcal{N}$ is no longer the fixed-point algebra under an $N$ action). We of course impose certain requirement (see the beginning of Chapter 12) to make sure that $\mathcal{N} \subseteq \mathcal{M}$ remains irreducible and of depth 2, and will investigate the resulting Kac algebra. As an underlying algebra a twisted crossed product naturally shows up (as expected). When a Kac algebra is equipped with a finite group action preserving every relevant structure, the ordinary crossed product becomes a Kac algebra in the natural way as was shown by De Cannière ([**11**]). The work in this chapter is a generalization of what was done in Chapter 4 and can be also regarded as a deformation theory of his construction. This consideration naturally gives us a notion of 2-cocycles of Kac algebras (the case of the dual of a compact group and related ergodic actions were studied in Wassermann's article [**73**]), and general discussions on them as well as basic examples are presented in Chapter 13. We will study 2-cocycles of the Kac-Paljutkin algebra as well as the group rings $\mathbf{C}(D_8), \mathbf{C}(D_{12}), \mathbf{C}(\mathfrak{A}_4)$. The subjects touched upon in these two chapters might be of independent interest, and seem to deserve further investigation. Close examination on twisted crossed products (i.e., deformed Kac algebras) $\mathbf{C}(D_{12}) \rtimes \mathbf{Z}_2$ and $\mathbf{C}(\mathfrak{A}_4) \rtimes \mathbf{Z}_2$ (in Chapter 13) brings us valuable information on Kac algebras of dimension 24 with intrinsic group of order 12, and will in fact enter into our classification procedure.

Our basic references on Kac algebras via the index theory are [**29, 48**] while basic facts on the index theory can be found in [**16**]. As was pointed out before Kac algebra structure is completely determined by a relevant multiplicative unitary as was shown by Baaj-Skandalis ([**1**]), and in the above two articles this multiplicative unitary was captured in terms of inclusion data via Cuntz algebra approach (see [**8**]).

**Acknowledgements**: The authors are grateful to Professor Akira Masuoka for informing them of recent works [**15, 17, 56, 58**] in the Hopf algebra context. The authors have also learned from him that Y. Kashina obtained classification of semisimple Hopf algebras of dimension 16 (over an algebraically closed field of characteristic 0).

This work is partially supported by Grant-in-Aid for Science Research, Ministry of Education, Science and Culture (No. 09304017).

CHAPTER 2

# Actions of matched pairs

Let $G$ be a finite group with subgroups $G_1, G_2$, and we assume that $G$ is a product group $G = G_1 \cdot G_2$. In such a case, we will call $(G_1, G_2)$ a matched pair. Throughout, let $\mathcal{R}$ be a factor.

DEFINITION 2.1.
*(i) A pair of actions $\alpha : G_1 \to Aut(\mathcal{R})$ and $\beta : G_2 \to Aut(\mathcal{R})$ is called an action of the matched pair $(G_1, G_2)$ if $\alpha, \beta$ give $G$-kernel (i.e., a homomorphism from $G$ to $Out(\mathcal{R}) = Aut(\mathcal{R})/Int(\mathcal{R})$.*
*(ii) An action $(\alpha, \beta)$ of the matched pair $(G_1, G_2)$ is outer if $\alpha, \beta$ are outer actions of $G_1, G_2$ respectively and the associated $G$-kernel is injective. Recall that this $G$-kernel or rather a map $G = G_1 \cdot G_2 \longrightarrow Aut(\mathcal{R})$ is called a near action in* [**32**].
*(iii) Two actions $(\alpha, \beta), (\alpha', \beta')$ are cocycle conjugate if there are an automorphism $\theta \in Aut(\mathcal{R})$, an $\alpha$-cocycle $\{u_{g_1}\}_{g_1 \in G_1}$, and a $\beta$-cocycle $\{v_{g_2}\}_{g_2 \in G_2}$ (i.e., $u_{g_1 g_1'} = u_{g_1}\alpha_{g_1}(u_{g_1'})$ and $v_{g_2 g_2'} = v_{g_2}\beta_{g_2}(v_{g_2'})$) satisfying*

$$\begin{aligned} \theta\alpha'_{g_1}\theta^{-1} &= Ad(u_{g_1})\alpha_{g_1}, \\ \theta\beta'_{g_2}\theta^{-1} &= Ad(v_{g_2})\beta_{g_2}. \end{aligned}$$

In this chapter we will study various cocycles arising from an outer action of a matched pair $(G_1, G_2)$ (in the semi-direct product case $G = N \rtimes H$) and introduce a certain cohomology group $H^2((N, H), \mathbf{T})$. (See [**21, 22**] for related results.) We will show that this cohomology group is a complete invariant for cocycle conjugacy classes of outer actions of $(N, H)$ when $\mathcal{R}$ is the hyperfinite $II_1$ factor (Theorem 2.5). In the next chapter the cohomology group $H^2((N, H), \mathbf{T})$ will be also identified with the one that naturally arises from the multiplicative unitary described in the appendix to [**1**].

**Remarks.**
1. When $(\alpha, \beta)$ is an outer action of the matched pair $(G_1, G_2)$, by the standard argument of bimodules or equivalently sectors we see that $\mathcal{R} \rtimes_\alpha G_1 \supseteq \mathcal{R}^{(G_2, \beta)}$ is an irreducible inclusion of factors of depth 2.
2. When $(\alpha, \beta)$ and $(\alpha', \beta')$ are cocycle conjugate, the two inclusions $\mathcal{R} \rtimes_\alpha G_1 \supseteq \mathcal{R}^{(\beta, G_2)}$ and $\mathcal{R} \rtimes_{\alpha'} G_1 \supseteq \mathcal{R}^{(\beta', G_2)}$ are conjugate.

In fact, let $w$ be the canonical implementation (see [**67**]) of $\theta$, we consider the unitary $U$ on $L^2(\mathcal{R}) \otimes \ell^2(G_1)$ defined by $(U\xi)(g_1) = u^*_{g_1^{-1}} w\xi(g_1)$. Then, $Ad(U)$ sends $\mathcal{R} \rtimes_{\alpha'} G_1$ onto $\mathcal{R} \rtimes_\alpha G_1$, and $\mathcal{R}$ onto itself in such a way that the restriction to $\mathcal{R}$ ($\subseteq \mathcal{R} \rtimes_{\alpha'} G_1$) is $\theta$. Therefore, $Ad(U)$ sends $\mathcal{R}^{(\beta', G_2)}$ onto the subfactor $\{x \in \mathcal{R} : Ad(v_{g_2})\beta_{g_2}(x) = x\}$. Since $\{v_{g_2}\}_{g_2 \in G_2}$ is a $\beta$-cocycle, one finds a unitary $V$ in $\mathcal{R}$ satisfying $v_{g_2} = V^*\beta_{g_2}(V)$ (see [**4, 6**]) and the above condition means

$\beta_{g_2}(VxV^*) = VxV^*$. Therefore, $Ad(VU)$ sends $\mathcal{R}^{(\beta',G_2)}$ onto $\mathcal{R}^{(\beta,G_2)}$, which proves the second statement.

## 1. Cocycles and cohomology group

In the rest we will assume that $G = N \rtimes H$ is a semi-direct product with $G_1 = N$ and $G_2 = H$. For $n \in N, h \in H$ we set

$$n^h = h^{-1}nh \ (\in N).$$

Note

$$(n_1n_2)^h = n_1^h n_2^h, \quad (n^{h_1})^{h_2} = n^{h_1h_2}.$$

When $(\alpha, \beta)$ is an action of the matched pair $(N, H)$, for each $(h, n) \in H \times N$ one finds a unitary $u(h, n)$ in $\mathcal{R}$ satisfying

$$\beta_h \alpha_{n^h} \beta_{h^{-1}} = Ad(u(h, n))\alpha_n.$$

(Also note that $u(h, N)$ is determined up to a scalar, which will be a source of a coboundary.) For each fixed $h \in H$, the above right side (and hence the left side) is an action of $N$. Thus, we find a scalar $\eta_h(n_1, n_2) \in \mathbf{T}$ satisfying

$$u(h, n_1n_2) = \eta_h(n_1, n_2)u(h, n_1)\alpha_{n_1}(u(h, n_2)).$$

Since

$$\begin{aligned}
\beta_{h_1h_2}\alpha_{n^{h_1h_2}}\beta_{(h_1h_2)^{-1}} &= \beta_{h_1}\left(\beta_{h_2}\alpha_{(n^{h_1})^{h_2}}\beta_{h_2^{-1}}\right)\beta_{h_1^{-1}} \\
&= \beta_{h_1}\left(Ad(u(h_2, n^{h_1}))\alpha_{n^{h_1}}\right)\beta_{h_1^{-1}} \\
&= Ad\left(\beta_{h_1}(u(h_2, n^{h_1}))\right)\beta_{h_1}\alpha_{n^{h_1}}\beta_{h_1^{-1}} \\
&= Ad\left(\beta_{h_1}(u(h_2, n^{h_1}))u(h_1, n)\right)\alpha_n,
\end{aligned}$$

we similarly find a scalar $\zeta_n(h_1, h_2) \in \mathbf{T}$ satisfying

$$u(h_1h_2, n)^* = \zeta_n(h_1, h_2)u(h_1, n)^*\beta_{h_1}(u(h_2, n^{h_1})^*).$$

Note that for each $h_1, h_2$, the function $n \in N \longrightarrow \zeta_n(h_1, h_2) \in \mathbf{T}$ is an element in $\mathbf{U}(\ell^\infty(N))$, the unitary group of $\ell^\infty(N)$.

LEMMA 2.2. *(i) For each $h \in H$, we have $\eta_h \in Z^2(N, \mathbf{T})$,*
*(ii) We have $\zeta \in Z^2(H, \mathbf{U}(\ell^\infty(N)))$ with the natural $H$-action on $\ell^\infty(N)$, that is,*

$$\zeta_n(h_1h_2, h_3)\zeta_n(h_1, h_2) = \zeta_n(h_1, h_2h_3)\zeta_{n^{h_1}}(h_2, h_3).$$

*(iii) The two cocycles $\{\eta_h\}_{h\in H}, \zeta$ are related by*

$$\frac{\eta_{h_1}(n_1, n_2)\eta_{h_2}(n_1^{h_1}, n_2^{h_1})}{\eta_{h_1h_2}(n_1, n_2)} = \frac{\zeta_{n_1n_2}(h_1, h_2)}{\zeta_{n_1}(h_1, h_2)\zeta_{n_2}(h_1, h_2)}.$$

PROOF. We compute

$$\begin{aligned}
u(h, (n_1n_2)n_3) &= \eta_h(n_1n_2, n_3)u(h, n_1n_2)\alpha_{n_1n_2}(u(h, n_3)) \\
&= \eta_h(n_1n_2, n_3)\left(\eta_h(n_1, n_2)u(h, n_1)\alpha_{n_1}(u(h, n_2)\right)\alpha_{n_1n_2}(u(h, n_3)), \\
u(h, n_1(n_2n_3)) &= \eta_h(n_1, n_2n_3)u(h, n_1)\alpha_{n_1}(u(h, n_2n_3)) \\
&= \eta_h(n_1, n_2n_3)u(h, n_1)\alpha_{n_1}\left(\eta_h(n_2, n_3)u(h, n_2)\alpha_{n_2}(u(h, n_3))\right).
\end{aligned}$$

From the above computations we conclude $\eta_h \in Z^2(N, \mathbf{T})$.

Similarly, we compute

$$\begin{aligned}
u(h_1(h_2h_3),n) &= \overline{\zeta_n(h_1,h_2h_3)}\beta_{h_1}(u(h_2h_3,n^{h_1}))u(h_1,n)\\
&= \overline{\zeta_n(h_1,h_2h_3)}\beta_{h_1}\left(\overline{\zeta_{n^{h_1}}(h_2,h_3)}\beta_{h_2}(u(h_3,(n^{h_1})^{h_2}))u(h_2,n^{h_1})\right)\\
&\qquad \times u(h_1,n),\\
u((h_1h_2)h_3,n) &= \overline{\zeta_n(h_1h_2,h_3)}\beta_{h_1h_2}(u(h_3,n^{h_1h_2}))u(h_1h_2,n)\\
&= \overline{\zeta_n(h_1h_2,h_3)}\beta_{h_1h_2}(u(h_3,n^{h_1h_2}))\\
&\qquad \times\overline{\zeta_n(h_1,h_2)}\beta_{h_1}(u(h_2,n^{h_1}))u(h_1,n).
\end{aligned}$$

From these, we see $\zeta \in Z^2(H, \mathbf{U}(\ell^\infty(N)))$.

To show (iii), we next compute

$$\begin{aligned}
u(h_1h_2,n_1n_2) &= \eta_{h_1h_2}(n_1,n_2)u(h_1h_2,n_1)\alpha_{n_1}(u(h_1h_2,n_2))\\
&= \eta_{h_1h_2}(n_1,n_2)\overline{\zeta_{n_1}(h_1,h_2)}\beta_{h_1}(u(h_2,n_1^{h_1}))u(h_1,n_1)\\
&\qquad \times\alpha_{n_1}\left(\overline{\zeta_{n_2}(h_1,h_2)}\beta_{h_1}(u(h_2,n_2^{h_1}))u(h_1,n_2)\right).
\end{aligned}$$

On the other hand, we can also compute the same quantity as follows:

$$\begin{aligned}
u(h_1h_2,n_1n_2) &= \overline{\zeta_{n_1n_2}(h_1,h_2)}\beta_{h_1}(u(h_2,(n_1n_2)^{h_1}))u(h_1,n_1n_2)\\
&= \overline{\zeta_{n_2n_2}(h_1,h_2)}\beta_{h_1}\left(\eta_{h_2}(n_1^{h_1},n_2^{h_1})u(h_2,n_1^{h_1})\alpha_{n_1^{h_1}}(u(h_2,n_2^{h_1}))\right)\\
&\qquad \times\eta_{h_1}(n_1,n_2)u(h_1,n_1)\alpha_{n_1}(u(h_1,n_2)).
\end{aligned}$$

Note that the unitaries $\beta_{h_1}(u(h_2,n_1^{h_1})), \alpha_{n_1}(u(h_1,n_2))$ appear in common in the above two computations. By noticing

$$\begin{aligned}
&u(h_1,n_1))\alpha_{n_1}\beta_{h_1}(u(h_2,n_2^{h_1}))\\
&\quad = Ad(u(h_1,n_1))\alpha_{n_1}\beta_{h_1}(u(h_2,n_2^{h_1}))u(h_1,n_1)\\
&\quad = \beta_{h_1}\alpha_{n_1^{h_1}}(u(h_2,n_2^{h_1}))u(h_1,n_1)
\end{aligned}$$

and comparing the scalars appearing in the preceding two computations, we get (iii). □

If we start from $u'(h,n) = \overline{\xi(h,n)}u(h,n)$ with $\xi(h,n) \in \mathbf{T}$ (instead of $u(h,n)$), the corresponding $\eta, \zeta$ are changed to

$$\begin{aligned}
\eta'_h(n_1,n_2) &= \eta_h(n_1,n_2)\overline{\xi(h,n_1n_2)}\xi(h,n_1)\xi(h,n_2),\\
\zeta'_n(h_1,h_2) &= \zeta_n(h_1,h_2)\xi(h_1h_2,n)\overline{\xi(h_1,n)\xi(h_2,n^{h_1})}.
\end{aligned}$$

Thus, we set

$$\begin{aligned}
(\partial^N\xi)_h(n_1,n_2) &= \overline{\xi(h,n_1n_2)}\xi(h,n_1)\xi(h,n_2),\\
(\partial^H\xi)_n(h_1,h_2) &= \xi(h_1h_2,n)\overline{\xi(h_1,n)\xi(h_2,n^{h_1})},
\end{aligned}$$

and define

DEFINITION 2.3.

$$\begin{aligned}
Z^2((N,H),\mathbf{T}) &= \{(\eta,\zeta);\ \eta = \{\eta_h\}_{h\in H}, \zeta \textit{ satisfy}\\
&\qquad\qquad (i), (ii), \textit{ and } (iii) \textit{ in Lemma 2.2}\},\\
B^2((N,H),\mathbf{T}) &= \{(\partial^N\xi, \partial^H\xi);\ \xi \textit{ is a } \mathbf{T}\textit{-valued function on } H\times N\},\\
H^2((N,H),\mathbf{T}) &= Z^2((N,H),\mathbf{T})/B^2((N,H),\mathbf{T}).
\end{aligned}$$

Related topics can be found in [**18, 37, 65**].

LEMMA 2.4. *(i) When* $(\alpha, \beta), (\alpha', \beta')$ *are cocycle conjugate outer actions, they give rise to the same element in* $H^2((N, H), \mathbf{T})$.
*(ii) Let* $(\alpha, \beta)$ *be an outer action, and we assume that unitaries* $\{w_h\}_{h\in H}$ *in* $\mathcal{R}$ *(not necessarily a* $\beta$*-cocycle) give rise to an* $H$*-action* $Ad(w_h)\beta_h$*. In this case* $(\alpha, \beta)$ *and* $(\alpha, Ad(w)\beta)$ *determine the same element in* $H^2((N, H), \mathbf{T})$
*(iii) Let* $(\alpha, \beta)$ *be an outer action, and we assume that unitaries* $\{v_n\}_{n\in N}$ *in* $\mathcal{R}$ *(not necessarily an* $\alpha$*-cocycle) satisfy*

$$v_{n_1 n_2} = \omega(n_1, n_2) v_{n_1} \alpha_{n_1}(v_{n_2})$$

*with* $\omega(n_1, n_2) \in \mathbf{T}$*. If* $(\eta, \zeta)$ *is associated with* $(\alpha, \beta)$*, then the pair* $(Ad(v)\alpha, \beta)$ *gives rise to* $(\eta', \zeta')$ *determined by*

$$\begin{aligned} \eta'_h(n_1, n_2) &= \eta_h(n_1, n_2)\omega(n_1^h, n_2^h)\overline{\omega(n_1, n_2)}, \\ \zeta' &= \zeta. \end{aligned}$$

PROOF. Assume $v_{n_1 n_2} = \omega(n_1, n_2) v_{n_1} \alpha_{n_1}(v_{n_2})$ (as in (iii)), and we set

$$\alpha_n^v = Ad(v_n)\alpha_n \quad \text{(a perturbed action of } N).$$

Notice

$$\begin{aligned} \beta_h \alpha^v_{n^h} \beta_{h^{-1}} &= Ad(\beta_h(v_{n^h}))\beta_h \alpha_{n^h} \beta_{h^{-1}} \\ &= Ad\left(\beta_h(v_{n^h}) u(h, n)\right) \alpha_n \\ &= Ad\left(\beta_h(v_{n^h}) u(h, n) v_n^*\right) \alpha_n^v. \end{aligned}$$

Hence the perturbation of $\alpha$ to $\alpha^v$ changes the unitary $u(h, n)$ to

$$u'(h, n) = \beta_h(v_{n^h}) u(h, n) v_n^*.$$

To see what the corresponding $\eta'$ is, we need to compute $u'(h, n_1 n_2)$. This unitary is equal to

$$\begin{aligned} &\beta_h\left(\omega(n_1^h, n_2^h) v_{n_1^h} \alpha_{n_1^h}(v_{n_2^h})\right) \\ \times\ &\eta_h(n_1, n_2) u(h, n_1) \alpha_{n_1}(u(h, n_2)) \\ \times\ &\overline{\omega(n_1, n_2)} \alpha_{n_1}(v_{n_2}^*) v_{n_1}^*. \end{aligned}$$

The product of the six unitaries appearing above is equal to

$$\begin{aligned} &\beta_h(v_{n_1^h}) u(h, n_1) \alpha_{n_1}\left(\beta_h(v_{n_2^h}) u(h, n_2) v_{n_2}^*\right) v_{n_1}^* \\ &\qquad\qquad \text{(thanks to } \beta_h \alpha_{n_1^h} = Ad(u(h, n_1))\alpha_n \beta_h) \\ = \ &\beta_h(v_{n_1^h}) u(h, n_1) v_{n_1}^* v_{n_1} \alpha_{n_1}\left(\beta_h(v_{n_2^h}) u(h, n_2) v_{n_2}^h\right) v_{n_1}^* \\ = \ &u'(h, n_1) \alpha_{n_1}^v(u'(h, n_2)) \end{aligned}$$

Therefore, by taking the product of the three scalars in the preceding expression, we observe

$$\eta'_h(n_1, n_2) = \eta_h(n_1, n_2)\omega(n_1^h, n_2^h)\overline{\omega(n_1, n_2)}. \tag{2.1}$$

The unitary $u'(h_1 h_2, n)$ is equal to

$$\beta_{h_1 h_2}(v_{n^{h_1 h_2}}) \times \overline{\zeta_n(h_1, h_2)} \beta_{h_1}(u(h_2, n^{h_1})) u(h_1, n) \times v_n^*.$$

The product of the unitaries here is equal to

$$\begin{aligned}\beta_{h_1}\left(\beta_{h_2}(v_{n^{h_1h_2}})u(h_2,n^{h_1})v^*_{n^{h_1}}\right)\times\beta_{h_1}(v_{n^{h_1}})u(h_1,n)v^*_n\\ =\beta_{h_1}(u'(h_2,n^{h_1}))u'(h_1,n).\end{aligned}$$

Therefore, we have seen that $\zeta$ is unchanged, i.e.,

$$\zeta'=\zeta. \tag{2.2}$$

On the other hand, (as in (ii)) we set

$$\begin{aligned}w_{h_1h_2}&=\mu(h_1,h_2)w_{h_1}\beta_{h_1}(w_{h_2})\quad(\text{with }\mu(\cdot,\cdot)\in\mathbf{T}),\\ \beta^w_h&=Ad(w_h)\beta_h,\ \text{a perturbed action of }H.\end{aligned}$$

Since

$$\begin{aligned}\beta^w_h\alpha_{n^h}\beta^w_{h^{-1}}&=Ad(w_h)\beta_h\alpha_{n^h}\beta_{h^{-1}}Ad(w^*_h)\\ &=Ad(w_hu(h,n))\alpha_nAd(w^*_h)=Ad\left(w_hu(h,n)\alpha_n(w^*_h)\right)\alpha_n,\end{aligned}$$

we can set

$$u''(h,n)=w_hu(h,n)\alpha_n(w^*_h).$$

We compute

$$\begin{aligned}u''(h,n_1n_2)&=w_h\times\eta_h(n_1,n_2)u(h,n_1)\alpha_{n_1}(u(h,n_2))\times\alpha_{n_1n_2}(w^*_h)\\ &=\eta_h(n_1,n_2)w_hu(h,n_1)\alpha_{n_1}(w^*_h)\alpha_{n_1}\left(w_hu(h,n_2)\alpha_{n_2}(w^*_h)\right)\\ &=\eta_h(n_1,n_2)u''(h,n_1)\alpha_{n_1}(u''(h,n_2)),\\ u''(h_1h_2,n)&=\mu(h_1,h_2)w_{h_1}\beta_{h_1}(w_{h_2})\zeta_n(h_1,h_2)\beta_{h_1}(u(h_2,n^{h_1}))u(h_1,n)\\ &\qquad\times\alpha_n\left(\overline{\mu(h_1,h_2)}\beta_{h_1}(w^*_{h_2})w^*_{h_1}\right)\\ &=\zeta_n(h_1,h_2)w_{h_1}\beta_{h_1}\left(w_{h_2}u(h_2,n^{h_1})\right)\\ &\qquad\times u(h_1,n)\alpha_n(\beta_{h_1}(w^*_{h_2}))\alpha_n(w^*_{h_1})\\ &=\zeta_n(h_1,h_2)w_{h_1}\beta_{h_1}\left(w_{h_2}u(h_2,n^{h_1})\right)\\ &\qquad\times\beta_{h_1}\alpha_{n^{h_1}}(w^*_{h_2})u(h_1,n)\alpha_n(w^*_{h_1})\\ &=\zeta_n(h_1,h_2)w_{h_1}\beta_{h_1}\left(w_{h_2}u(h_2,n^{h_1})\alpha_{n^{h_1}}(w^*_{h_2})\right)w^*_{h_1}\\ &\qquad\times w_{h_1}u(h_1,n)\alpha_n(w^*_{h_1})\\ &=\zeta_n(h_1,h_2)\beta^w_{h_1}(u''(h_2,n^{h_1}))u''(h_1,n).\end{aligned}$$

Therefore, we conclude that the perturbation of $\beta$ to $\beta^w$ does not change the cocycles $\eta,\zeta$, that is,

$$\eta''=\eta,\ \zeta''=\zeta. \tag{2.3}$$

When $(\alpha,\beta),(\alpha',\beta')$ are cocycle conjugate, we have $\omega(n_1,n_2)=\mu(h_1,h_2)=1$ in the above computations (and $\theta=id$ can be assumed). Therefore, (2.1), (2.2), (2.3) imply (i). On the other hand, (ii) follows from (2.3) while (iii) follows from (2.1), (2.2). □

Recall that $\alpha,\beta$ give rise a $G$-kernel (with $G=N\rtimes H$). Namely, we set

$$\theta_{nh}=\alpha_n\beta_h$$

(and $nh \in G \to [\theta_{nh}] \in Out(\mathcal{R})$ is a homomorphism). Since

$$\begin{aligned}\theta_{n_1h_1}\theta_{n_2h_2} &= \alpha_{n_1}\beta_{h_1}\alpha_{n_2}\beta_{h_2} = \alpha_{n_1}\beta_{h_1}\alpha_{n_2}\beta_{h_1^{-1}}\beta_{h_1h_2}\\ &= \alpha_{n_1}Ad(u(h_1, n_2^{h_1^{-1}}))\alpha_{n_2^{h_1^{-1}}}\beta_{h_1h_2}\\ &= Ad\left(\alpha_{n_1}(u(h_1, n_2^{h_1^{-1}}))\right)\alpha_{n_1n_2^{h_1^{-1}}}\beta_{h_1h_2},\end{aligned}$$

we have

$$\theta_{g_1}\theta_{g_2} = Ad(U(g_1, g_2))\theta_{g_1g_2}$$

with

$$U(n_1h_1, n_2h_2) = \alpha_{n_1}(u(h_1, n_2^{h_1^{-1}})).$$

We next determine a 3-cocycle associated with the $G$-kernel $\theta$ (and the above unitaries $U(\cdot,\cdot)$), i.e., $\omega(\cdot,\cdot,\cdot) \in \mathbf{T}$ satisfying

$$\theta_{g_1}(U(g_2, g_3))U(g_1, g_2g_3) = \omega(g_1, g_2, g_3)U(g_1, g_2)U(g_1g_2, g_3).$$

Since $n_2h_2n_3h_3 = n_2n_3^{h_2^{-1}}h_2h_3$ and $(n_2n_3^{h_2^{-1}})^{h_1^{-1}} = n_2^{h_1^{-1}}n_3^{h_2^{-1}h_1^{-1}}$, we compute

$$\begin{aligned}&\theta_{n_1h_1}(U(n_2h_2, n_3h_3))U(n_1h_1, n_2h_2n_3h_3)\\ =\ &\alpha_{n_1}\beta_{h_1}\alpha_{n_2}\left(u(h_2, n_3^{h_2^{-1}})\right)\alpha_{n_1}\left(u(h_1, n_2^{h_1^{-1}}n_3^{h_2^{-1}h_1^{-1}})\right)\\ =\ &\alpha_{n_1}\Big(\beta_{h_1}\alpha_{n_2}(u(h_2, n_3^{h_2^{-1}}))\\ &\qquad\times\eta_{h_1}(n_2^{h_1^{-1}}n_3^{h_2^{-1}h_1^{-1}})u(h_1, n_2^{h_1^{-1}})\alpha_{n_2^{h_1^{-1}}}(u(h_1, n_3^{h_2^{-1}h_1^{-1}}))\Big).\end{aligned}$$

On the other hand, since $n_1h_1n_2h_2 = n_1n_2^{h_1^{-1}}h_1h_2$, we compute

$$\begin{aligned}&U(n_1h_1, n_2h_2)U(n_1h_1n_2h_2, n_3h_3)\\ =\ &\alpha_{n_1}(u(h_1, n_2^{h_1^{-1}}))\alpha_{n_1n_2^{h_1^{-1}}}(u(h_1h_2, n_3^{h_2^{-1}h_1^{-1}}))\\ =\ &\alpha_{n_1}\left(u(h_1, n_2^{h_1^{-1}})\alpha_{n_2^{h_1^{-1}}}\left(\overline{\zeta_{n_3^{h_2^{-1}h_1^{-1}}}(h_1, h_2)}\beta_{h_1}(u(h_2, n_3^{h_2^{-1}}))u(h_1, n_3^{h_2^{-1}h_1^{-1}})\right)\right)\\ =\ &\overline{\zeta_{n_3^{h_2^{-1}h_1^{-1}}}(h_1, h_2)}\\ &\qquad\times\alpha_{n_1}\left(u(h_1, n_2^{h_1^{-1}})\alpha_{n_2^{h_1^{-1}}}\beta_{h_1}(u(h_2, n_3^{h_2^{-1}}))\alpha_{n_2^{h_1^{-1}}}(u(h_1, n_3^{h_2^{-1}h_1^{-1}}))\right)\\ =\ &\overline{\zeta_{n_3^{h_2^{-1}h_1^{-1}}}(h_1, h_2)}\\ &\qquad\times\alpha_{n_1}\left(\beta_{h_1}\alpha_{n_2}(u(h_2, n_3^{h_2^{-1}}))u(h_1, n_2^{h_1^{-1}})\alpha_{n_2^{h_1^{-1}}}(u(h_1, n_3^{h_2^{-1}h_1^{-1}}))\right).\end{aligned}$$

Note that the same unitaries are around. Therefore, by comparing the coefficients we conclude that the 3-cocycle $\omega$ is

$$\omega(n_1h_1, n_2h_2, n_3h_3) = \eta_{h_1}(n_2^{h_1^{-1}}, n_3^{(h_1h_2)^{-1}})\zeta_{n_3^{(h_1h_2)^{-1}}}(h_1, h_2).$$

## 2. Cocycle conjugacy

Our goal is to prove

THEOREM 2.5. *Let $\mathcal{R}$ be the hyperfinite $II_1$ factor, and assume that $G = N \rtimes H$ is a semi-direct product. Then, $H^2((N,H),\mathbf{T})$ is a complete invariant for cocycle conjugacy classes of outer actions of $(N,H)$. Furthermore, every element in $H^2((N,H),\mathbf{T})$ can be realized.*

LEMMA 2.6. *(i) The map*

$$\varphi : (\eta,\zeta) \in Z^2((N,H),\mathbf{T}) \to \omega \in Z^3(G,\mathbf{T}),$$

*where $\omega$ is defined as above, is a well-defined homomorphism satisfying*

$$\varphi(B^2((N,H),\mathbf{T})) \subseteq B^3(G,\mathbf{T}).$$

*(ii) The kernel of the induced map $[\varphi] : H^2((N,H),\mathbf{T}) \to H^3(G,\mathbf{T})$ is*

$$\{[(\eta,\zeta)] \in H^2((N,H),\mathbf{T});\ \eta_h(n_1,n_2) = \mu(n_1^h,n_2^h)\overline{\mu(n_1,n_2)} \\ \text{for some } \mu \in Z^2(N,\mathbf{T}) \text{ and } \zeta = 1\}.$$

PROOF. We begin by checking the 3-cocycle property. By noticing

$$n_1h_1n_2h_2 = n_1n_2^{h_1^{-1}}h_1h_2,\ n_2h_2n_3h_3 = n_2n_3^{h_2^{-1}}h_2h_3,$$
$$n_3h_3n_4h_4 = n_3n_4^{h_3^{-1}}h_3h_4,$$

from the definition of $\omega$ we have

$$\begin{aligned}
&\omega(n_2h_2,n_3h_3,n_4h_4)\overline{\omega(n_1h_1n_2h_2,n_3h_3,n_4h_4)}\omega(n_1h_1,n_2h_2n_3h_3,n_4h_4)\\
&\qquad\times \overline{\omega(n_1h_1,n_2h_2,n_3h_3n_4h_4)}\omega(n_1h_1,n_2h_2,n_3h_3)\\
&= \eta_{h_2}(n_3^{h_2^{-1}},n_4^{(h_2h_3)^{-1}})\zeta_{n_4^{(h_2h_3)^{-1}}}(h_2,h_3)\\
&\times \overline{\eta_{h_1h_2}(n_3^{(h_1h_2)^{-1}},n_4^{(h_1h_2h_3)^{-1}})\zeta_{n_4^{(h_1h_2h_3)^{-1}}}(h_1h_2,h_3)}\\
&\times \eta_{h_1}(n_2^{h_1^{-1}}n_3^{(h_1h_2)^{-1}},n_4^{(h_1h_2h_3)^{-1}})\zeta_{n_4^{(h_1h_2h_3)^{-1}}}(h_1,h_2h_3)\\
&\times \overline{\eta_{h_1}(n_2^{h_1^{-1}},n_3^{(h_1h_2)^{-1}}n_4^{(h_1h_2h_3)^{-1}})\zeta_{n_3^{(h_1h_2)^{-1}}n_4^{(h_1h_2h_3)^{-1}}}(h_1,h_2)}\\
&\times \eta_{h_1}(n_2^{h_1^{-1}},n_3^{(h_1h_2)^{-1}})\zeta_{n_3^{(h_1h_2)^{-1}}}(h_1,h_2).
\end{aligned}$$

By the cocycle property for $\eta_{h_1}$ (Lemma 2.2,(i)), the product of the three $\eta_{h_1}$ factors is equal to

$$\eta_{h_1}(n_3^{(h_1h_2)^{-1}},n_4^{(h_1h_2h_3)^{-1}}).$$

Also, by the cocycle property for $\zeta$ (Lemma 2.2,(ii)), the product of the first three $\zeta$ factors above is equal to

$$\zeta_{n_4^{(h_1h_2h_3)^{-1}}}(h_1,h_2).$$

Therefore, the product of the ten factors in the preceding computation can be rewritten as that of six factors consisting of the above two and the following four remaining factors (from the ten factors):

$$\eta_{h_2}(n_3^{h_2^{-1}},n_4^{(h_2h_3)^{-1}})\overline{\eta_{h_1h_2}(n_3^{(h_1h_2)^{-1}},n_4^{(h_1h_2h_3)^{-1}})}$$
$$\times\zeta_{n_3^{(h_1h_2)^{-1}}}(h_1,h_2)\overline{\zeta_{n_3^{(h_1h_2)^{-1}}n_4^{(h_1h_2h_3)^{-1}}}(h_1,h_2)}.$$

Thanks to the relation between $\eta$ and $\zeta$ (Lemma 2.2,(iii))), the product of the six factors is 1, and hence $\omega$ is indeed a 3-cocycle.

Next we assume $(\eta, \zeta) = (\partial^N \xi, \partial^H \xi)$. Then, the corresponding $\omega$ is

$$\begin{aligned} &\overline{\xi(h_1, (n_2 n_3^{h_2^{-1}})^{h_1^{-1}})}\xi(h_1, n_2^{h_1^{-1}})\xi(h_1, n_3^{(h_1h_2)^{-1}}) \\ &\times \xi(h_1h_2, n_3^{(h_1h_2)^{-1}})\overline{\xi(h_1, n_3^{(h_1h_2)^{-1}})\xi(h_2, n_3^{h_2^{-1}})} \\ &\quad = \overline{\xi(h_2, n_3^{h_2^{-1}})}\xi(h_1h_2, n_3^{(h_1h_2)^{-1}})\overline{\xi(h_1, (n_2n_3^{h_2^{-1}})^{h_1^{-1}})}\xi(h_1, n_2^{h_1^{-1}}) \end{aligned}$$

from the definitions of $\omega(\cdot,\cdot,\cdot)$ and $(\partial^N\xi, \partial^H\xi)$. If we set

$$\xi'(n_1h_1, n_2h_2) = \xi(h_1, n_2^{h_1^{-1}}),$$

then the above product is

$$\overline{\xi'(n_2h_2, n_3h_3)}\xi'(n_1h_1n_2h_2, n_3h_3)\overline{\xi'(n_1h_1, n_2h_2n_3h_3)}\xi'(n_1h_1, n_2h_2),$$

which means that $\omega$ in the present case is a coboundary.

To show the second statement, let us assume that $\omega$ is a coboundary. Hence, there is a $\mathbf{T}$-valued function $\mu$ on $(H \rtimes N) \times (H \rtimes N)$ satisfying

$$\begin{aligned} \omega(n_1h_1, n_2h_2, n_3h_3) &= \eta_{h_1}(n_2^{h_1^{-1}}, n_3^{(h_1h_2)^{-1}})\zeta_{n_3^{(h_1h_2)^{-1}}}(h_1, h_2) \\ &= \mu(n_1h_1, n_2h_2)\mu(n_1n_2^{h_1^{-1}}h_1h_2, n_3h_3) \\ &\qquad \overline{\times\mu(n_2h_2, n_3h_3)\mu(n_1h_1, n_2n_3^{h_2^{-1}}h_2h_3)}. \end{aligned}$$

By changing $n_2$ and $n_3$ to $n_2^{h_1}$ and $n_3^{h_1h_2}$ respectively, we have

$$\begin{aligned} (2.4) \qquad & \eta_{h_1}(n_2, n_3)\zeta_{n_3}(h_1, h_2) \\ &= \mu(n_1h_1, n_2^{h_1}h_2)\mu(n_1n_2h_1h_2, n_3^{h_1h_2}h_3) \\ &\qquad \overline{\times\mu(n_1h_1, n_2^{h_1}n_3^{h_1}h_2h_3)\mu(n_2^{h_1}h_2, n_3^{h_1h_2}h_3)} \\ &= \nu(n_1, n_2, h_1, h_2)\nu(n_1n_2, n_3, h_1h_2, h_3) \\ &\qquad \overline{\times\nu(n_1, n_2n_3, h_1, h_2h_3)\nu(n_2^{h_1}, n_3^{h_1}, h_2, h_3)} \end{aligned}$$

with

$$\nu(n_1, n_2, h_1, h_2) = \mu(n_1h_1, n_2^{h_1}h_2).$$

The above product (2.4) is 1 if either all $h$'s or all $n$'s are $e$ because of

$$\omega(n_1, n_2, n_3) = \omega(h_1, h_2, h_3) = 1$$

(note that $\theta\,|_N = \alpha$ and $\theta\,|_H = \beta$ are actions). This means that if we set

$$\begin{aligned} \omega_1(n_1, n_2) &= \nu(n_1, n_2, e, e), \\ \omega_2(h_1, h_2) &= \nu(e, e, h_1, h_2), \end{aligned}$$

then they are 2-cocycles on $N$ and $H$ respectively. We set

$$\xi(h, n) = \nu(e, n, h, e),$$

and define $\nu_0(n_1, n_2, h_1, h_2)$ by

$$\nu(n_1, n_2, h_1, h_2) = \omega_1(n_1, n_2)\omega_2(h_1, h_2)\xi(h_1, n_2)\nu_0(n_1, n_2, h_1, h_2).$$

Notice that we may and do assume

$$(2.5) \qquad \nu_0(e, e, h_1, h_2) = \nu_0(n_1, n_2, e, e) = \nu_0(e, n, h, e) = 1$$

(after renormalization $\cdots$ Note that cocycles satisfy $\omega_1(n_1,e)=\omega_1(e,n_2)=\omega_1(e,e)$ and $\omega_2(h_1,e)=\omega_2(e,h_2)=\omega_2(e,e)$). From (2.4) we get

$$\begin{aligned}
&\eta_{h_1}(n_2,n_3)\zeta_{n_3}(h_1,h_2)\\
&= \nu(n_1,n_2,h_1,h_2)\nu(n_1n_2,n_3,h_1h_2,h_3)\overline{\nu(n_1,n_2n_3,h_1,h_2h_3)\nu(n_2^{h_1},n_3^{h_1},h_2,h_3)}\\
&\qquad\qquad \times \xi(h_1,n_3)\overline{\xi(h_1,n_3)}\\
&= \overline{\omega_1(n_2^{h_1}n_3^{h_1})}\omega_1(n_2,n_3)(\partial^N\xi)_{h_1}(n_2,n_3)(\partial^H\xi)_{n_3}(h_1,h_2)\\
&\qquad\qquad \times \frac{\nu_0(n_1,n_2,h_1,h_2)\nu_0(n_1n_2,n_3,h_1h_2,h_3)}{\nu_0(n_1,n_2n_3,h_1,h_2h_3)\nu_0(n_2^{h_1},n_3^{h_1}h_2,h_3)}.
\end{aligned}$$

In fact, by looking at the second and third variables in the above four $\nu$'s and $\xi(h_1,n_3)\overline{\xi(h_1,n_3)}$, we get $(\partial^N\xi)_{h_1}(n_2,n_3)(\partial^H\xi)_{n_3}(h_1,h_2)$. Also, by looking at the first two and the last two variables, we have used the cocycle properties

$$\omega_1(n_1,n_2)\omega_1(n_1n_2,n_3)\overline{\omega_1(n_1,n_2n_3)}=\omega_1(n_2,n_3),$$
$$\omega_2(h_1,h_2)\omega_2(h_1h_2,h_3)\overline{\omega_2(h_1,h_2h_3)}=\omega_2(h_2,h_3).$$

Hence, by setting

$$\begin{aligned}
\eta'_h(n_1,n_2) &= \eta_h(n_1,n_2)\omega_1(n_1^h,n_2^h)\overline{\omega_1(n_1,n_2)(\partial^N\xi)_h(n_1,n_2)},\\
\zeta'_n(h_1,h_2) &= \zeta_n(h_1,h_2)\overline{(\partial^H\xi)_n(h_1,h_2)},
\end{aligned}$$

we have shown

$$\eta'_{h_1}(n_2,n_3)\zeta'_{n_3}(h_1,h_2)=\frac{\nu_0(n_1,n_2,h_1,h_2)\nu_0(n_1n_2,n_3,h_1h_2,h_3)}{\nu_0(n_1,n_2n_3,h_1,h_2h_3)\nu_0(n_2^{h_1},n_3^{h_1},h_2,h_3)}. \tag{2.6}$$

In (2.6), we set

$$\begin{aligned}
&n_1=e, h_2=h_3=e,\\
&n_1=n_2=e, h_3=e,\\
&n_3=e, h_1=h_2=e,\\
&n_2=n_3=e, h_1=e,\\
&n_1=n_3=e, h_2=e.
\end{aligned}$$

Then, in each case by (2.5) we get

$$\eta'_{h_1}(n_2,n_3)=\nu_0(n_2,n_3,h_1,e), \tag{2.7}$$
$$\zeta'_{n_3}(h_1,h_2)=\overline{\nu_0(e,n_3,h_1,h_2)}, \tag{2.8}$$
$$\nu_0(n_1,n_2,e,h_3)=\nu_0(n_1n_2,e,e,,h_3)\overline{\nu_0(n_2,e,e,h_3)}, \tag{2.9}$$
$$\nu_0(n_1,e,h_2,h_3)=\nu_0(n_1,e,e,h_2h_3)\overline{\nu_0(n_1,e,e,h_2)}, \tag{2.10}$$
$$\nu_0(n_2,e,h_1,h_3)=\nu_0(e,n_2,h_1,h_3)\nu_0(n_2^{h_1},e,e,h_3). \tag{2.11}$$

Thanks to (2.8), (2.11), and (2.10), we get

$$\begin{aligned}
\zeta'_{n_2}(h_1,h_3) &= \nu_0(n_2^{h_1},e,e,h_3)\overline{\nu_0(n_2,e,h_1,h_3)}\\
&= \nu_0(n_2^{h_1},e,e,h_3)\nu_0(n_2,e,e,h_1)\overline{\nu_0(n_2,e,e,h_1h_3)}.
\end{aligned}$$

This equation means that, by setting

$$\xi'(h,n)=\nu_0(n,e,e,h),$$

we have

$$\zeta'_n(h_1,h_2)=\overline{(\partial^H\xi')_n(h_1,h_2)}.$$

We next set $n_2 = e, h_1 = h_3 = e$ in (2.6) to get

$$\nu_0(n_1, n_3, e, h_2) = \nu_0(n_1, e, e, h_2)\nu_0(n_1, n_3, h_2, e).$$

The formulas (2.7), (2.9) together with this identity show

$$\begin{aligned} \eta'_{h_2}(n_1, n_2) &= \nu_0(n_1, n_2, e, h_2)\overline{\nu_0(n_1, e, e, h_2)} \\ &= \nu_0(n_1 n_2, e, e, h_2)\overline{\nu_0(n_1, e, e, h_2)\nu_0(n_2, e, e, h_2)} \\ &= \overline{(\partial^N \xi')_{h_2}(n_1, n_2)} \end{aligned}$$

with the above $\xi'$. Therefore, we conclude

$$\begin{aligned} \eta_h(n_1, n_2) &= \overline{\omega_1(n_1^h, n_2^h)}\omega_1(n_1, n_2)(\partial^N \xi)_h(n_1, n_2)\overline{(\partial^N \xi')_h(n_1, n_2)}, \\ \zeta_n(h_1, h_2) &= (\partial^H \xi)_n(h_1, h_2)\overline{(\partial^H \xi')_n(h_1, h_2)}, \end{aligned}$$

and the lemma has been proved. □

**Proof of Theorem 2.5 (Existence).** For a given $(\eta, \zeta) \in Z^2((N, H), \mathbf{T})$, we set $\omega = \varphi((\eta, \zeta)) \in Z^3(G, \mathbf{T})$ (Lemma 2.6,(i)). By the Jones theorem [**35**] there exists a $G$-kernel (together with its lifting $G \to Aut(\mathcal{R})$ and unitaries) realizing $\omega$. Since $\omega(n_1, n_2, n_3) = \omega(h_1, h_2, h_3) = 1$, the obstructions of the restrictions to the subgroups $N, H$ vanish and hence (after inner perturbation) one may choose a lifting $\theta : G \to Aut(\mathcal{R})$ in such a way that $\alpha = \theta\,|_N, \beta = \theta\,|_H$ are actions of $N, H$ respectively. By Lemma 2.6,(ii) the cocycles $(\eta', \zeta')$ attached to $(\alpha, \beta)$ is given by

$$\begin{aligned} \eta'_h(n_1, n_2) &= \eta_h(n_1, n_2)\overline{\mu(n_1^h, n_2^h)}\mu(n_1, n_2) \\ \zeta'_n(h_1, h_2) &= \zeta_n(h_1, h_2) \end{aligned}$$

with some $\mu \in Z^2(N, \mathbf{T})$ (up to a coboundary in $B^2((N, H), \mathbf{T})$). Then one can find a family $\{v_n\}_{n \in N}$ of unitaries in $\mathcal{R}$ satisfying

$$v_{n_1 n_2} = \mu(n_1, n_2)v_{n_1}\alpha_{n_1}(v_{n_2})$$

(see [**68**]). Thanks to Lemma 2.4,(iii), the perturbed action $(Ad(v)\alpha, \beta)$ has the the desired invariant $[(\eta, \zeta)] \in H^2((N, H), \mathbf{T})$, and the existence has been proved. □

To show the uniqueness, we need

LEMMA 2.7. *Let $\theta$ be an outer action of $G = N \bowtie H$ on the hyperfinite $II_1$ factor $\mathcal{R}$, and we set $\alpha^\circ = \theta\,|_N$ and $\beta^\circ = \theta\,|_H$. Let $\omega^1 \in Z^2(N, \mathbf{T})$ and $\omega^2 \in Z^2(H, \mathbf{T})$ be cocycles. Assume that there is a $\mathbf{T}$-valued function $\xi$ on $H \times N$ satisfying*

$$\begin{aligned} \omega^1(n_1^h, n_2^h)\overline{\omega^1(n_1, n_2)} &= \xi(h, n_1 n_2)\overline{\xi(h, n_1)\xi(h, n_2)}, \\ \xi(h_1 h_2, n) &= \xi(h_1, n)\xi(h_1, n^{h_1}). \end{aligned}$$

*Then there exist an automorphism $\nu \in Aut(\mathcal{R})$ and unitaries $\{v_n\}_{n \in N}, \{w_h\}_{h \in H}$ in $\mathcal{R}$ satisfying*

$$\begin{aligned} v_{n_1}\alpha^\circ_{n_1}(v_{n_2}) &= \omega^1(n_1, n_2)v_{n_1, n_2}, \\ w_{h_1}\beta^\circ_{h_1}(w_{h_2}) &= \omega^2(h_1, h_2)w_{h_1, h_2}, \\ \nu\alpha^\circ_n\nu^{-1} &= Ad(v_n)\alpha^\circ_n, \\ \nu\beta^\circ_h\nu^{-1} &= Ad(w_h)\beta^\circ_h. \end{aligned}$$

PROOF. Let $(\pi', \mathcal{H})$ be a projective representation of $N$ corresponding to $\omega^1$:

$$\pi'(n_1)\pi'(n_2) = \omega^1(n_1, n_2)\pi'(n_1 n_2) \text{ and } \dim \mathcal{H} < \infty.$$

We set $\pi^h(n) = \xi(h, n)\pi'(n^h)$ so that

$$\begin{aligned}\pi^h(n_1)\pi^h(n_2) &= \xi(h, n_1)\xi(h, n_2)\omega^1(n_1^h, n_2^h)\pi'((n_1 n_2)^h) \\ &= \omega^1(n_1, n_2)\xi(h, n_1 n_2)\pi'((n_1 n_2)^h) \quad \text{(by the assumption)} \\ &= \omega^1(n_1, n_2)\pi^h(n_1 n_2).\end{aligned}$$

We set

$$\pi(n) = \oplus_{h \in H} \pi^h(n),$$

a unitary acting on $\mathcal{H} \otimes \ell^2(H)$. On the other hand, let $\pi(h)$ (a unitary on the same Hilbert space) be $id_{\mathcal{H}}$ tensored by the left regular representation of the group $H$. We claim

$$\pi(h^{-1})\pi(n)\pi(h) = \xi(h, n)\pi(n^h). \tag{2.12}$$

In fact, for a vector $a \otimes \delta_k \in \mathcal{H} \otimes \ell^2(H)$, we compute

$$\begin{aligned}\pi(h^{-1})\pi(n)\pi(h)(a \otimes \delta_k) &= \pi(h^{-1})\pi(n)(a \otimes \delta_{hk}) \\ &= \pi(h^{-1})(\pi^{hk}(n)a \otimes \delta_{hk}) \\ &= \pi^{hk}(n)a \otimes \delta_k \\ &= \xi(hk, n)(\pi'(n^{hk})a \otimes \delta_k) \\ &= \xi(hk, n)\overline{\xi(k, n^h)}(\pi^k(n^h)a \otimes \delta_k) \\ &= \xi(h, n)(\pi^k(n^h)a \otimes \delta_k) \quad \text{(by the assumption)} \\ &= \xi(h, n)\pi(n^h)(a \otimes \delta_k).\end{aligned}$$

Let $(\sigma, \mathcal{K})$ be a (finite-dimensional) projective representation of $H$ corresponding to $\omega^2$:

$$\sigma(h_1)\sigma(h_2) = \omega^2(h_1, h_2)\sigma(h_1 h_2).$$

Let $v_n, w_h$ be the unitaries on $L^2(\mathcal{R}) \otimes (\mathcal{H}) \otimes \ell^2(H)) \otimes \mathcal{K}$ defined by

$$\begin{aligned}v_n &= 1 \otimes \pi(n) \otimes 1, \\ w_h &= 1 \otimes \pi(h) \otimes \sigma(h).\end{aligned}$$

Since the products of $\pi(n)$'s and $\sigma(h)$'s produce $\omega^1$ and $\omega^2$ respectively, we easily observe

$$\begin{aligned}v_{n_1}(\alpha^\circ_{n_1} \otimes id \otimes id)(v_{n_2}) &= v_{n_1}v_{n_2} = \omega^1(n_1, n_2)v_{n_1 n_2}, \\ w_{h_1}(\beta^\circ_{h_1} \otimes id \otimes id)(w_{h_2}) &= w_{h_1}w_{h_2} = \omega^2(h_1, h_2)w_{h_1 h_2}.\end{aligned}$$

Therefore, $Ad(v_n) \cdot (\alpha^\circ_n \otimes id \otimes id)$ and $Ad(w_h) \cdot (\beta^\circ_h \otimes id \otimes id)$ give rise to outer actions on $\mathcal{R} \otimes B(\mathcal{H} \otimes \ell^2(H)) \otimes B(\mathcal{K})$ of the groups $N, H$ respectively. Due to (2.12) (and the fact that that starting $\theta$ was an action), they actually determine an outer action of the group $G = N \rtimes H$. Therefore, we get an automorphism $\nu \in Aut(\mathcal{R})$ from the Jones theorem ([**35**]) stating that all outer actions (of a finite group) of $\mathcal{R}$ are conjugate. □

**Proof of Theorem 2.5 (Uniqueness).** Assume $[(\eta, \zeta)] \in H^2((N, H), \mathbf{T})$, and let $(\alpha, \beta)$ be an outer action of $(N, H)$ realizing $(\eta, \zeta)$. We set

$$\tilde{\alpha} = \alpha \otimes \alpha^\circ, \ \tilde{\beta} = \beta \otimes \beta^\circ$$

with $\alpha^\circ, \beta^\circ$ as in Lemma 2.7. Note that this tensoring does not change $\eta, \zeta$, and $(\tilde{\alpha}, \tilde{\beta})$ will become a "model" action.

Let us assume that $(\alpha', \beta')$ has the same invariant $[(\eta, \zeta)]$. Since $G$-kernels determined by $(\tilde{\alpha}, \tilde{\beta}), (\alpha', \beta')$ have the same invariant (Lemma 2.6,(i)) and conjugate by [**35**]. Hence, we have an automorphism $\mu$ and unitaries $\{v'_n\}_{n\in N}, \{w'_h\}_{h\in H}$ satisfying

$$\begin{aligned} \mu\tilde{\alpha}_n\mu^{-1} &= Ad(v'_n)\alpha'_n, \\ \mu\tilde{\beta}_h\mu^{-1} &= Ad(w'_h)\beta'_h. \end{aligned}$$

Notice that the right sides and $\tilde{\alpha}, \tilde{\beta}$ are actions, and we have

$$\begin{aligned} v'_{n_1n_2} &= \omega^1(n_1, n_2)v'_{n_1}\alpha'_{n_1}(v'_{n_2}) \\ w'_{h_1h_2} &= \omega^2(h_1, h_2)w'_{h_1}\beta'_{h_1}(w'_{h_2}) \end{aligned}$$

with $\omega^1 \in Z^2(N, \mathbf{T})$ and $\omega^2 \in Z^2(H, \mathbf{T})$ (see Appendix §3 at the end of the chapter). Since the $G$-kernels determined by $(\alpha', \beta'), (Ad(v')\alpha', Ad(w')\beta')$ have the same invariant (in $H^3(G, \mathbf{T})$), Lemma 2.4,(iii) and Lemma 2.6,(ii) guarantee

$$\begin{aligned} \omega^1(n_1^h, n_2^h)\overline{\omega^1(n_1, n_2)} &= \xi(h, n_1n_2)\overline{\xi(h, n_1)\xi(h, n_2)} \\ 1 &= \overline{\xi(h_1h_2, n)}\xi(h_1, n)\xi(h_2, n^{h_1}) \end{aligned}$$

with some $\xi$. By Lemma 2.7 one finds $\nu$, $\{v_n\}_{n\in N}, \{w_h\}_{h\in H}$ satisfying

$$\begin{aligned} (id\otimes\nu)\tilde{\alpha}_n(id\otimes\nu)^{-1} &= Ad(1\otimes v_n)\tilde{\alpha}_n, \\ (id\otimes\nu)\tilde{\beta}_h(id\otimes\nu)^{-1} &= Ad(1\otimes w_h)\tilde{\beta}_h, \\ (1\otimes v_{n_1n_2}) &= \omega^1(n_1, n_2)(1\otimes v_{n_1})\tilde{\alpha}_{n_1}(1\otimes v_{n_2}), \\ (1\otimes w_{h_1h_2}) &= \omega^2(h_1, h_2)(1\otimes w_{h_1})\tilde{\beta}_{h_1}(1\otimes w_{h_2}). \end{aligned}$$

Now it is clear that $(\alpha', \beta')$ is cocycle conjugate to $(\tilde{\alpha}, \tilde{\beta})$ (see Appendix §3), and the theorem has been proved. □

Finally some remarks are in order.

1. The invariant (in $H^2((N, H), \mathbf{T})$) of $(\alpha, \beta)$ is trivial if and only if there exist an outer action $\theta$ of $G = N \rtimes H$ and an $\theta\,|_N$-cocycle $\{w_n\}_{n\in N}$ satisfying

$$\begin{aligned} \alpha_n &= Ad(w_n)\theta_n, \\ \beta_h &= \theta_h. \end{aligned}$$

In fact, assume that $(\alpha, \beta)$ has the trivial invariant. Let $\chi$ be an outer action of $G = N \rtimes H$. Since $(\chi\,|_N, \chi\,|_H)$ has the trivial invariant and cocycle conjugate to $(\alpha, \beta)$ by Theorem 2.5, we get

$$\begin{aligned} \mu\alpha_n\mu^{-1} &= Ad(u_n)\chi_n, \\ \mu\beta_h\mu^{-1} &= Ad(v_h)\chi_h \end{aligned}$$

with unitary cocycles $u_n$ and $v_h$. Notice that $u_n = U\chi_n(U^*), v_h = V\chi_h(V^*)$ for some unitaries $U, V$. Therefore, we have

$$\begin{aligned} \mu\beta_h\mu^{-1} &= U\chi_h(U^*\cdot U)U^*, \\ \mu\alpha_n\mu^{-1} &= V\chi_n(V^*\cdot V)V^* \\ &= WU\chi_n(U^*W^*\cdot WU)U^*W^* \end{aligned}$$

with the unitary $W = VU^*$. By setting

$$\begin{aligned}\theta_g &= \mu^{-1}\left(U\chi_g(U^*\mu(\cdot)U)U^*\right)\\ w_n &= \mu^{-1}(W)\theta_n(\mu^{-1}(W^*)),\end{aligned}$$

we have $\alpha_n = Ad(w_n)\theta_n$ and $\beta_h = \theta_h$.

Conversely, in this case we have

$$\beta_h\alpha_{n^h}\beta_{h^{-1}} = Ad(\theta_h(w_{n^h}))\theta_n = Ad(\theta_h(w_{n^h})w_n^*)\alpha_n.$$

Therefore, we can take

$$u(h,n) = \theta_h(w_{n^h})w_n^*.$$

It is straight-forward to see that the corresponding $\eta, \zeta$ are both identically 1 due to the cocycle property of $\{w_n\}_{n\in N}$.

2. Those cohomology classes with a representative satisfying $\eta = 1$ can be identified with $H^2(H, Hom(N,\mathbf{T}))$. In fact, $\eta = 1$ means

$$\zeta_{n_1n_1}(h_1,h_2) = \zeta_{n_1}(h_1,h_2)\zeta_{n_2}(h_1,h_2), \text{ i.e., } \zeta_\cdot(h_1,h_2)\in Hom(N,\mathbf{T}),$$

(Lemma 2.2,(iii)) and also $\partial^N\xi = 1$ means

$$\xi(h,n_1n_2) = \xi(h,n_1)\xi(h,n_2).$$

Assume that $H^2(N,\mathbf{T})$ vanishes. Then, for a given $[(\eta,\zeta)] \in H^2((N,H),\mathbf{T})$, we can find $\xi(\cdot,\cdot)$ satisfying

$$\eta_h(n_1,n_2) = \xi(h,n_1n_2)\overline{\xi(h,n_1)\xi(h,n_2)}.$$

Thus, by changing $(\eta,\zeta)$ to $(\eta(\partial^N\xi),\zeta(\partial^H\xi))$, one can always assume $\eta = 1$ up to a coboundary. Therefore, in this case we conclude that $H^2((N,H),\mathbf{T})$ is isomorphic to $H^2(H, Hom(N,\mathbf{T}))$, i.e., it is parameterized by the extensions

$$1 \longrightarrow Hom(N,\mathbf{T}) \longrightarrow K \longrightarrow H \longrightarrow 1.$$

3. (Normalization of cocycles) One can always assume that the following quantities are 1 up to a coboundary (see Appendix §4):

$$\eta_e(n_1,n_2),\ \eta_h(n_1,e),\ \eta_h(e,n_2),\ \eta_h(n,n^{-1}),\ \zeta_e(h_1,h_2),\ \zeta_n(n_1,e),\ \zeta_n(e,n_2).$$

## 3. Appendix Iterated perturbation

Let $\alpha,\beta,\gamma$ be actions of a finite group (on a factor $\mathcal{R}$). Assume $\mu\alpha_g\mu^{-1} = Ad(u_g)\beta_g$ with unitaries $\{u_g\}_{g\in G}$ in $\mathcal{R}$ satisfying

$$u_{g_1g_2} = \omega^1(g_1,g_2)u_{g_1}\beta_{g_1}(u_{g_2}).$$

At first we notice $\omega^1 \in Z^2(G,\mathbf{T})$. In fact, we compute

$$\begin{aligned}\overline{\omega^1(g_2,g_3)\omega^1(g_1,g_2g_3)} &= u_{g_1}\overline{\omega^1(g_2,g_3)}\beta_{g_1}(u_{g_2g_3})u^*_{g_1g_2g_3}\\ &= u_{g_1}\beta_{g_1}\left(u_{g_2}\beta_{g_2}(u_{g_3})u^*_{g_2g_3}\right)\beta_{g_1}(u_{g_2g_3})u^*_{g_1g_2g_3}\\ &= u_{g_1}\beta_{g_1}(u_{g_2})\beta_{g_1g_2}(u_{g_3})u^*_{g_1g_2g_3} \qquad \text{(after canceling } \beta_{g_1}(u_{g_2g_3}))\\ &= u_{g_1}\beta_{g_1}(u_{g_2})u^*_{g_1g_2}\times u_{g_1g_2}\beta_{g_1g_2}(u_{g_3})u_{g_1g_2g_3}\\ &= \overline{\omega^1(g_1,g_2)\omega^1(g_1g_2,g_3)}.\end{aligned}$$

Since $\beta_g = Ad(u_g^*)\mu\alpha_g\mu^{-1} = \mu Ad(\mu^{-1}(u_g^*))\alpha_g\mu^{-1}$, we have

$$\mu^{-1}\beta_g\mu = Ad(w_g')\alpha_g \qquad \text{with } w_g' = \mu^{-1}(u_g^*).$$

The family $\{w'_g\}_{g\in G}$ of unitaries satisfies

$$\begin{aligned} w'_{g_1g_2} &= \mu^{-1}(u^*_{g_1g_2}) = \overline{\omega^1(g_1,g_2)}\mu^{-1}(\beta_{g_1}(u^*_{g_2}))\mu^{-1}(u^*_{g_1}) \\ &= \overline{\omega^1(g_1,g_2)}\mu^{-1}(u^*_{g_1}) \times \mu^{-1}\left(Ad(u_{g_1})\beta_{g_1}(u^*_{g_2})\right) \\ &= \overline{\omega^1(g_1,g_2)}w'_{g_1}\mu^{-1}\left(\mu\alpha_{g_1}\mu^{-1}(u^*_{g_2})\right) \\ &= \overline{\omega^1(g_1,g_2)}w'_{g_1}\alpha_{g_1}(w'_{g_2}). \end{aligned}$$

Next we assume $\nu\beta_g\nu^{-1} = Ad(v_g)\gamma_g$ with

$$v_{g_1g_2} = \omega^2(g_1,g_2)v_{g_1}\gamma_{g_1}(v_{g_2}).$$

Then,

$$(\nu\mu)\alpha_g(\nu\mu)^{-1} = \nu Ad(u_g)\beta_g\nu^{-1} = Ad(\nu(u_g))\nu\beta_g\nu^{-1} = Ad(w''_g)\gamma_g$$

with $w''_g = \nu(u_g)v_g$. The family $\{w''_g\}_{g\in G}$ of unitaries satisfies

$$\begin{aligned} w''_{g_1g_2} &= \nu(u_{g_1g_2})v_{n_1n_2} = \omega^1(g_1,g_2)\omega^2(g_1,g_2)\nu\left(u_{g_1}\beta_{g_1}(u_{g_2})\right)v_{g_1}\gamma_{g_1}(v_{g_2}) \\ &= \omega^1(g_1,g_2)\omega^2(g_1,g_2)\nu(u_{g_1})v_{g_1} \times Ad(v^*_{g_1})\nu\beta_{g_1}(u_{g_2}) \times \gamma_{g_1}(v_{g_2}) \\ &= \omega^1(g_1,g_2)\omega^2(g_1,g_2)\nu(u_{g_1})v_{g_1} \times \gamma_{g_1}\nu(u_{g_2}) \times \gamma_{g_1}(v_{g_2}) \\ &= \omega^1(g_1,g_2)\omega^2(g_1,g_2)w''_{g_1}\gamma_{g_1}(w''_{g_2}). \end{aligned}$$

## 4. Appendix Normalization of cocycles

In the cocycle equation for $\zeta$ (Lemma 2.2,(ii)) we set $h_1 = 1, h_3 = 1$, or $h_1^{-1} = h_2 = h_3^{-1} = h$. In each case we get

$$\begin{aligned} \zeta_n(e,h) &= \zeta_n(e,e), \\ \zeta_n(hh',e) &= \zeta_{n^h}(h',e), \\ \zeta_{n^{h^{-1}}}(h,h^{-1}) &= \zeta_n(h^{-1},h) \times \frac{\zeta_n(e,h^{-1})}{\zeta_n(h^{-1},e)}. \end{aligned}$$

The second equation in particular says $\zeta_n(h,e) = \zeta_{n^h}(e,e)$.

We set $\xi(h,n) = \zeta_n(e,e)$ so that

$$(\partial^H\xi)_n(e,e) = \zeta_n(e,e)\overline{\zeta_n(e,e)\zeta_n(e,e)} = \overline{\zeta_n(e,e)}.$$

Thus, after perturbation by the coboundary associated to $\xi$, we may and do assume $\zeta_n(e,e) = 1$. This means

$$\zeta_n(h,e) = \zeta_n(e,h) = 1 \text{ and } \zeta_{n^{h^{-1}}}(h,h^{-1}) = \zeta_n(h^{-1},h).$$

Also the fundamental relation between $\eta$ and $\zeta$ (Lemma 2.2,(iii)) says

$$\eta_e(n_1,n_2) = 1.$$

We next set $\xi(h,n) = \overline{\eta_h(e,e)}$. We get $(\partial^N\xi)_h(e,e) = \overline{\eta_h(e,e)}$ as above, and hence after this perturbation we may and do assume $\eta_h(e,e) = 1$. Notice

$$(\partial^H\xi)_n(h_1,h_2) = \frac{\xi(h_1h_2,n)}{\xi(h_1,n)\xi(h,n^{h_1})} = \frac{\eta_{h_1}(e,e)\eta_{h_2}(e,e)}{\eta_{h_1h_2}(e,e)} = \overline{\zeta_e(h_1,h_2)}.$$

Therefore, $(\partial^H\xi)_n(h,e) = (\partial^H\xi)_n(e,h) = 1$, and this second perturbation preserves the property $\zeta_n(h,e) = \zeta_n(e,h) = 1$.

In the cocycle equation

$$\eta_h(n_1,n_2)\eta_h(n_1n_2,n_3) = \eta_h(n_1n_2,n_3)\eta_h(n_2,n_3)$$

(Lemma 2.2,(i)) we set $n_2 = e$ or $n_1 = n_2^{-1} = n_3 = n$ and get

$$\eta_h(n,e) = \eta_h(e,n) = \eta_h(e,e), \tag{2.13}$$

$$\eta_h(n,n^{-1}) = \eta_h(n^{-1},n). \tag{2.14}$$

Since $\eta_h(e,e) = 1$, by (2.13) we have

$$\eta_h(n,e) = \eta_h(e,n) = 1.$$

Also Lemma 2.2,(iii) says

$$\zeta_e(h_1,h_2) = 1.$$

Finally, we define

$$\xi(h,n) = \sqrt{\eta_h(n,n^{-1})},$$

one of square roots (and $\sqrt{1} = 1$). Since $\xi(h,n) = \xi(h,n^{-1})$ by (2.14) and $\xi(h,e) = 1$, we have $(\partial^N \xi)_h(n,n^{-1}) = \overline{\eta_h(n,n^{-1})}$. Therefore, after this perturbation we may and do assume

$$\eta_h(n,n^{-1}) = \eta_h(n^{-1},n) = 1.$$

It is easy to see $(\partial^N \xi)_h(n,e) = (\partial^N \xi)_h(e,n) = 1$, and hence the property $\eta_h(n,e) = \eta_h(e,n) = 1$ is still preserved. Notice

$$(\partial^H \xi)_n(h_1,h_2) = \frac{\sqrt{\eta_{h_1h_2}(n,n^{-1})}}{\sqrt{\eta_{h_1}(n,n^{-1})}\sqrt{\eta_{h_2}(n^{h_1},(n^{-1})^{h_1})}}.$$

Since $\xi(e,n) = 1$, we have $(\partial^H \xi)_n(e,e) = 1$ and the perturbation preserves the property $\zeta_n(e,e) = 1$

CHAPTER 3

# Cocycles attached to the pentagon equation

We assume that a finite group $G$ is a semi-direct product $G = N \rtimes H$. Let $e$ be the unit in $G$, and $n, n', n_1, n_2, \cdots$ (resp. $h, h', h_1, h_2, \cdots$) will denote elements in $N$ (resp. $N$). We will employ the same notations as in Chapter 2 (such as $n^h = h^{-1}nh$ ($\in N$) and so on). As in the appendix to the Baaj-Skandalis article [**1**], let $v$ be the map (from $G \times G$ to itself) defined by

$$v(n_1h_1, n_2h_2) = (n_1n_2h_1, n_2^{h_1}h_1^{-1}h_2).$$

Then we have $v_{23}v_{13}v_{12} = v_{12}v_{23}$. Hence, if we set

$$(V\xi)(n_1h_1, n_2h_2) = \xi(v(n_1h_1, n_2h_2))$$

(which is a unitary operator on $\ell^2(G) \otimes \ell^2(G) \cong \ell^2(G \times G)$), then $V$ satisfies the pentagon equation

$$V_{12}V_{13}V_{23} = V_{23}V_{12}.$$

In this chapter we study cocycles arising from the above fundamental unitary $V$ and the resulting cohomology group (which is denoted by $H^2_V((N, H), \mathbf{T})$). We will show that this cohomology group is the same as $H^2((N, H), \mathbf{T})$ introduced in the previous chapter (Theorem 3.6).

## 1. Cocycles

Let $\theta(\cdot, \cdot)$ be a $\mathbf{T}$-valued function on $G \times G$, and we set

$$(V^\theta\xi)(n_1h_1, n_2h_2) = \theta(n_1h_1, n_2h_2)\xi(v(n_1h_1, n_2h_2)).$$

It is straight-forward to see that $V^\theta$ satisfies the pentagon equation if and only if

$$\theta(n_1h_1, n_2h_2)\theta(n_1n_2h_1, n_3h_3)\theta(n_2^{h_1}h_1^{-1}h_2, n_3^{h_1}h_1^{-1}h_3) = \theta(n_2h_2, n_3h_3)\theta(n_1h_1, n_2n_3h_2). \tag{3.1}$$

We set

$$\begin{aligned} \tilde{\eta}_h(n, n') &= \theta(nh, n')\overline{\theta(h, n')}, \\ \tilde{\zeta}_n(h, h') &= \theta(h, nhh')\overline{\theta(h, hh')}, \\ \tilde{\theta}(h, h') &= \theta(h, h'). \end{aligned}$$

The following cocycle properties will be checked:

$$\tilde{\eta}_h(n_1, n_2)\tilde{\eta}_h(n_1n_2, n_3) = \tilde{\eta}_h(n_1, n_2n_3)\tilde{\eta}_h(n_2, n_3), \tag{3.2}$$

$$\tilde{\zeta}_n(h_1, h_2)\tilde{\zeta}_n(h_1h_2, h_3) = \tilde{\zeta}_n(h_1, h_2h_3)\tilde{\zeta}_{n^{h_1}}(h_2, h_3), \tag{3.3}$$

$$\tilde{\theta}(h_1, h_3)\tilde{\theta}(h_1^{-1}h_2, h_1^{-1}h_3) = \tilde{\theta}(h_2, h_3). \tag{3.4}$$

Furthermore, $\tilde{\eta}_\cdot, \tilde{\zeta}_\cdot$ are related as follows:

$$\frac{\tilde{\eta}_h(n,n')(h\cdot\tilde{\eta}_{h'})(n,n')}{\tilde{\eta}_{hh'}(n,n')} = \frac{\tilde{\zeta}_{nn'}(h,h')}{\tilde{\zeta}_n(h,h')\tilde{\zeta}_{n'}(h,h')}. \tag{3.5}$$

Here, $h\cdot\tilde{\eta}_{n'}$ means

$$(h\cdot\tilde{\eta}_{h'})(n,n') = \tilde{\eta}_{h'}(n^h, n'^h).$$

Then, $\theta(\cdot,\cdot)$ can be recovered from $\tilde{\eta}_\cdot, \tilde{\zeta}_\cdot, \tilde{\theta}$ in the following way:

$$\theta(nh, n'h') = \eta_h(n,n')\tilde{\zeta}_{n'}(h, h^{-1}h')\tilde{\theta}(h,h'). \tag{3.6}$$

Actually, we now prove

THEOREM 3.1. *There is a one-to-one correspondence between $\theta$'s satisfying* (3.1) *and triples $\{\tilde{\eta}_\cdot, \tilde{\zeta}_\cdot, \tilde{\theta}\}$ of cocycles* ((3.2), (3.3), (3.4)) *satisfying the relation* (3.5).

We start from the equation (3.1). By setting $n_1 = h_3 = 1$, we get

$$\theta(h_1, n_2h_2)\theta(n_2h_1, n_3)\theta(n_2^{h_1}h_1^{-1}h_2, n_3^{h_1}h_1^{-1}) = \theta(n_2h_2, n_3)\theta(h_1, n_2n_3h_2).$$

After changing $n_2^{h_1}, h_1^{-1}h_2, n_3^{h_1}, h_1^{-1}$ to $n, h, , n', h'$ respectively, we solve the above for the third factor in the left side, and we have

$$\theta(nh, n'h') = \frac{\theta(h'^{-1}, (nn')^{h'}h'^{-1}h)}{\theta(h'^{-1}, n^{h'}h'^{-1}h)} \times \frac{\theta(n^{h'}h'^{-1}h, h'^{h'})}{\theta(n^{h'}h'^{-1}, h'^{h'})}. \tag{3.7}$$

By further specializing $h = n' = e$, we have

LEMMA 3.2. *We have $\theta(n,h) = 1$ for $n \in N, h \in H$.*

We have the following invariance:

LEMMA 3.3. (*i*) $\theta(n, n'h') = \theta(n, n')$, (*ii*) $\theta(nh, h') = \theta(h, h')$.

PROOF. (i) By setting $n_1 = h_1 = e$ in (3.1), we get

$$\theta(n_2, n_3h_3) = \frac{\theta(e, n_2n_3h_2)}{\theta(e, n_2h_2)}.$$

Since the right (resp. left) side does not depend upon $h_3$ (resp. $h_2$), we have the desired result.
(ii) By setting $n_2 = n_3 = e, h_3 = h_2$ in (3.1), we get

$$\theta(n_1h_1, h_2) = \frac{\theta(h_2, h_2)}{\theta(h_1^{-1}h_2, h_1^{-1}h_2)}.$$

Since the right side does not depend upon $n_1$, we are done. □

By setting $n_2 = h_2 = e$ in (3.1), we get

$$\theta(n_1h_1, e)\theta(n_1h_1, n_3h_3)\theta(h_1^{-1}, n_3^{h_1}h_1^{-1}h_3) = \theta(e, n_3h_3)\theta(n_1h_1, n_3).$$

Notice $\theta(n_1h_1, e) = \theta(h_1, e)$ and $\theta(e, n_3h_3) = \theta(e, n_3)$ by Lemma 3.3. Therefore, by changing $n_1, h_1, n_3, h_3$ to $n, h, n', h'$, we have

$$\theta(nh, n'h') = \frac{\theta(nh, n')}{\theta(h, e)} \times \frac{\theta(e, n')}{\theta(h^{-1}, n'^h h^{-1}h')}. \tag{3.8}$$

By setting $n_1 = n_2 = e$ (and $n_3 = n'$) in (3.1), we get

$$\theta(h_1, h_2)\theta(h_1, n'h_3)\theta(h_1^{-1}h_2, n'^{h_1}h_1^{-1}h_3) = \theta(h_2, n'h_3)\theta(h_1, n'h_2), \tag{3.9}$$

or equivalently (by changing $h_2, h_3$ to $h_1h_2, h_1h_2h_3$),

$$\theta(h_1, h_1h_2)\theta(h_1, n'h_1h_2h_3)\theta(h_2, n'^{h_1}h_2h_3) = \theta(h_1h_2, n'h_1h_2h_3)\theta(h_1, n'h_1h_2).$$

By setting $h_1 = h, h_2 = h^{-1}, h_3 = h'$ here, we get

$$\theta(h, e)\theta(h, n'h')\theta(h^{-1}, n'^hh^{-1}h') = \theta(e, n'h')\theta(h, n') = \theta(e, n')\theta(h, n')$$

thanks to Lemma 3.3,(i). Solving this for $\theta(h^{-1}, n'^hh^{-1}h')$ and substituting the result to (3.8), we get

$$\theta(nh, n'h') = \frac{\theta(nh, n')\theta(h, n'h')}{\theta(h, n')}. \tag{3.10}$$

By setting $h_3 = e, n_1 = n_2 = e$ (and $n_3 = n$) in (3.1) (i.e., set $h_3 = e$ in (3.9)), we get

$$\theta(h_1, h_2)\theta(h_1, n)\theta(h_1^{-1}h_2, n^{h_1}h_1^{-1}) = \theta(h_2, n)\theta(h_1, nh_2).$$

Hence, by setting $h = h_1^{-1}h_2, n' = n^{h_1}, h' = h_1^{-1}$, we have

$$\theta(h, n'h') = \frac{\theta(h'^{-1}h, n'^{h'})\theta(h'^{-1}, n'^{h'}h'^{-1}h)}{\theta(h'^{-1}, h'^{-1}h)\theta(h'^{-1}, n'^{h'})}.$$

Substituting this to (3.10), we get

$$\theta(nh, n'h') = \frac{\theta(nh, n')}{\theta(h, n')} \times \frac{\theta(h'^{-1}h, n'^{h'})\theta(h'^{-1}, n'^{h'}h'^{-1}h)}{\theta(h'^{-1}, h'^{-1}h)\theta(h'^{-1}, n'k')}. \tag{3.11}$$

By comparing (3.7)and (3.11), we easily get

$$\frac{\theta(h'^{-1}, (nn')^{h'}h'^{-1}h)}{\theta(h'^{-1}, n^{k'}h'^{-1}h)\theta(h'^{-1}, n'^{h'}h'^{-1}h)}$$
$$= \frac{\theta(nh, n')\theta(n^{h'}h'^{-1}, n'^{h'})}{\theta(n^{h'}h'^{-1}h, n'^{h'})} \times \frac{\theta(h'^{-1}h, n'^{h'})}{\theta(h, n')\theta(h'^{-1}, n'^{h'})} \times \frac{1}{\theta(h'^{-1}, h'^{-1}h)}.$$

Notice that the above right side can be rewritten as

$$\frac{\{\theta(nh, n')\overline{\theta(h, n')}\}\{\theta(n^{h'}h'^{-1}, n'^{h'})\overline{\theta(h'^{-1}, n'^{h'})}\}}{\{\theta(n^{h'}h'^{-1}h, n'^{h'})\overline{\theta(h'^{-1}h, n'^{h'})}\}} \times \frac{1}{\theta(h'^{-1}, h'^{-1}h)}.$$

Recall

$$\begin{aligned} \tilde{\zeta}_n(h, h') &= \theta(h, nhh')\overline{\theta(h, hh')}, \\ \tilde{\eta}_h(n, n') &= \theta(nh, n')\overline{\theta(h, n')}. \end{aligned}$$

Therefore, the right side is actually

$$\frac{\tilde{\eta}_{h'^{-1}}(n^{h'}, n'^{h'})\tilde{\eta}_h(n, n')}{\tilde{\eta}_{h'^{-1}h}(n^{h'}, n'^{h'})} \times \frac{1}{\theta(h'^{-1}, h'^{-1}h)}.$$

On the other hand, the left side together with the above $\theta(h'^{-1}, h'^{-1}h)$ gives us $\tilde{\zeta}$'s. In this way we conclude

$$\frac{\tilde{\zeta}_{(nn')^{h'}}(h'^{-1}, h)}{\tilde{\zeta}_{n^{h'}}(h'^{-1}, h)\tilde{\zeta}_{n'^{h'}}(h'^{-1}, h)} = \frac{\tilde{\eta}_{h'^{-1}}(n^{h'}, n'^{h'})\tilde{\eta}_h(n, n')}{\tilde{\eta}_{h'^{-1}h}(n^{h'}, n'^{h'})}.$$

This is exactly (3.5) because the second factor in the numerator of the right side is $(h'^{-1} \cdot \eta_h)(n^{h'}, n'^{h'})$.

We next prove the cocycle properties (3.2) and (3.3). Recall (3.9) and set $n' = 1$ there:

$$\theta(h_1, h_2)\theta(h_1, h_3)\theta(h_1^{-1}h_2, h_1^{-1}h_3) = \theta(h_2, h_3)\theta(h_1, h_2).$$

Divide (3.9) by the above. It is not difficult to observe that the result (after canceling $\theta(h_1, h_2)$) is exactly the cocycle property (3.3).

We set $h_2 = h_3 = e$ in (3.1) and get

$$\theta(n_1h_1, n_2)\theta(n_1n_2h_1, n_3)\theta(n_2^{h_1}h_1^{-1}, n_3^{h_1}h_1^{-1}) = \theta(n_2, n_3)\theta(n_1h_1, n_2n_3). \tag{3.12}$$

By setting $n_1 = e$ in (3.12) and solving this for the third factor in the left side, we have

$$\theta(n_2^{h_1}h_1^{-1}, n_3^{h_1}h_1^{-1}) = \frac{\theta(n_2, n_3)\theta(h_1, n_2n_3)}{\theta(h_1, n_2)\theta(n_2h_1, n_3)}.$$

By substituting this back to (3.12), we get

$$\theta(n_1h_1, n_2)\theta(n_1n_2h_1, n_3) \times \frac{\theta(n_2, n_3)\theta(h_1, n_2n_3)}{\theta(h_1, n_2)\theta(n_2h_1, n_3)} = \theta(n_2, n_3)\theta(n_1h_1, n_2n_3),$$

or equivalently (by canceling $\theta(n_2, n_3)$),

$$\frac{\theta(n_1h_1, n_2)}{\theta(h_1, n_2)} \times \theta(n_1n_2h_1, n_3) = \frac{\theta(n_1h_1, n_2n_3)}{\theta(h_1, n_2n_3)} \times \theta(n_2h_1, n_3).$$

Dividing the both sides by $\theta(h_1, n_3)$, we get

$$\tilde{\eta}_{h_1}(n_1, n_2)\tilde{\eta}_{h_1}(n_1n_2, n_3) = \tilde{\eta}_{h_1}(n_1, n_2n_3)\tilde{\eta}_{h_1}(n_2, n_3),$$

which is exactly (3.2).

Finally, it is elementary to see that $\tilde{\theta}(h, h') = \theta(h, h')$ satisfies (3.4). (Set $n_1 = n_2 = n_3 = e$ in (3.1)) Hence, we have shown a half of Theorem 3.1.

**Remarks.**
(i) If one further assumes $\theta(h, n'h') = \theta(h, h')$ for $n' \in N$ and $h, h' \in H$ (compare this with the invariance in Lemma 3.3), then $\tilde{\zeta}_n = 1$ from the definition and (3.5) means

$$\tilde{\eta}_h(n, n')(h \cdot \tilde{\eta}_{h'})(n, n') = \tilde{\eta}_{hh'}(n, n').$$

Notice that this is exactly the situation analyzed in the appendix of [**1**]. The above "$H$-equivariant" 2-cocycle easily changes the underlying algebra structure of the dual Kac algebra $\hat{A}$, and for example the non-trivial 8-dimensional Kac algebra ([**38**]) can be captured in this way. (Detailed discussions will be presented in Chapter 7,§2.)
(ii) If one requires $\theta(nh, n') = \theta(h, n')$ for $n, n' \in N$ and $h \in H$ instead, then $\tilde{\eta}_k = 1$ and hence (3.5) means

$$\tilde{\zeta}_{nn'}(h, h') = \tilde{\zeta}_n(h, h')\tilde{\zeta}_{n'}(h, h'),$$

that is, $n \in H \to \tilde{\zeta}_n(h, h') \in \mathbf{T}$ is a character (for each $h, h' \in H$). The presence of this $\tilde{\zeta}_\cdot$ also changes Kac algebra structure. In fact, the effect of the above character appears when one determines the group structure among group-like elements of the Kac algebra $A$, and an abundance of examples will be worked out in Chapter 7.

To show the remaining half of Theorem 3.1, let us assume that $\tilde{\eta}, \tilde{\zeta}, \tilde{\theta}$ (with the cocycle properties (3.2),(3.3),(3.4)) satisfy the relation (3.5). The equation (3.10) means

$$\theta(nh, n'h') = \frac{\theta(nh, n')}{\theta(h, n')} \times \frac{\theta(h, n'h')}{\theta(h, h')} \times \theta(h, h').$$

Motivated by this, we set

$$\theta(nh, n'h') = \tilde{\eta}_h(n, n')\tilde{\zeta}_{n'}(h, h^{-1}h')\tilde{\theta}(h, h').$$

We have to show that this $\theta$ satisfies (3.1). It is straight-forward to see that the "$\tilde{\theta}$-part" alone obviously satisfies (3.1) thanks to (3.4). Therefore, to this end we may and do assume $\tilde{\theta} = 1$.

Notice that we have to show

$$\begin{aligned}
&\tilde{\zeta}_{n_2}(h_1, h_1^{-1}h_2)\tilde{\eta}_{h_1}(n_1, n_2)\tilde{\zeta}_{n_3}(h_1, h_1^{-1}h_3)\tilde{\eta}_{h_1}(n_1n_2, n_3) \\
&\qquad \times \tilde{\zeta}_{n_3^{h_1}}(h_1^{-1}h_2, h_2^{-1}h_3)\tilde{\eta}_{h_1^{-1}h_2}(n_2^{h_1}, n_3^{h_1}) \\
&= \tilde{\zeta}_{n_3}(h_2, h_2^{-1}h_3)\tilde{\eta}_{h_2}(n_2, n_3) \\
&\qquad \times \tilde{\zeta}_{n_2n_3}(h_1, h_1^{-1}h_2)\tilde{\eta}_{h_1}(n_1, n_2n_3).
\end{aligned}$$

The relation (3.5) says

$$\tilde{\zeta}_{n_2n_3}(h_1, h_1^{-1}h_2) = \tilde{\zeta}_{n_2}(h_1, h_1^{-1}h_2)\tilde{\zeta}_{n_3}(h_1, h_1^{-1}h_2) \times \frac{\tilde{\eta}_{h_1}(n_2, n_3)\tilde{\eta}_{h_1^{-1}h_2}(n_2^{h_1}, n_3^{h_1})}{\tilde{\eta}_{h_2}(n_2, n_3)}.$$

Notice the three $\tilde{\eta}_{h_1}$ terms in the former equation (to be checked) correspond to the single $\tilde{\eta}_{h_1}$ term in the latter thanks to the cocycle property (3.2). Also notice that $\tilde{\eta}_{h_2}, \tilde{\eta}_{h_1^{-1}h_2}$ and $\tilde{\zeta}_{n_2}$ appear in common in the both equations. Hence, by finding $\tilde{\eta}_{h_2h_3}$ from the former equation and comparing the result with the latter, we observe that the following relation should be checked:

$$\tilde{\zeta}_{n_3}(h_1, h_1^{-1}h_3) \times \frac{\tilde{\zeta}_{n_3^{h_1}}(h_1^{-1}h_2, h_2^{-1}h_3)}{\tilde{\zeta}_{n_3}(h_2, h_2^{-1}h_3)} = \tilde{\zeta}_{n_3}(h_1, h_1^{-1}h_2).$$

However, this is just the cocycle property (3.3) for $\tilde{\zeta}_\cdot$. Therefore, we have shown (3.1), and Theorem 3.1 has been proved.

## 2. Cohomology group

Let $a(\cdot)$ be a $\mathbf{T}$-valued function on $G$. Of course the multiplication operator $(U\xi)(g) = a(g)\xi(g)$ is a unitary operator on $\ell^2(G)$, and we compute

$$\begin{aligned}
((U \otimes U)V^\theta(U \otimes U)^*\xi)(g_1, g_2) &= a(g_1)a(g_2)(V^\theta(U \otimes U)^*)\xi)(g_1, g_2) \\
&= \theta(g_1, g_2)a(g_1)a(g_2)((U \otimes U)^*\xi)(v(g_1, g_2)) \\
&= \theta(g_1, g_2)t(g_1, g_2)\overline{t(v(g_1, g_2))}\xi(v(g_1, g_2))
\end{aligned}$$

with $t(g_1, g_2) = a(g_1)a(g_2)$. Obviously $(U \otimes U)V^\theta(U \otimes U)^*$ is a multiplicative unitary, and we have $(U \otimes U)V^\theta(U \otimes U)^* = V^{\theta'}$ with

$$\theta'(g_1, g_2) = \theta(g_1, g_2)t(g_1, g_2)\overline{t(v(g_1, g_2))}.$$

The unitary $U \otimes U$ coming from $a(\cdot)$ of course does not change relevant Kac algebras, and we get the notion of a coboundary in the present setting. More precisely, a cocycle (in the sense of (3.1)) of the form

$$\theta'(n_1h_1, n_2h_2) = a(n_1h_1)a(n_2h_2)\overline{a(n_1n_2h_1)a(n_2^{h_1}h_1^{-1}h_2)} \tag{3.13}$$

(for some $\mathbf{T}$-valued function $a(\cdot)$ on $G$) is called a coboundary.

Here is a normalization result whose proof is quite standard and will be given in Appendix §3 at the end of the chapter.

PROPOSITION 3.4. *Up to a coboundary we can assume that our cocycle satisfies*

$$\tilde{\theta}(h,h') = 1,$$
$$\tilde{\eta}_e(n,n') = \tilde{\eta}_h(n,e) = \tilde{\eta}_h(e,n') = \tilde{\eta}_h(n,n^{-1}) = 1,$$
$$\tilde{\zeta}_e(h,h') = \tilde{\zeta}_n(h,e) = \tilde{\zeta}_n(e,h') = 1$$

*for each $n, n' \in N$ and $h, h' \in H$.*

Let $Z^2$ be the cocycles of the multiplicative unitary $V$ (i.e., $\theta$'s satisfying (3.1)), and $B^2$ be the coboundaries defined by (3.13). We define

DEFINITION 3.5.

$$H^2_V((N,H),\mathbf{T}) = Z^2/B^2.$$

THEOREM 3.6. *The cohomology group $H^2_V((N,H),\mathbf{T})$ arising from $V$ in Definition 3.5 is isomorphic to $H^2((N,H),\mathbf{T})$ in Definition 2.3.*

PROOF. For $(\eta,\zeta) \in Z^2((N,H),\mathbf{T})$, we define $\theta = \pi(\eta,\zeta)$ by

$$\theta(nh,n'h') = \eta_h(n,n')\zeta_{n'}(h,h^{-1}h').$$

We already know that $\theta$ satisfies (3.1) (the second half of the proof of Theorem 3.1), and hence $\pi$ is a map from $Z^2((N,H),\mathbf{T})$ to $Z^2$. When $(\eta,\zeta) \in B^2((N,K),\mathbf{T})$ (i.e., $\eta_h = (\partial^N\xi)_h, \zeta_n = (\partial^H\xi)_n$ for some $\mathbf{T}$-valued function $\xi$ on $H \times N$), we compute the corresponding $\theta$ as follows:

$$\begin{aligned}\theta(nh,n'h') &= \xi(h,n)\xi(h,n')\overline{\xi(h,nn')} \times \xi(h',n')\overline{\xi(h,n')\xi(h^{-1}h',n'^h)} \\ &= \xi(h,n)\overline{\xi(h,nn')}\xi(h',n')\overline{\xi(h^{-1}h',n'^h)}.\end{aligned}$$

This is a coboundary in $B^2$ (with $a(nh) = \xi(h,n)$). Thus, $\pi(B^2((N,H),\mathbf{T})) \subseteq B^2$, and we have the induced map $[\pi]: H^2((N,H),\mathbf{T}) \to Z^2/B^2$.

The map $[\pi]$ is surjective. In fact, for a given $\theta' \in Z^2$, we may and do assume $\theta'(h,h') = 1$ up to a coboundary in $B^2$ thanks to Proposition 3.4. We set

$$\begin{aligned}\eta_h(n,n') &= \theta'(nh,n')\overline{\theta'(h,n')}, \\ \zeta_n(h,h') &= \theta'(h,nhh').\end{aligned}$$

Then, $(\eta,\zeta)$ belongs to $Z^2((H,K),\mathbf{T})$ (the first half of the proof of Theorem 3.1), and $\theta = \pi(\eta,\zeta)$ is computed by

$$\begin{aligned}\theta(nh,n'h') &= \eta_h(n,n')\zeta_{n'}(h,h^{-1}h') \\ &= \theta'(nh,n')\overline{\theta'(h,n')}\theta'(h,n'h') \\ &= \theta'(nh,n'h')\end{aligned}$$

thanks to (3.10).

It remains to show the injectivity of $[\pi]$. Take $[(\eta,\zeta)] \in H^2((N,H),\mathbf{T})$ from the kernel of $[\pi]$. We may and do assume that $(\eta,\zeta)$ is a normalized cocycle. Then $\theta = \pi(\eta,\zeta)$ satisfies

$$\theta(h,h') = \eta_h(e,e)\zeta_e(h,h^{-1}h') = 1. \tag{3.14}$$

Also notice

$$\theta(nh,n')\overline{\theta(h,n')} = \eta_h(n,n')\zeta_{n'}(h,h^{-1})\overline{\eta_h(e,n')\zeta_{n'}(h,h^{-1})} = \eta_h(n,n'), \tag{3.15}$$

$$\theta(h,nhh') = \eta_h(e,n)\zeta_n(h,h') = \zeta_n(h,h'). \tag{3.16}$$

Since $[(\eta,\zeta)]$ is in the kernel of $[\pi]$, there exists a $\mathbf{T}$-valued function $a(\cdot)$ on $G$ satisfying

$$\theta(n_1h_1,n_2h_2) = a(n_1h_1)a(n_2h_2)\overline{a(n_1n_2h_1)a(n_2^{h_1}h_1^{-1}h_2)}.$$

Since $\theta(h_1,h_2) = a(h_2)\overline{a(h_1^{-1}h_2)}$, (3.14) means that $a(h)(= e^{i\theta_0})$ is independent of $h \in H$. Notice that (3.15),(3.16) imply

$$\begin{aligned}
\eta_h(n_1,n_2) &= a(n_1h)a(n_2)\overline{a(n_1n_2h)a(n_2^hh^{-1})} \times \overline{a(h)a(n_2)}a(n_2h)a(n_2^hh^{-1}) \\
&= e^{-i\theta_0}a(n_1h)a(n_2h)\overline{a(n_1n_2h)}, \\
\zeta_n(h_1,h_2) &= a(h_1)a(nh_1h_2)\overline{a(nh_1)a(n^{h_1}h_2)} \\
&= e^{i\theta_0}a(nh_1h_2)\overline{a(nh_1)a(n^{h_1}h_2)}.
\end{aligned}$$

Therefore, $\eta = \partial^N\xi, \zeta = \partial^H\xi$ with $\xi(h,n) = e^{-i\theta_0}a(nh)$, and hence $(\eta,\zeta)$ falls into $B^2((N,H),\mathbf{T})$, i.e., $[(\eta,\zeta)] = 0 \in H^2((H,K),\mathbf{T})$. □

## 3. Appendix Normalization of cocycles

We prove Proposition 3.4 here. At first we notice

$$\begin{aligned}
\tilde{\eta}_h(n,e) &= \theta(nh,e)\overline{\theta(h,e)} = 1 \quad \text{(by Lemma 3.3, (ii))}, \\
\tilde{\eta}_h(e,n) &= \theta(h,n)\overline{\theta(h,n)} = 1, \\
\tilde{\zeta}_e(h_1,h_2) &= \theta(h_1,h_1h_2)\overline{\theta(h_1,h_1h_2)} = 1.
\end{aligned}$$

By using the first two equations and (3.2), we immediately get

$$\tilde{\eta}_h(n,n^{-1}) = \tilde{\eta}_h(n^{-1},n).$$

By setting $n_1 = n_2 = n_3 = e$ and $h_1 = h_2 = h_3 = e$ in the cocycle equation (3.1), we get

$$\begin{aligned}
\theta(h_1,h_3)\theta(h_1^{-1}h_2,h_1^{-1}h_3) &= \theta(h_2,h_3), \\
\theta(n_1,n_2)\theta(n_1n_2,n_3) &= \theta(n_1,n_2n_3).
\end{aligned}$$

Therefore, we get

$$\theta(n,n') = \frac{\theta(e,nn')}{\theta(e,n)}, \tag{3.17}$$

$$\theta(h,h') = \frac{\theta(h'^{-1}h,e)}{\theta(h'^{-1},e)}, \tag{3.18}$$

and in particular we have $\theta(n,e) = \theta(e,h) = 1$.

With $a(nh) = \overline{\theta(h^{-1},e)}$ the coboundary $\theta'(h_1 \cdot k_1, h_2 \cdot k_2)$ (see (3.13)) becomes

$$\overline{\theta(h_1^{-1},e)\theta(h_2^{-1},e)}\theta(h_1^{-1},e)\theta(h_2^{-1}h_1,e) = \frac{\theta(h_2^{-1}h_1,e)}{\theta(h_2^{-1},e)} = \theta(h_1,h_2)$$

by (3.18). Therefore, up to a coboundary we can assume $\tilde{\theta}(h_1,h_2) = \theta(h_1,h_2) = 1$. On the other hand, with $a(nh) = \overline{\theta(1,n)}$, we get

$$\theta'(n_1,n_2) = \overline{\theta(e,n_1)\theta(e,n_2)}\theta(e,n_1n_2)\theta(e,n_2) = \theta(n_1,n_2)$$

thanks to (3.17). It is straight-forward to see that this $\theta'$ (arising from $a(bh) = \overline{\theta(1,n)}$) still satisfies $\theta'(h_1, h_2) = 1$.

The arguments so far mean that up to a coboundary we may and do assume the invariance

$$\theta(n, n') = \theta(h, h') = 1.$$

In particular, we have

$$\tilde{\eta}_e(n_1, n_2) = 1 \quad \text{and} \quad \tilde{\zeta}_n(e, e) = 1.$$

From the cocycle equation (3.3), we observe

$$\begin{aligned} \tilde{\zeta}_n(e, h) &= \tilde{\zeta}_n(e, e), \\ \tilde{\zeta}_n(hh', e) &= \tilde{\zeta}_{n^h}(h', e), \\ \tilde{\zeta}_{n^{h^{-1}}}(h, h^{-1}) &= \tilde{\zeta}_n(h^{-1}, h) \times \frac{\tilde{\zeta}_n(e, h^{-1})}{\tilde{\zeta}_n(h^{-1}, e)}. \end{aligned}$$

(set $h_1 = e, h_3 = e$ and $h_1^{-1} = h_2 = h_3^{-1} = h$ respectively in (3.3)) From the first two equations and $\tilde{\zeta}_n(e, e) = 1$, we get

$$\tilde{\zeta}_n(h, e) = \tilde{\zeta}_n(e, h) = 1,$$

and the third equation means

$$\tilde{\zeta}_{n^{h^{-1}}}(h, h^{-1}) = \tilde{\zeta}_n(h^{-1}, h). \tag{3.19}$$

Let $a(nh)$ $(= a(n^{-1}h))$ be one of square roots of $\tilde{\eta}_h(n, n^{-1}) = \tilde{\eta}_h(n^{-1}, n)$ and $\theta'(n_1h_1, n_2h_2)$ be the coboundary defined by (3.13). Then, the $\tilde{\eta}$-cocyle

$$\tilde{\eta}'_h(n, n') = \theta'(nh, n')\overline{\theta'(h, n')}$$

arising from this $\theta'$ satisfies

$$\begin{aligned} \tilde{\eta}'_h(n, n^{-1}) &= \theta'(nh, n^{-1})\overline{\theta'(h, n^{-1})} \\ &= a(nh)a(n^{-1})\overline{a(h)a((n^{-1})^h h^{-1})} \\ &\qquad\qquad \times \overline{a(h)a(n^{-1})}a(n^{-1}h)a((n^{-1})^h h^{-1}) \\ &= a(nh)^2 \times \overline{a(h)^2}. \end{aligned}$$

But, since $a(h) = \sqrt{\tilde{\eta}_h(e, e)} = 1$, we conclude

$$\tilde{\eta}'_h(n, n^{-1}) = a(nh)^2 = \tilde{\eta}_h(n, n^{-1}).$$

Note that we still have $\theta'(n, n') = \theta'(h, h') = 1$ because of $\tilde{\eta}_h(e, e) = \tilde{\eta}_e(h, h') = 1$. Hence, (up to a coboundary) we can further assume

$$\tilde{\eta}_h(n, n^{-1}) = 1.$$

## 4. Appendix Multiplicative unitary $V^\theta$

We present a slightly different expression for the multiplicative unitary $V^\theta$, and for convenience we assume that our cocycle is a normalized one in Proposition 3.4.

For $f \in \ell^\infty(N)$ and $h \in H$, we define the operator $A_{f,h}$ (acting on $\ell^2(G)$) by

$$(A_{f,h}\xi)(n'h') = f(n')\tilde{\zeta}_{n'}(h, h^{-1}h')\xi(n'^h h^{-1}h')$$

with $\xi \in \ell^2(G)$ and $n'h' \in G$. Also, for $n \in N$ and $h \in H$, we define the operator $B_{n,h}$ (acting on $\ell^2(G)$) by

$$(B_{n,h}\xi)(n'h') = \delta_{h,h'}\tilde{\eta}_h(n', n)\xi(n'nh')$$

($\delta_{h,h'}$ is the Kronecker symbol). Notice $(A_{f,h}\xi)(n'hh') = f(n')\tilde{\zeta}_{n'}(h,h')\xi(n'^h h')$. Therefore, in terms of matrix units $E_{g_1,g_2}$, we have the following expression:

$$A_{f,h} = \sum_{n',h'} f(n')\tilde{\zeta}_{n'}(h,h')E_{n'hh',n'^h h'}.$$

Similarly, we have

$$B_{n,h} = \sum_{n',h'} \delta_{h,h'}\tilde{\eta}_h(n',n)E_{n'h',n'nh'} = \sum_{n'} \tilde{\eta}_h(n',n)E_{n'h,n'nh}.$$

Let $\delta_n \in \ell^\infty(N)$ be the delta function. For a vector $\xi \in \ell^2(G\times G) = \ell^2(G)\otimes \ell^2(G)$ we compute

$$\begin{aligned}
&\left(\left(\sum_{n\in N,h\in H} B_{n,h}\otimes A_{\delta_n,h}\right)\xi\right)(n_1h_1,n_2h_2)\\
&= \sum_{n\in N,h\in H} \delta_{h,h_1}\tilde{\eta}_h(n_1,n)\delta_n(n_2)\tilde{\zeta}_{n_2}(h,h^{-1}h_2)\xi(n_1nh_1,n_2^h h^{-1}h_2)\\
&= \tilde{\eta}_{h_1}(n_1,n_2)\tilde{\zeta}_{n_2}(h_1,h_1^{-1}h_2)\xi(n_1n_2h_1,n_2^{h_1}h_1^{-1}h_2).
\end{aligned}$$

Therefore, (3.6) shows that the multiplicative unitary $V^\theta$ is given by

$$V^\theta = \sum_{n\in N,h\in H} B_{n,h}\otimes A_{\delta_n,h}.$$

Let us check the product rules for $A$'s and $B$'s. We compute

$$\begin{aligned}
(A_{f,h_1}A_{g,h_2}\xi)(nh) &= f(n)\tilde{\zeta}_n(h_1,h_1^{-1}h)(A_{g,h_2}\xi)(n^{h_1}h_1^{-1}h)\\
&= f(n)\tilde{\zeta}_n(h_1,h_1^{-1}h)\\
&\qquad\times g(n^{h_1})\tilde{\zeta}_{n^{h_1}}(h_2,h_2^{-1}h_1^{-1}h)\times\xi(h^{h_1h_2}(h_1h_2)^{-1}h)\\
&= \tilde{\zeta}_n(h_1,h_2)(A_{fg^{h_1^{-1}},h_1h_2}\xi)(nh).
\end{aligned}$$

Here, $g^{h_1^{-1}} = g(h_1^{-1}\cdot h_1) \in \ell^\infty(N)$, and $fg^{h_1^{-1}}$ means the product in the abelian algebra $\ell^\infty(N)$. In fact, (3.3) shows

$$\tilde{\zeta}_n(h_1,h_1^{-1}h)\tilde{\zeta}_{n^{h_1}}(h_2,h_2^{-1}h_1^{-1}h) = \tilde{\zeta}_n(h_1,h_2)\tilde{\zeta}_n((h_1h_2),(h_1h_2)^{-1}h).$$

Thus, we have shown

$$A_{f,h_1}A_{g,h_2} = A_{\tilde{\zeta}.(h_1,h_2)fg^{h_1^{-1}},h_1h_2}$$

with the product $\tilde{\zeta}.(h_1,h_2)fg^{h_1^{-1}} \in \ell^\infty(N)$.

On the other hand, due to the presence of the Kronecker symbol in the definition of $B$, we easily see $B_{n_1,h_1}B_{n_2,h_2} = 0$ unless $h_1 = h_2$. When $h_1 = h_2$, we compute

$$\begin{aligned}
(B_{n_1,h_1}B_{n_2,h_1}\xi)(n'h') &= \delta_{h_1,h'}\tilde{\eta}_{h_1}(n',n_1)(B_{n_2,h_1}\xi)(n'n_1h')\\
&= \delta_{h_1,h'}\tilde{\eta}_{h_1}(n',n_1)\tilde{\eta}_{h_1}(n'n_1,n_2)\xi(n'n_1n_2h')\\
&= \delta_{h_1,h'}\tilde{\eta}_{h_1}(n_1,n_2)\tilde{\eta}_{h_1}(n',n_1n_2)\xi(n'n_1n_2h')\\
&= \tilde{\eta}_{h_1}(n_1,n_2)(B_{n_1n_2,h_1}\xi)(n'h'),
\end{aligned}$$

where the third equation comes from the cocycle property (3.3) for $\tilde{\eta}$. Computations so far show the following product rule for $B$'s:

$$B_{n_1,h_1}B_{n_2,h_2} = \delta_{h_1,h_2}\eta_{h_1}(n_1,n_2)B_{n_1n_2,h_1}.$$

We now determine the adjoints of $A_{f,h}$ and $B_{n,h}$. It is straight-forward to check $A^*_{f,h} = A_{\tilde{f}_h,h^{-1}}$ with $\tilde{f}_h(n) = \overline{\tilde{\zeta}_n(h^{-1},h)f^h(n)}$. In fact, we compute

$$\begin{aligned}(\xi, A_{f,h_1}\xi') &= \sum_{n,h} \xi(nh)\overline{f(n)\tilde{\zeta}_n(h_1,h_1^{-1}h)\xi'(n^{h_1}h_1^{-1}h)} \\ &= \sum_{n,h} \xi(n^{h_1^{-1}}h_1h)\overline{f(n^{h_1^{-1}})\tilde{\zeta}_{n^{h_1^{-1}}}(h_1,h)\xi'(nh)}\end{aligned}$$

by changing at first $h$ to $h_1h$ and then $n$ to $n^{h_1^{-1}}$. We also have

$$\overline{\tilde{\zeta}_{n^{h_1^{-1}}}(h_1,h)} = \tilde{\zeta}_n(h_1^{-1},h_1h)\overline{\tilde{\zeta}_n(e,h)\tilde{\zeta}_n(h_1^{-1},h_1)} = \tilde{\zeta}_n(h_1^{-1},h_1h)\overline{\tilde{\zeta}_n(h_1^{-1},h_1)}$$

by (3.3), and the above formula for $A^*_{f,h}$ is now obvious.

We have $A^*_{f,e} = A_{\overline{f},e}$ and each $A_{1,h}$ ($1 \in \ell^\infty(N)$ is the identity function) is a unitary. It is also straight-forward to check the following covariance:

$$A_{1,h}A_{f,e}A^*_{1,h} = A_{f^{h^{-1}},e} \quad A^*_{1,h}A_{f,e}A_{1,h} = A_{f^h,e}.$$

For example, $A^*_{1,h} = A_{g,h^{-1}}$ with $g(n) = \overline{\tilde{\zeta}_n(h^{-1},h)}$, and hence $A_{1,h}A^*_{1,h} = A_{f,e}$ with

$$f(n) = \tilde{\zeta}_n(h,h^{-1})g^{h^{-1}}(n) = \tilde{\zeta}_n(h,h^{-1})\overline{\tilde{\zeta}_{n^h}(h^{-1},h)} = 1$$

thanks to (3.19), which shows that $A_{1,h}$ is indeed a unitary.

We can check $B^*_{n,h} = B_{n^{-1},h}$. In fact, we compute

$$\begin{aligned}(\xi, B_{n,h}\xi') &= \sum_{n',h'} \xi(n'h')\delta_h(h')\overline{\tilde{\eta}_h(n',n)\xi'(n'nh')} \\ &= \sum_{n',h'} \xi(n'n^{-1}h')\delta_h(h')\overline{\tilde{\eta}_h(n'n^{-1},n)\xi'(n'h')}\end{aligned}$$

by changing $n'$ to $n'n^{-1}$. Hence, we have

$$\begin{aligned}(B^*_{n,h}\xi)(n'h') &= \delta_h(h')\overline{\tilde{\eta}_h(n'n^{-1},n)}\xi(n'n^{-1}h') \\ &= \overline{\tilde{\eta}_h(n'n^{-1},n)\tilde{\eta}_h(n',n^{-1})} \times \delta_h(h')\tilde{\eta}_h(n',n^{-1})\xi(n'n^{-1}h') \\ &= \overline{\tilde{\eta}_h(n'n^{-1},n)\tilde{\eta}_h(n',n^{-1})}(B_{n^{-1},h}\xi)(n'h').\end{aligned}$$

However, the cocycle equation (3.2) shows

$$\tilde{\eta}_h(n'n^{-1},n)\tilde{\eta}_h(n',n^{-1}) = \tilde{\eta}_h(n',e)\tilde{\eta}_h(n^{-1},n) = 1.$$

The discussions so far show that the product rules for $A$'s and $B$'s are exactly the ones in a twisted crossed product ([**76**]) and a twisted group ring.

CHAPTER 4

# Multiplicative unitary

Let $\alpha$ be an injective kernel of a semi-direct product $G = N \rtimes H$ on a properly infinite factor $\mathcal{R}$ such that its restrictions to the subgroups $N$ and $H$ are actions and they are related by

$$\alpha_h \alpha_{n^h} \alpha_{h^{-1}} = Ad(u(h,n))\alpha_n$$

with $n^h = h^{-1}nh$ (see the beginning of Chapter 2).

In this chapter we capture the multiplicative unitary of the relevant Kac algebra via sector technique (Theorem 4.6), and our basic references for this method are [**25, 46, 47**]. (A more general case will be considered in Chapter 12.) We fix endomorphisms $\rho_1$ and $\rho_2$ of $\mathcal{R}$ satisfying

$$\mathcal{R}^{(N,\alpha)} = \rho_1(\mathcal{R}) \quad \text{and} \quad \mathcal{R}^{(H,\alpha)} = \rho_2(\mathcal{R})$$

respectively. We set

$$\sigma = \bar{\rho}_1 \rho_2$$

so that $\mathcal{R} \supseteq \sigma(\mathcal{R})$ is conjugate to the irreducible depth 2 inclusion $\mathcal{R} \rtimes_\alpha H \supseteq \mathcal{R}^{(N,\alpha)}$ of index $\#N \times \#H = \#G$.

Let us briefly recall what was done in [**29, 48**] (see also [**8**]). We consider the following space of intertwiners:

$$\mathcal{H} = (\sigma, \sigma\bar{\sigma}\sigma).$$

This is a $\#G$-dimensional Hilbert space in $\mathcal{R}$ due to the depth 2 condition. Let $\{W_i\}$ be an orthonormal basis consisting of isometries. Then,

$$R = \sum_i \sigma\bar{\sigma}(W_i)W_i^*$$

belongs to $\mathcal{H}^2\mathcal{H}^{*2}$ ($\cong B(\mathcal{H}) \otimes B(\mathcal{H})$) and the composition $V = RF$ with the the flip $F = \sum_{i,j} W_i W_j W_i^* W_j^*$ is a multiplicative unitary (acting on $\mathcal{H}$). Thus, we get a Kac algebra $\mathcal{A}$ (together with the natural action of the dual Kac algebra $\hat{\mathcal{A}}$ on $\mathcal{R}$), and it is known that the subfactor $\sigma(\mathcal{R})$ is the fixed-point algebra by this action.

## 1. Preliminaries

LEMMA 4.1. *There exist isometries $\{S_n\}_{n\in N}$ and $\{T_h\}_{h\in H}$ in $\mathcal{R}$ satisfying*

$$\rho_1\bar{\rho}_1 = \sum_{n\in N} S_n \alpha_n(x) S_n^*,$$

$$\rho_2\bar{\rho}_2 = \sum_{h\in H} T_h \alpha_h(x) T_h^*,$$

*and* $\alpha_{n_1}(S_{n_2}) = S_{n_1 n_2}$, $\alpha_{h_1}(T_{h_2}) = T_{h_1 h_2}$.

PROOF. We choose an isometry $S_e$ from the one-dimensional space $(id, \rho_1\bar{\rho}_1)$ of intertwiners, and set

$$S_n = \alpha_n(S_e)$$

(so that $\alpha_{n_1}(S_{n_2}) = S_{n_1 n_2}$). Since $S_e x = \rho_1\bar{\rho}_1(x)S_e$, by applying $\alpha_n$ to the both sides we see

$$S_n\alpha_n(x) = \alpha_n\rho_1\bar{\rho}_1(x)S_n = \rho_1\bar{\rho}_1(x)S_n,$$

i.e., $S_n \in (\alpha_n, \rho_1\bar{\rho}_1)$, because $\rho_1\bar{\rho}_1(x)$ falls into $\mathcal{R}^{(N,\alpha)}$. Since $\rho_1\bar{\rho}_1 = \oplus_n\alpha_n$, we have $\sum_{n\in N} S_nS_n^* = 1$. Isometries $\{T_h\}_{h\in H}$ can be constructed in the analogous way. □

LEMMA 4.2. *There exist a unitary representation $\{\lambda_n\}_{n\in N}$ in $\mathcal{R}$ satisfying the relations*

$$\bar{\rho}_1\alpha_n = Ad(\lambda_n)\bar{\rho}_1 \text{ and } \rho_1(\lambda_{n_1^{-1}})S_{n_2} = S_{n_2 n_1}.$$

PROOF. Since $\mathcal{R}$ is a crossed product of $\bar{\rho}_1(\mathcal{R})$ by an $N$-action, one finds unitaries $\lambda_n$ in $\mathcal{R}$ satisfying $Ad(\lambda_n)\bar{\rho}_1 = \bar{\rho}_1\alpha_n$. In fact, we have $\mathcal{R} = \bar{\rho}_1(\mathcal{R}) \rtimes N$, and let $\lambda_n$'s be the canonical unitaries in $\mathcal{R}$ corresponding to the $N$-action on $\bar{\rho}_1(\mathcal{R})$. Since $Ad(\lambda_n)\bar{\rho}_1(\mathcal{R}) = \bar{\rho}_1(\mathcal{R})$, one finds an automorphism $\theta \in Aut(\mathcal{R})$ with $Ad(\lambda_n)\bar{\rho}_1 = \bar{\rho}_1\theta$ as an endomorphism. By applying $\rho_1$ to the both sides, we see that $\theta = \alpha_{n'}$ (up to inner perturbation) for some $n'$. From $\alpha_{n'} = \bar{\rho}_1^{-1}Ad(\lambda_n)\bar{\rho}_1$, we see that $n'$ is uniquely determined by $n$. In this way we have the map: $n \to n'$, and it is obviously an injective homomorphism of $N$. By changing indices and adjusting $\lambda_n$ by a unitary in $\bar{\rho}_1(\mathcal{R})$ if necessary, we get $Ad(\lambda_n)\bar{\rho}_1 = \bar{\rho}_1\alpha_n$ as an endomorphism.

For each $n \in N$, one computes

$$\begin{aligned}
\rho_1(\lambda_{n^{-1}})S_e\alpha_n(x) &= \rho_1(\lambda_{n^{-1}})\rho_1\bar{\rho}_1\alpha_n(x)S_e \\
&= \rho_1\left(\lambda_{n^{-1}}\bar{\rho}_1\alpha_n(x)\right)S_e \\
&= \rho_1\left(Ad(\lambda_{n^{-1}})\bar{\rho}_1\alpha_n(x)\lambda_{n^{-1}}\right)S_e \\
&= \rho_1\left(\bar{\rho}_1(x)\lambda_{n^{-1}}\right)S_e \\
&= \rho_1\bar{\rho}_1(x)\rho_1(\lambda_{n^{-1}})S_e.
\end{aligned}$$

Therefore, $\rho_1(\lambda_{n^{-1}})S_e$ belongs to the one-dimensional space $(\alpha_n, \rho_1\bar{\rho}_1)$, and (after changing a phase) we may and do assume that $\lambda_n$ satisfies $\rho_1(\lambda_{n^{-1}})S_e = S_n$. We have

$$S_{n_1n_2} = \alpha_{n_1}(S_{n_2}) = \alpha_{n_1}\left(\rho_1(\lambda_{n_2^{-1}}\right)S_e) = \rho_1(\lambda_{n_2^{-1}})S_{n_1}$$

because of $\mathcal{R}^{(N,\alpha)} = \rho_1(\mathcal{R})$. It remains to show that $n \to \lambda_n$ is a representation, but it follows from

$$\begin{aligned}
&\rho_1(\lambda_{(n_1n_2)^{-1}})S_e = S_{n_1n_2} \\
&\quad = \rho_1(\lambda_{n_2^{-1}})S_{n_1} = \rho_1(\lambda_{n_2^{-1}})\rho_1(\lambda_{n_1^{-1}})S_e \in (\alpha_{n_1n_2}, \rho_1\bar{\rho}_1).
\end{aligned}$$

□

## 2. Orthonormal basis and multiplicative unitary

To know the multiplicative unitary $V$, we at first compute $\sigma\bar{\sigma}\rho_1 = \bar{\rho}_1\rho_2\bar{\rho}_2\rho_1\bar{\rho}_1$. By Lemma 4.1 and Lemma 4.2, we have

$$\begin{aligned}
\sigma\bar{\sigma}\bar{\rho}_1(x) &= \sum_{h,n} \bar{\rho}_1\left(T_h\alpha_h(S_n\alpha_n(x)S_n^*)T_h^*\right) \\
&= \sum_{h,n} \bar{\rho}_1\left(T_h\alpha_h(S_n)\alpha_h\alpha_n(x)\alpha_h(S_n^*)T_h^*\right) \\
&= \sum_{h,n} \bar{\rho}_1\left(T_h\alpha_h(S_n)Ad(u(h,n^{h^{-1}}))\alpha_{n^{h^{-1}}}\alpha_h(x)\alpha_h(S_n^*)T_h^*\right) \\
&= \sum_{h,n} \bar{\rho}_1\left(T_h\alpha_h(S_{n^h})u(h,n)\right) \times \bar{\rho}_1\alpha_n\alpha_h(x) \times \bar{\rho}_1\left(u(h,n)^*\alpha_h(S_{n^h}^*)T_h^*\right) \\
&= \sum_{h,n} \bar{\rho}_1\left(T_h\alpha_h(S_{n^h})u(h,n)\right)\lambda_n \times \bar{\rho}_1\alpha_h(x) \\
&\qquad\qquad \times\lambda_n^*\bar{\rho}_1\left(u(h,n)^*\alpha_h(S_{n^h}^*)T_h^*\right).
\end{aligned}$$

Here, on the fourth line we have changed $n$ to $n^h$. For each $n \in N$ and $h \in H$, we set

$$W(n,h) = \bar{\rho}_1\left(T_h\alpha_h(S_{n^h})u(h,n)\right)\lambda_n,$$

which is obviously an isometry in $\mathcal{R}$. The computations so far mean

$$\sigma\bar{\sigma}\bar{\rho}_1(x) = \sum_{h,n} W(n,h)\bar{\rho}_1\alpha_h(x)W(n,h)^*. \tag{4.1}$$

Observe that $W(n,h)$'s have orthogonal ranges summing up to 1. By changing $x$ to $\rho_2(x)$ in the above equation (note that $\bar{\rho}_1\alpha_h\rho_2 = \bar{\rho}_1\rho_2 = \sigma$ because of $\mathcal{R}^{(H,\alpha)} = \rho_2(\mathcal{R})$), we have

PROPOSITION 4.3. *The family $\{W(n,h)\}_{(n,h)\in N\times H}$ of isometries defined above is an orthonormal basis for the Hilbert space $\mathcal{H} = (\sigma, \sigma\bar{\sigma}\sigma)$.*

We next compute $\sigma\bar{\sigma}(W(n,h))$. This is equal to

$$\begin{aligned}
&\sigma\bar{\sigma}\bar{\rho}_1(T_h\alpha_h(S_{n^h})u(h,n)) \times \sigma\bar{\sigma}(\lambda_n) \\
&\quad= \sum_{m,k} W(m,k)\bar{\rho}_1\alpha_k\left(T_h\alpha_h(S_{n^h})u(h,n)\right)W(m,k)^* \times \sigma\bar{\sigma}(\lambda_n) \quad \text{(by (4.1))} \\
&\quad= \sum_{m,k} W(m,k)\bar{\rho}_1\alpha_k\left(T_h\alpha_h(S_{n^h})u(h,n)\right) \\
&\qquad\qquad \times \lambda_m^*\bar{\rho}_1\left(u(k,m)^*\alpha_k(S_{m^k}^*)T_k^*\rho_2\bar{\rho}_2\rho_1(\lambda_n)\right) \\
&\qquad\qquad \text{(by the definition of } W(m,k)^* \text{ and } \sigma\bar{\sigma} = \bar{\rho}_1\rho_2\bar{\rho}_2\rho_1) \\
&\quad= \sum_{m,k} W(m,k)\bar{\rho}_1\left(T_{kh}\alpha_{kh}(S_{n^h})\alpha_k(u(h,n))\right) \\
&\quad\quad \times \lambda_m^*\bar{\rho}_1\left(u(k,m)^*\alpha_k(S_{m^k}^*\rho_1(\lambda_n))T_k^*\right) \quad \text{(by } T_k^*\rho_2\bar{\rho}_2(x) = \alpha_k(x)T_k^*).
\end{aligned}$$

Notice

$$W(n^{k^{-1}},kh) = \bar{\rho}_1(T_{kh}\alpha_{kh}(S_{n^h})u(kh,n^{k^{-1}}))\lambda_{n^{k^{-1}}} \text{ and } \rho_1(\lambda_{n^{-1}})S_{m^k} = S_{m^k n}.$$

Thus, we further compute

$$\begin{aligned}
&\sigma\bar{\sigma}(W(n,h)) \\
&= \sum_{m,k} W(m,k)W(n^{k^{-1}},kh)\lambda^*_{n^{k^{-1}}}\bar{\rho}_1\left(u(kh,n^{k^{-1}})^*\alpha_k(u(h,n))\right) \\
&\qquad \times \lambda^*_m\bar{\rho}_1\left(u(k,m)^*\alpha_k(S^*_{m^kn})T^*_k\right) \\
&= \sum_{m,k} \zeta_{n^{k-1}}(k,h)W(m,k)W(n^{k^{-1}},kh)\lambda^*_{n^{k-1}}\bar{\rho}_1\left(u(k,n^{k^{-1}})\right) \\
&\qquad \times \lambda^*_m\bar{\rho}_1\left(u(k,m)^*\alpha_k(S^*_{m^kn})T^*_k\right) \quad \text{(by the definition of } \zeta) \\
&= \sum_{m,k} \zeta_{n^{k-1}}(k,h)W(m,k)W(n^{k^{-1}},kh)\lambda^*_{mn^{k-1}}\bar{\rho}_1\left(\alpha_m(u(k,n^{k^{-1}})^*)\right) \\
&\qquad \times \bar{\rho}_1\left(u(k,m)^*\alpha_k(S^*_{m^kn})T^*_k\right).
\end{aligned}$$

Here, the last equality follows from Lemma 4.2:

$$\bar{\rho}_1\left(u(k,n^{k^{-1}})^*\right)\lambda^*_m = \lambda^*_m Ad(\lambda_m)\bar{\rho}_1\left(u(k,n^{k^{-1}})^*\right) = \lambda^*_m\bar{\rho}_1\left(\alpha_m(u(k,n^{k^{-1}})^*)\right).$$

The product of the two $\bar{\rho}_1$ terms in the preceding expression together with $\lambda^*_{mn^{k-1}}$ in front can be combined to

$$\begin{aligned}
&\lambda^*_{mn^{k-1}}\bar{\rho}_1\left(\alpha_m(u(k,n^{k^{-1}})^*)u(k,m)^*\alpha_k(S^*_{m^kn})T^*_k\right) \\
&\quad = \eta_k(m,n^{k^{-1}})\lambda^*_{mn^{k-1}}\bar{\rho}_1\left(u(k,mn^{k^{-1}})^*\alpha_k(S^*_{m^kn})T^*_k\right) \\
&\quad = \eta_k(m,n^{k^{-1}})W(mn^{k^{-1}},k)^*
\end{aligned}$$

by the definition of $\eta$, and hence we conclude

$$\sigma\bar{\sigma}(W(n,h)) = \sum_{m,k}\zeta_{n^{k-1}}(k,h)\eta_k(m,n^{k^{-1}})W(m,k)W(n^{k^{-1}},kh)W(mn^{k^{-1}},k)^*.$$

We thus have shown

$$\begin{aligned}
R &= \sum_{n,h}\sigma\bar{\sigma}(W(n,h))W(n,h)^* \\
&= \sum_{n,m,h,k}\zeta_{n^{k-1}}(k,h)\eta_k(m,n^{k^{-1}}) \\
&\qquad\qquad \times W(m,k)W(n^{k^{-1}},kh)W(mn^{k^{-1}},k)^*W(n,h)^*.
\end{aligned}$$

Composing this with the flip

$$F = \sum_{n_1,n_2,h_1,h_2} W(n_1,h_1)W(n_2,k_2)W(n_1,h_1)^*W(n_2,k_2)^*,$$

we get

$$\begin{aligned}
V &= \sum_{n,m\in N;h,k\in H}\zeta_{n^{k-1}}(k,h)\eta_k(m,n^{k^{-1}}) \\
&\qquad\qquad \times W(m,k)W(n^{k^{-1}},kh)W(n,h)^*W(mn^{k^{-1}},k)^*.
\end{aligned}$$

Therefore, by changing the variable $n$ to $n^k$ and then switching $h$ and $k$, we have

THEOREM 4.4. *The multiplicative unitary $V$ corresponding to the irreducible depth 2 inclusion $\mathcal{R} \supseteq \sigma(\mathcal{R})$ is given by*

$$V = \sum_{n,m \in N; h,k \in H} \zeta_n(h,k)\eta_h(m,n)W(m,h)W(n,hk)W(n^h,k)^* W(mn,h)^*$$

*in terms of the basis in Proposition 4.3.*

We set

$$\begin{aligned} \rho_h^\eta(n) &= \sum_m \eta_h(m,n)W(m,h)W(mn,h)^*, \\ \lambda^\zeta(h) &= \sum_{n,k} \zeta_n(h,k)W(n,hk)W(n^h,k)^*. \end{aligned}$$

We also set

$$\delta_{n'} = \sum_{k'} W(n',k')W(n',k')^* \quad \text{(the projection onto the “}n'\text{-th” component).}$$

Since $\delta_n \lambda^\zeta(h) = \sum_k \zeta_n(h,k)W(n,hk)W(n^h,k)^*$, we conclude

$$V = \sum_{(n,h) \in N \times H} \rho_h^\eta(n) \otimes \delta_n \lambda^\zeta(h).$$

under the standard identification $\mathcal{H}^2\mathcal{H}^{*2} \cong B(\mathcal{H}) \otimes B(\mathcal{H})$ in the Cuntz algebra theory.

We compute

$$\begin{aligned} &\lambda^\zeta(h_1)\lambda^\zeta(h_2) \\ &= \sum_{n,n';k,k'} \zeta_n(h_1,k)\zeta_{n'}(h_2,k')W(n,h_1k)W(n^{h_1},k)^*W(n',h_2k')W(n^{h_2},k')^* \\ &= \sum_{n,k} \zeta_n(h_1,k)\zeta_{n^{h_1}}(h_2,h_2^{-1}k)W(n,h_1k)W(n^{h_1h_2},h_2^{-1}k)^* \\ &= \sum_{n,k} \zeta_n(h_1,h_2k)\zeta_{n^{h_1}}(h_2,k)W(n,h_1h_2k)W(n^{h_1h_2},k)^*. \end{aligned}$$

Here, the second equality follows from the fact that we get non-zero terms only when $n' = n^{h_1}$ and $k' = h_2^{-1}k$, and to get the third we have changed $k$ to $h_2k$. Therefore, the cocycle identity $\zeta_n(h_1,h_2k)\zeta_{n^{h_1}}(h_2,k) = \zeta_n(h_1,h_2)\zeta_n(h_1h_2,k)$ shows

$$\begin{aligned} \lambda^\zeta(h_1)\lambda^\zeta(h_2) &= \sum_{n,k} \zeta_n(h_1,h_2)\zeta_n(h_1h_2,k)W(n,h_1h_2k)W(n^{h_1h_2},k)^* \\ &= \sum_n \zeta_n(h_1,h_2)\delta_n\lambda^\zeta(h_1h_2). \end{aligned}$$

Note that $\rho_{k_1}^\eta(n_1)\rho_{k_2}^\eta(n_2) = 0$ when $k_1 \neq k_2$ and that $\rho_{k_1}^\eta(n)$ is a matrix (at the “position $k_1$”) corresponding to the (twisted) regular representation of $N$. Therefore, we also conclude

$$\rho_{k_1}^\eta(n_1)\rho_{k_2}^\eta(n_2) = \delta_{k_1,k_2}\eta_{k_1}(n_1,n_2)\rho_{k_1}^\eta(n_1n_2)$$

with the Kronecker symbol $\delta_{k_1,k_2}$.

CHAPTER 5

# Kac algebra structure

A multiplicative unitary $V$ contains complete information on the corresponding Kac algebra $\mathcal{A}$. Indeed, every relevant quantity can be expressed in terms of $V$ as was shown in [**1**] (see below). For the multiplicative unitary $V$ obtained in Chapter 3 or Chapter 4 (in Chapter 3 the notation $V^\theta$ was used instead to emphasize the presence of a "cocycle" $\theta$), we will explicitly write down the corresponding Kac algebra $\mathcal{A}$ and its dual Kac algebra $\hat{\mathcal{A}}$ (Theorem 5.1).

Let $\delta_g$ be the element in $\ell^2(G)$ defined by $\delta_g(g') = \delta_{g,g'}$, and $\omega$ $(= \omega_{g_1,g_2})$ in $B(\ell^2(G))_*$ is defined by

$$\omega = Tr((\delta_{g_1} \otimes \delta^c_{g_2})\cdot)$$

with the rank-one operator $\delta_{g_1} \otimes \delta^c_{g_2}$. We begin by summarizing what is shown in [**1**].

- The Kac algebra $\mathcal{A}$ and the dual Kac algebra $\hat{\mathcal{A}}$ are recovered from the multiplicative unitary $V$ via slice maps. Namely, since $V$ is a unitary acting on $\ell^2(G)\otimes\ell^2(G)$, $(\omega\otimes id)(V)$ and $(id\otimes\omega)(V)$ are operators on $\ell^2(G)$. These operators generate $\mathcal{A}$ and $\hat{\mathcal{A}}$ respectively (as linear spaces).
- The duality between $\mathcal{A}$ and $\hat{\mathcal{A}}$ is described by the following pairing between $A = (\omega \otimes id)(V)$ and $B = (id \otimes \omega')(V)$:

$$\mathcal{A} \times \hat{\mathcal{A}} \ni (A, B) \mapsto < A, B >= (\omega \otimes \omega')(V) \in \mathbf{C}.$$

- The coproducts $\Gamma : \mathcal{A} \to \mathcal{A} \otimes \mathcal{A}$ and $\hat{\Gamma} : \hat{\mathcal{A}} \to \hat{\mathcal{A}} \otimes \hat{\mathcal{A}}$ are given as follows:

$$\begin{aligned} \Gamma(A) &= V(A \otimes 1)V^*, \\ \hat{\Gamma}(B) &= V^*(1 \otimes B)V. \end{aligned}$$

- Finally, the antipodes of $\mathcal{A}, \hat{\mathcal{A}}$ are given by

$$\begin{aligned} \kappa :\ & (\omega \otimes id)(V) \in \mathcal{A} \longrightarrow (\omega \otimes id)(V^*) \in \mathcal{A}, \\ \hat{\kappa} :\ & (id \otimes \omega)(V) \in \hat{\mathcal{A}} \longrightarrow (id \otimes \omega)(V^*) \in \hat{\mathcal{A}}. \end{aligned}$$

For a vector $\xi$ in $\ell^2(G)$, we compute

$$\begin{aligned} ((\omega \otimes id)(V)\xi)(g) &= ((\omega \otimes id)(V)\xi, \delta_g) \\ &= Tr\left(((\omega \otimes id)(V)\xi) \otimes \delta^c_g\right) \\ &= Tr\left(((\omega \otimes id)(V))\,(\xi \otimes \delta^c_g)\right) \\ &= (Tr \otimes Tr)\left(((\delta_{g_1} \otimes \delta^c_{g_2}) \otimes 1)\, V\, \left(1 \otimes (\xi \otimes \delta^c_g)\right)\right) \\ &= (Tr \otimes Tr)\left(V\left((\delta_{g_1} \otimes \delta^c_{g_2}) \otimes (\xi \otimes \delta^c_g)\right)\right) \end{aligned}$$

$$\begin{aligned} &= (Tr \otimes Tr)\left(V\left((\delta_{g_1} \otimes \xi) \otimes (\delta_{g_2} \otimes \delta_g)^c\right)\right) \\ &= (V(\delta_{g_1} \otimes \xi), (\delta_{g_2} \otimes \delta_g)) \\ &= (V(\delta_{g_1} \otimes \xi))(g_2, g). \end{aligned}$$

Here, $(\cdot, \cdot)$ means an inner product on the relevant Hilbert space (i.e., $\ell^2(G)$ or $\ell^2(G) \otimes \ell^2(G)$), and $(\delta_{g_1} \otimes \delta_{g_2}^c) \otimes (\xi \otimes \delta_g^c) = (\delta_{g_1} \otimes \xi) \otimes (\delta_{g_2} \otimes \delta_g)^c$ is a rank-one operator on $\ell^2(G) \otimes \ell^2(G)$. Symmetric computations also show

$$((id \otimes \omega)(V)\xi)(g) = (V(\xi \otimes \delta_{g_1}))(g, g_2).$$

Recall that in the present setting the multiplicative unitary $V$ is given by

$$\begin{aligned} (V\xi)(n_1h_1, n_2h_2) &= \theta(n_1h_1, n_2h_2)\xi(n_1n_2h_1, n_2^{h_1}h_1^{-1}h_2) \\ &= \tilde{\eta}_{h_1}(n_1, n_2)\tilde{\zeta}_{n_2}(h_1, h_1^{-1}h_2) \times \xi(n_1n_2h_1, n_2^{h_1}h_1^{-1}h_2) \end{aligned}$$

(see Chapters 3 and 4). Therefore, when $g = nh, g_i = n_ih_i$ $(i = 1, 2)$, we have

$$((\omega \otimes id)(V)\xi)(nh) = \theta(n_2h_2, nh)(\delta_{g_1} \otimes \xi)(n_2nh_2, n^{h_2}h_2^{-1}h)), \tag{5.1}$$

$$((id \otimes \omega)(V)\xi)(nh) = \theta(nh, n_2h_2)(\xi \otimes \delta_{g_1})(nn_2h, n_2^hh^{-1}h_2). \tag{5.2}$$

Notice that the first (resp. second) quantity can be non-zero only when

$$n_1 = n_2n \quad \text{and} \quad h_1 = h_2 \tag{5.3}$$

$$(\text{resp.} \quad n_1 = n_2^h \quad \text{and} \quad h_1 = h^{-1}h_2) \tag{5.4}$$

due to the presence of $\delta_{g_i}$ and $g_i = n_ih_i$.

Let us assume $h_2 = h_1$ (because we get zero otherwise). Then, the operator (5.1) becomes

$$((\omega \otimes id)(V)\xi)(nh) = \delta_{n_2^{-1}n_1}(n)\tilde{\eta}_{h_1}(n_2, n)\tilde{\zeta}_n(h_1, h_1^{-1}h)\xi(k_1^{-1}hk_1 \cdot k_1^{-1}k).$$

Notice that the presence of the delta function $\delta_{n_2^{-1}n_1}$ above is due to the first condition of (5.3). For given $n'$ and $h'$, we choose $n_1 = e, n_2 = n'^{-1}$ (so that $\delta_{n_2^{-1}n_1} = \delta_{n'}$), and $h_1 = h_2 = h'$. Then, the above equation means

$$\begin{aligned} A_{\delta_{n'},h'} &= \frac{1}{\tilde{\eta}_{h'}(n'^{-1}, n')}(\omega \otimes id)(V) = (\omega \otimes id)(V) \\ &\text{with} \quad \omega = Tr((\delta_{h'} \otimes \delta^c_{n'^{-1}h'})\cdot). \end{aligned} \tag{5.5}$$

On the other hand, thanks to (5.4), the operator (5.2) is given by

$$\begin{aligned} ((id \otimes \omega)(V)\xi)(nh) &= \tilde{\eta}_h(n, n_2)\tilde{\zeta}_{n_2}(h, h^{-1}h_2) \\ &\quad \times \delta_{h_2h_1^{-1}}(h)\chi(h)\xi(nn_2h), \end{aligned}$$

where $\chi$ means the characteristic function of the subset $\{h;\ n_1 = n_2^h\}$ in $H$. For given $n'$ and $h'$, we choose $n_1 = n'^{h'}, h_1 = h'^{-1}, n_2 = n'$, and $h_2 = 1$ (so that $\delta_{h_2h_1^{-1}} = \delta_{h'}$ and $n_1 = n_2^{h'}$). Then, from the above equation we observe

$$\begin{aligned} B_{n',h'} &= \frac{1}{\tilde{\zeta}_{n'}(h', h'^{-1})}(id \otimes \omega')(V) \\ &\text{with} \quad \omega' = Tr((\delta_{n'^{h'}h'^{-1}} \otimes \delta^c_{n'})\cdot). \end{aligned} \tag{5.6}$$

The computations so far show $A_{\delta_n,h} \in \mathcal{A}$ and $B_{n,h} \in \hat{\mathcal{A}}$. Since $\dim \mathcal{A} = \dim \hat{\mathcal{A}} = \#G$, $A$'s and $B$'s generate $\mathcal{A}$ and $\hat{\mathcal{A}}$ respectively. We will actually show

THEOREM 5.1. *(i) The Kac algebra $\mathcal{A}$ is*

$$\mathcal{A} = \ell^\infty(N) \rtimes_{\tilde{\zeta}} H,$$

*the twisted crossed product (with the natural action of $H$ on $\ell^\infty(N) : f \to f^{h^{-1}} = f(h^{-1} \cdot h)$) by the cocycle $\tilde{\zeta}$ (see* [**76**]*), and the dual Kac algebra $\hat{\mathcal{A}}$ is*

$$\hat{\mathcal{A}} = \oplus_{h \in H} \mathbf{C}_{\tilde{\eta}_h}(N),$$

*the direct sum of the twisted group rings of $N$ by the cocycles $\tilde{\eta}_h$.*
*(ii) The duality between $\mathcal{A}$ and $\hat{\mathcal{A}}$ is given by*

$$\mathcal{A} \times \hat{\mathcal{A}} \ni (A_{\delta_n,h}, B_{n',h'}) \mapsto \delta_{n,n'}\delta_{h,h'}.$$

*(iii) The coproduct and the antipode of $\mathcal{A}$ are given by*

$$\begin{aligned} \Gamma(A_{\delta_n,h}) &= \sum_{n'n''=n} \tilde{\eta}_h(n',n'')(A_{\delta_{n'},h} \otimes A_{\delta_{n''},h}), \\ \kappa(A_{\delta_n,h}) &= \tilde{\zeta}_n(h,h^{-1}) A_{\delta_{(n^{-1})^h},h^{-1}}. \end{aligned}$$

*(iv) The coproduct and the antipode of $\hat{\mathcal{A}}$ are given by*

$$\begin{aligned} \hat{\Gamma}(B_{n,h}) &= \sum_{h'h''=h} \tilde{\zeta}_n(h',h'')(B_{n,h'} \otimes B_{(n^{-1})^{h'},h''}), \\ \hat{\kappa}(B_{n,h}) &= \overline{\tilde{\zeta}_n(h,h^{-1})} B_{(n^{-1})^h,h^{-1}}. \end{aligned}$$

From the above formula for the antipode $\kappa$ of $\mathcal{A}$ one easily observes $\kappa(A_{\delta_n,e}) = A_{\delta_{n^{-1}},e}$ and $\kappa(A_{1,h}) = A_{1,h}^*$ (for $A_{1,h} = \sum_{n \in N} A_{\delta_n,h}$).

Let us begin with the duality between $\mathcal{A}$ and $\hat{\mathcal{A}}$. With the $\omega$ in (5.5) (but for $A_{\delta_n,h}$ instead of $A_{\delta_{n'},h'}$) and the $\omega'$ in (5.6), we compute

$$\begin{aligned} &(\omega \otimes \omega')(V) \\ &= (Tr \otimes Tr)((\delta_h \otimes \delta_{n'^{h'}h'^{-1}}) \otimes (\delta_{n^{-1}h} \otimes \delta_{n'})^c)(V)) \\ &= (V(\delta_h \otimes \delta_{n'^{h'}h'^{-1}}))(n^{-1}h, n') \\ &= \tilde{\eta}_h(n^{-1}, n')\tilde{\zeta}_{n'}(h,h^{-1})(\delta_h \otimes \delta_{n'^{h'}h'^{-1}})(n^{-1}n'h, n'^h h^{-1}). \end{aligned}$$

Notice that $e = n^{-1}n', n'^{h'} = n'^h$, and $h'^{-1} = h^{-1}$ if and only if $n = n'$ and $h = h'$. Therefore, the above quantity is equal to

$$\delta_{n,n'}\delta_{h,h'}\tilde{\eta}_h(n^{-1},n')\tilde{\zeta}_{n'}(h,h^{-1}) = \delta_{n,n'}\delta_{h,h'}\tilde{\zeta}_{n'}(h,h^{-1})$$

thanks to the normalization $\tilde{\eta}_h(n^{-1},n) = 1$. Hence, (by recalling the coefficient in (5.6)) the pairing between $A_{\delta_n,h}$ and $B_{n',h'}$ is given by

$$< A_{\delta_n,h}, B_{n',h'} >= \frac{1}{\tilde{\zeta}_{n'}(h',h'^{-1})} \times \delta_{n,n'}\delta_{h,h'}\tilde{\zeta}_{n'}(h,h^{-1}) = \delta_{n,n'}\delta_{h,h'}.$$

The next lemma is straight-forward.

LEMMA 5.2. *The adjoint of $V$ is given by*

$$\begin{aligned} (V^*\xi)(n_1h_1, n_2h_2) &= \overline{\tilde{\eta}_{h_1}(n_1(n_2^{-1})^{h_1^{-1}}, n_2^{h_1^{-1}})\tilde{\zeta}_{n_2^{h_1^{-1}}}(h_1,h_2)} \\ &\qquad \times \xi(n_1(n_2^{-1})^{h_1^{-1}}h_1, n_2^{h_1^{-1}}h_1h_2). \end{aligned}$$

Recall $(\Gamma(A_{f,h}))(n_1h_1, n_2h_2) = (V(A_{f,h}\otimes 1)V^*)(n_1h_1, n_2h_2)$, which is equal to

$$\begin{aligned}
&\tilde\eta_{h_1}(n_1,n_2)\tilde\zeta_{n_2}(h_1,h_1^{-1}h_2)((A_{f,h}\otimes 1)(V)^*\xi)(n_1n_2h_1, n_2^{h_1}h_1^{-1}h_2)\\
&\quad = \tilde\eta_{h_1}(n_1,n_2)\tilde\zeta_{n_2}(h_1,h_1^{-1}h_2)f(n_1n_2)\tilde\zeta_{n_1n_2}(h,h^{-1}h_1)\\
&\qquad\qquad \times ((V)^*\xi)((n_1n_2)^h h^{-1}h_1, n_2^{h_1}h_1^{-1}h_2)\\
&\quad = \tilde\eta_{h_1}(n_1,n_2)\tilde\zeta_{n_2}(h_1,h_1^{-1}h_2)f(n_1n_2)\tilde\zeta_{n_1n_2}(h,h^{-1}h_1)\\
&\qquad\qquad \times \overline{\tilde\eta_{h^{-1}h_1}(n_1^{h_1}, n_2^h)\tilde\zeta_{n_2^h}(h^{-1}h_2, h_1^{-1}h_2)}\\
&\qquad\qquad \times \xi(n_1^h h^{-1}h_1, n_2^h h^{-1}h_2).
\end{aligned}$$

By the cocycle equation of $\tilde\zeta$ and the fundamental relation between $\tilde\zeta$ and $\tilde\eta$, the product of the third and second $\tilde\zeta$ factors in the last expression is

$$\begin{aligned}
&\overline{\tilde\zeta_{n_2^h}(h^{-1}h_2, h_1^{-1}h_2)}\tilde\zeta_{n_1n_2}(h,h^{-1}h_1)\\
&\quad = \tilde\zeta_{n_2}(h,h^{-1}h_2)\overline{\tilde\zeta_{n_2}(h_1,h_1^{-1}h_2)\tilde\zeta_{n_2}(h,h^{-1}h_1)}\tilde\zeta_{n_1n_2}(h,h^{-1}h_1)\\
&\quad = \tilde\zeta_{n_2}(h,h^{-1}h_2)\overline{\tilde\zeta_{n_2}(h_1,h_1^{-1}h_2)}\\
&\qquad\qquad \times \tilde\zeta_{n_1}(h,h^{-1}h_1) \times \frac{\tilde\eta_h(n_1,n_2)(h\cdot\tilde\eta_{h^{-1}h_1})(n_1,n_2)}{\tilde\eta_{h_1}(n_1,n_2)}.
\end{aligned}$$

Observe that $\overline{\tilde\zeta_{n_2}(h_1,h_1^{-1}h_2)}$, $(h\cdot\tilde\eta_{h^{-1}h_1})(n_1,n_2) = \tilde\eta_{h^{-1}h_1}(n_1^h, n_2^h)$, and $\frac{1}{\tilde\eta_{h_1}(n_1,n_2)}$ here cancel with those in the preceding expression and disappear. Consequently, we have

$$\begin{aligned}
&(\Gamma(A_{f,h})\xi)(n_1h_1, n_2h_2)\\
&\quad = f(n_1n_2)\tilde\eta_h(n_1,n_2)\tilde\zeta_{n_1}(h,h^{-1}h_1)\tilde\zeta_{n_2}(h,h^{-1}h_2)\\
&\qquad\qquad \times \xi(n_1^h h^{-1}h_1, n_2^h h^{-1}h_2).
\end{aligned}$$

Hence, we have shown

$$\Gamma(A_{f,h}) = \sum_{n',n''} f(n'n'')\tilde\eta_h(n',n'')(A_{\delta_{n'},h}\otimes A_{\delta_{n''},h}),$$

where the sum is taken over all $(n',n'')$'s in $N\times N$, and this is exactly the formula in the theorem.

We now compute

$$(\hat\Gamma(B_{n,h}))(n_1h_1, n_2h_2) = (V^*(1\otimes B_{n,h})V)(n_1h_1, n_2h_2).$$

It is easy to see that this quantity is equal to

$$\begin{aligned}
(5.7)\qquad &\overline{\tilde\eta_{h_1}(n_1(n_2^{-1})^{h_1^{-1}}, n_2^{h_1^{-1}})\tilde\zeta_{n_2^{h_1^{-1}}}(h_1,h_2)}\\
&\times\delta_h(h_1h_2)\tilde\eta_h(n_2^{h_1^{-1}}, n)\\
&\times\tilde\eta_{h_1}(n_1(n_2^{-1})^{h_1^{-1}}, n_2^{h_1^{-1}}n)\tilde\zeta_{n_2^{h_1^{-1}}n}(h_1,h_2)\\
&\times\xi(n_1nh_1, n_2n^{h_1}h_2).
\end{aligned}$$

The product of the two $\tilde\eta_{h_1}$ factors in (5.7) is

$$(5.8)\qquad \tilde\eta_{h_1}(n_1,n)\overline{\tilde\eta_{h_1}(n_2^{h_1^{-1}}, n)}$$

by the cocycle equation while the product of the two $\tilde{\zeta}$ factors in (5.7) is

$$\begin{aligned}(5.9)\qquad &\tilde{\zeta}_n(h_1,h_2)\times\frac{\tilde{\eta}_{h_1}(n_2^{h_1^{-1}},n)(h_1\cdot\tilde{\eta}_{h_2})(n_2^{h_1^{-1}},n)}{\tilde{\eta}_{h_1h_2}(n_2^{h_1^{-1}},n)}\\ &=\tilde{\zeta}_n(h_1,h_2)\times\frac{\tilde{\eta}_{h_1}(n_2^{h_1^{-1}},n)\tilde{\eta}_{h_2}(n_2,n^{h_1})}{\tilde{\eta}_{h_1h_2}(n_2^{h_1^{-1}},n)}\end{aligned}$$

due to the fundamental relation between $\tilde{\zeta}$ and $\tilde{\eta}$. After canceling $\tilde{\eta}_{h_1}(n_2^{h_1^{-1}},n)$ in (5.8) and (5.9), (5.7) is equal to

$$\delta_h(h_1h_2)\tilde{\zeta}_n(h_1,h_2)\times\frac{\tilde{\eta}_h(n_2^{h_1^{-1}},n)}{\tilde{\eta}_{h_1h_2}(n_2^{h_1^{-1}},n)}\times\tilde{\eta}_{h_1}(n_1,n)\tilde{\eta}_{h_2}(n_2,n^{h_1})\xi(n_1nh_1,n_2n^{h_1}h_2).$$

Note that the above quantity can be non-zero only when $h=h_1h_2$ due to the presence of $\delta_h(h_1h_2)$. When $h=h_1h_2$, the above $\tilde{\eta}_h$ and $\tilde{\eta}_{h_1h_2}$ cancel out and we conclude

$$\begin{aligned}(\hat{\Gamma}(B_{n,h}))(n_1h_1,n_2h_2) &= \delta_h(h_1h_2)\tilde{\zeta}_n(h_1,h_2)\tilde{\eta}_{h_1}(n_1,n)\tilde{\eta}_{h_2}(n_2,n^{h_1})\\ &\qquad\times\xi(n_1nh_1,n_1n^{h_1}h_2),\end{aligned}$$

which is what we wanted.

Finally, we deal with the antipodes of $\mathcal{A},\hat{\mathcal{A}}$. With the $\omega$ in (5.5) (but for $A_{\delta_n,h}$ instead of $A_{\delta_{n'},h'}$), we compute

$$\begin{aligned}((\omega\otimes id)(V^*)\xi)(n'h') &= ((\omega\otimes id)(V^*)\xi,\delta_{n'h'})\\ &= Tr\left(((\omega\otimes id)(V^*)\xi)\otimes\delta^c_{n'h'}\right)\\ &= Tr\left(((\omega\otimes id)(V^*))\,(\xi\otimes\delta^c_{n'h'})\right)\\ &= (Tr\otimes Tr)\left(((\delta_h\otimes\delta^c_{n^{-1}h})\otimes(\xi\otimes\delta^c_{n'h'}))\,V^*\right)\\ &= (Tr\otimes Tr)\left(((\delta_h\otimes\xi)\otimes(\delta_{n^{-1}h}\otimes\delta_{n'h'})^c)\,V^*\right)\\ &= (V^*(\delta_h\otimes\xi))(n^{-1}h,n'h').\end{aligned}$$

By Lemma 5.2, this is equal to

$$\begin{aligned}&\overline{\tilde{\eta}_h(n^{-1}(n'^{-1})^{h^{-1}},n'^{h^{-1}})\tilde{\zeta}_{n'^{h^{-1}}}(h,h')}\\ &\qquad\times(\delta_h\otimes\xi)(n^{-1}(n'^{-1})^{h^{-1}}h,n'^{h^{-1}}hh')\\ &=\delta_{(n^{-1})^h}(n')\overline{\tilde{\zeta}_{n^{-1}}(h,h')}\xi(n^{-1}hh').\end{aligned}$$

Indeed, $h=n^{-1}(n'^{-1})^{h^{-1}}h$ if and only if $n'=(n^{-1})^h$ (and we have also used the normalization $\tilde{\eta}_h(e,n^{-1})=1$). On the other hand, recall

$$(A_{\delta_{(n^{-1})^h},h^{-1}})(n'h')=\delta_{(n^{-1})^h}(h')\tilde{\zeta}_{(n^{-1})^h}(h^{-1},hh')\xi(n^{-1}hh').$$

Therefore, we observe

$$\begin{aligned}(\kappa(A_{\delta_n,h})\xi)(n'h') &= \overline{\tilde{\zeta}_{n^{-1}}(h,h')\tilde{\zeta}_{(n^{-1})^h}(h^{-1},hh')}(A_{\delta_{(n^{-1})^h},h^{-1}}\xi)(n'h')\\ &= \overline{\tilde{\zeta}_{n^{-1}}(1,hh')\tilde{\zeta}_{n^{-1}}(h,h^{-1})}(A_{\delta_{(n^{-1})^h},h^{-1}}\xi)(n'h')\\ &= \overline{\tilde{\zeta}_{n^{-1}}(n,n^{-1})}(A_{\delta_{(n^{-1})^h},h^{-1}}\xi)(n'h')\\ &= \tilde{\zeta}_h(n,n^{-1})(A_{\delta_{(n^{-1})^h},h^{-1}}\xi)(n'h').\end{aligned}$$

Here, the second equation follows from the cocycle equation for $\tilde{\zeta}$ while the fourth follows from the fact that the normalization $\tilde{\eta}_{\cdot}(h, h^{-1}) = 1$ implies

$$\tilde{\zeta}_n(h, h^{-1})\tilde{\zeta}_{n^{-1}}(h, h^{-1}) = \tilde{\zeta}_e(h, h^{-1}) = 1.$$

We next compute the antipode $\hat{\kappa}$ of $\hat{\mathcal{A}}$. With the $\omega'$ in (5.6) we have

$$(\hat{\kappa}(B_{n',h'})\xi)(nh) = \frac{1}{\tilde{\zeta}_{n'}(h', h'^{-1})} \times ((id \otimes \omega')(V^*)\xi)(nh). \tag{5.10}$$

By similar computations as before, the second factor in the above right side is easily seen to be equal to

$$((V^*)(\xi \otimes \delta_{n'^{h'}h'^{-1}}))(nh, n').$$

Thanks to Lemma 5.2 and the definition of $B$, this quantity is equal to

$$\begin{aligned}
&\overline{\tilde{\eta}_h(n(n'^{-1})^{h^{-1}}, n'^{h^{-1}})} \times (\xi \otimes \delta_{n'^{h'}h'^{-1}})(n(h'^{-1})^{h^{-1}}h, n'^{h^{-1}}h)\\
&= \overline{\tilde{\eta}_{h'^{-1}}(n(n'^{-1})^{h'}, n'^{h'})} \times \delta_{h'^{-1}}(h)\xi(nn'^{h'}h'^{-1})\\
&\qquad\qquad (\text{since } n'^{h'}h'^{-1} = n'^{h^{-1}}h \text{ if and only if } h = h'^{-1})\\
&= \overline{\tilde{\eta}_{h'^{-1}}(n(n'^{-1})^{h'}, n'^{h'})\tilde{\eta}_{h'^{-1}}(n, (n'^{-1})^{h'})} \times (B_{(n'^{-1})^{h'},h'^{-1}}\xi)(nh).
\end{aligned}$$

However, the cocycle equation for $\tilde{\eta}$ and its normalization implies that the product of the two $\tilde{\eta}_{h'^{-1}}$ factors here is

$$\overline{\tilde{\eta}_{h'^{-1}}(n, e)\tilde{\eta}_{h'^{-1}}((n'^{-1})^{h'}, n'^{h'})} = 1.$$

Therefore, by recalling the coefficient $\frac{1}{\tilde{\zeta}_{n'}(h', h'^{-1})} = \overline{\tilde{\zeta}_{n'}(h', h'^{-1})}$ appearing in the right side of (5.10), we have shown the formula for $\hat{\kappa}$ in the theorem.

CHAPTER 6

# Group-like elements

An element $U$ ($\neq 0$) in a Kac algebra is called a group-like element if $\Gamma(U) = U \otimes U$ is satisfied. Such an element is automatically a unitary and corresponds to a central rank-one projection in the dual Kac algebra. The group of all group-like elements in a Kac algebra $\mathcal{A}$ is denoted by $G(\mathcal{A})$, the intrinsic group of $\mathcal{A}$. Here, the intrinsic groups $G(\mathcal{A})$ and $G(\hat{\mathcal{A}})$ are determined for the Kac algebra $\mathcal{A}$ and its dual $\hat{\mathcal{A}}$ described in the previous chapters (Theorem 6.1 and Theorem 6.2). This information is quite useful when one deals with concrete examples.

Let $p$ be a central rank-one (as an operator on $\ell^2(G)$) projection in $\mathcal{A}$. Being a rank-one projection, $p$ is of the form $\xi_0 \otimes \xi_0^c$ with a unit vector $\xi_0 \in \ell^2(G)$. We set

$$U = (id \otimes Tr(p\cdot))V \in \hat{\mathcal{A}}.$$

We claim

$$(U\xi, \xi) = (V(\xi \otimes \xi_0), (\xi \otimes \xi_0)). \tag{6.1}$$

In fact, we compute

$$\begin{aligned} ((id \otimes Tr(p\cdot))V\xi, \xi) &= Tr\left(((id \otimes Tr(p\cdot)V)\,\xi) \otimes \xi^c\right) \\ &= Tr\left(((id \otimes Tr(p\cdot))V)\,(\xi \otimes \xi^c)\right) \\ &= (Tr \otimes Tr)\left((1 \otimes p)\,V\,((\xi \otimes \xi^c) \otimes 1)\right) \\ &= (Tr \otimes Tr)\left(V\,(1 \otimes p)\,((\xi \otimes \xi^c) \otimes 1)\right) \\ &= (Tr \otimes Tr)\left(V\,((\xi \otimes \xi^c) \otimes p)\right), \end{aligned}$$

and we have

$$(\xi \otimes \xi^c) \otimes p = (\xi \otimes \xi^c) \otimes (\xi_0 \otimes \xi_0^c) = (\xi \otimes \xi_0) \otimes (\xi \otimes \xi_0)^c$$

(a rank-one projection on $\ell^2(G) \otimes \ell^2(G)$). Thanks to $\hat{\Gamma}(U) = V^*(1 \otimes U)V$ and (6.1), we compute

$$\begin{aligned} &(\hat{\Gamma}(U)(\xi_1 \otimes \xi_2), (\xi_1 \otimes \xi_2)) \\ &\quad = ((1 \otimes U)V(\xi_1 \otimes \xi_2), V(\xi_1 \otimes \xi_2)) \\ &\quad = (V_{23}V_{12}(\xi_1 \otimes \xi_2 \otimes \xi_0), V_{12}(\xi_1 \otimes \xi_2 \otimes \xi_0)). \end{aligned}$$

Recall the pentagon equation $V_{23}V_{12} = V_{12}V_{13}V_{23}$. Since $V_{12}$ is a unitary, the above quantity is equal to

$$(V_{13}V_{23}(\xi_1 \otimes \xi_2 \otimes \xi_0), (\xi_1 \otimes \xi_2 \otimes \xi_0)).$$

Since $V \in \hat{\mathcal{A}} \otimes \mathcal{A}$ and $p$ is central in $\mathcal{A}$, the two operators $V$ and $1 \otimes p$ $(= 1 \otimes (\xi_0 \otimes \xi_0^c))$ commute. Therefore, $V_{23}(\xi_1 \otimes \xi_2 \otimes \xi_0)$ is (a linear combination of vectors) of the form: $\xi_1 \otimes \xi_2' \otimes \xi_0$, and the above quantity is equal to

$$\begin{aligned}(V_{13}(\xi_1 \otimes \xi_2' \otimes \xi_0), (\xi_1 \otimes \xi_2 \otimes \xi_0))\\ &= ((U \otimes 1)(\xi_1 \otimes \xi_2'), (\xi_1 \otimes \xi_2))\\ &= ((U \otimes 1 \otimes 1)(\xi_1 \otimes \xi_2' \otimes \xi_0), (\xi_1 \otimes \xi_2 \otimes \xi_0))\\ &= ((U \otimes 1 \otimes 1)V_{23}(\xi_1 \otimes \xi_2 \otimes \xi_0), (\xi_1 \otimes \xi_2 \otimes \xi_0))\\ &= ((U \otimes 1)(1 \otimes U)(\xi_1 \otimes \xi_2), (\xi_1 \otimes \xi_2)).\end{aligned}$$

Here, we have used (6.1) to get the first and fourth equalities. Hence we have shown $\hat{\Gamma}(U) = (U \otimes 1)(1 \otimes U) = U \otimes U$, and $U$ is indeed a group-like element in $\hat{\mathcal{A}}$.

On the other hand, let $q = \xi_0 \otimes \xi_0^c$ be a central rank-one projection in $\hat{\mathcal{A}}$, and we set

$$U = (Tr(q\cdot) \otimes id)V \in \mathcal{A}.$$

Symmetric arguments show

$$\begin{aligned}(U\xi, \xi) &= (V(\xi_0 \otimes \xi), (\xi_0 \otimes \xi)),\\ \Gamma(U) &= U \otimes U,\end{aligned} \tag{6.2}$$

and hence $U$ is a group-like element in $\mathcal{A}$.

At first we determine the rank-one central projections in the dual Kac algebra

$$\hat{\mathcal{A}} = \oplus_{h \in H} \mathbf{C}_{\tilde{\eta}_h}(N).$$

The twisted group ring $\mathbf{C}_{\tilde{\eta}_h}(N)$ has a rank-one central projection if and only if $\tilde{\eta}_h \in B^2(N, \mathbf{T})$. In fact, if $\tilde{\eta}_h$ is a coboundary, then the twisted group ring is isomorphic to the ordinary group ring $\mathbf{C}(N)$ and hence possesses such a projection. On the other hand, let us assume that $p$ is a rank-one central projection in $\mathbf{C}_{\tilde{\eta}_h}(N)$. Then, $B_{n,h}p = pB_{n,h} = f(n)p$ with a scalar $f(n) \in \mathbf{T}$. By hitting $p$ to $B_{n_1,h}B_{n_2,h} = \tilde{\eta}_h(n_1, n_2)B_{n_1n_2,h}$, we get $f(n_1)f(n_2) = \tilde{\eta}_h(n_1, n_2)f(n_1n_2)$, i.e., $\tilde{\eta}_h$ is a coboundary. We set

$$H_0 = \{h \in H;\ \tilde{\eta}_h \in B^2(N, \mathbf{T})\}.$$

Notice that this is a subgroup in $H$ because of the fundamental relation between $\tilde{\zeta}$ and $\tilde{\eta}$ and the fact that the map

$$(n_1, n_2) \in N \times N \longrightarrow \frac{\tilde{\zeta}_{n_1}(h_1, h_2)\tilde{\zeta}_{n_2}(h_1, h_2)}{\tilde{\zeta}_{n_1n_2}(h_1, h_2)} \in \mathbf{T}$$

itself sits in $B^2(N, \mathbf{T})$ (for each fixed $h_1, h_2$). Assume $h \in H_0$, that is, we have

$$\tilde{\eta}_h(n_1, n_2) = \xi(h, n_1n_2)\overline{\xi(h, n_1)\xi(h, n_2)}$$

for some $\mathbf{T}$-valued function $\xi(h, \cdot)$ on $N$. We extend $\xi(\cdot, \cdot)$ to a $\mathbf{T}$-valued function on $H \times N$ (for example by just setting $\xi(h, n) = 1$ for $h \in H \setminus H_0$). Then, by changing $(\tilde{\eta}, \tilde{\zeta})$ to $(\tilde{\eta}(\partial^N \xi), \tilde{\zeta}(\partial^H \xi))$, we may and do assume

$$\tilde{\eta}_h(n_1, n_2) = 1 \text{ when } h \in H_0. \tag{6.3}$$

The rank-one central projections in $\hat{\mathcal{A}}$ are then given by

$$p_{\chi,h} = \frac{1}{\#N} \sum_{n \in N} \chi(n^{-1})B_{n,h}\ (\in \mathbf{C}_{\tilde{\eta}_h}(N) = \mathbf{C}(N)).$$

with $h \in H_0$ and $\chi \in Hom(N, \mathbf{T})$, the characters of $N$.

Secondly, we determine the rank-one central projections in the Kac algebra

$$\mathcal{A} = \ell^\infty(H) \rtimes_{\tilde{\zeta}} K.$$

Let $N' = \{n \in N;\ n^h = n \quad \text{for each } h \in H\}$. Let $\mathcal{O} \subseteq N$ be an orbit under the $H$-action containing more than two points. A rank-one projection in $\ell^\infty(\mathcal{O}) \rtimes_{\tilde{\zeta}} H$ is of the form

$$p = \frac{1}{\#H} A_{\delta_n, e} + \cdots$$

with some $n \in \mathcal{O}$. By taking $h \in H$ such that $n^h \neq n$, we see that $p$ does not commute with $A_{1,h}$ and cannot be central. Thus, rank-one central projections have to come from orbits $\{n\}$ $(n \in N')$. Let us assume $n \in N'$. Then, $A_{\delta_n,h}$'s $(h \in H)$ generate the algebra isomorphic to the twisted group ring $\mathbf{C}_{\tilde{\zeta}_n(\cdot,\cdot)}(H)$. This algebra admits a rank-one central projection if and only if $\tilde{\zeta}_n(\cdot,\cdot) \in B^2(H, \mathbf{T})$ (as was seen before). We thus set

$$N_0 = \{n \in N;\ n^h = n \quad \text{for each } h \in H \text{ and } \quad \tilde{\zeta}_n(\cdot,\cdot) \in B^2(H, \mathbf{T})\},$$

which is a subgroup in $H$ as before. We also renormalize $\tilde{\zeta}_n$ in such a way that

$$\tilde{\zeta}_n(h_1, h_2) = 1 \ (n \in N_0). \tag{6.4}$$

The rank-one central projections in $\mathcal{A}$ are

$$q_{n,\chi} = \frac{1}{\#H} \sum_{h \in H} \chi(h) A_{\delta_n, h}$$

with $n \in N_0$ and $\chi \in Hom(H, \mathbf{T})$, the characters on $H$.

To determine a group-like element in $\mathcal{A}$ (corresponding to $p_{\chi,h} \in \hat{\mathcal{A}}$), we need a unit vector $\xi_{\chi,h} \in \ell^2(G)$ $(h \in H_0, \chi \in Hom(N, \mathbf{T}))$ satisfying $p_{\chi,h} = \xi_{\chi,h} \otimes \xi^c_{\chi,h}$. For example,

$$\xi_{\chi,h}(n'h') = \delta_{h,h'} \times \frac{1}{\sqrt{\#N}} \chi(n')$$

does the job. Indeed, we compute

$$\begin{aligned}(p_{\chi,h}\xi_{\chi,h})(n'h') &= \frac{1}{\#N} \sum_{n \in N} \chi(n^{-1})(B_{n,h}\xi_{\chi,h})(n'h') \\ &= \frac{1}{\#N} \sum_{n \in N} \chi(n^{-1}) \times \delta_{h,h'}\xi_{\chi,h}(n'nh') = \frac{1}{(\#N)^{3/2}} \sum_{n \in N} \delta_{h,h'}\chi(n^{-1})\chi(n'n) \\ &= \delta_{h,h'} \times \frac{1}{\sqrt{\#N}} \chi(n') = \xi_{\chi,h}(n'h').\end{aligned}$$

Hence, thanks to (6.2) (with $\xi_0 = \xi_{\chi,h}$), the corresponding group-like element $U_{\chi,h} \in \mathcal{A}$ is determined by

$$(U_{\chi,h}\xi_1, \xi_2) = (V(\xi_{\chi,h} \otimes \xi_1), (\xi_{\chi,h} \otimes \xi_2)).$$

Recalling the definition of $V$, we compute

$$\begin{aligned}(U_{\chi,h}\xi_1,\xi_2) &= \sum_{n_1,h_1,n_2,h_2} \tilde{\zeta}_{n_2}(h_1,h_1^{-1}h_2)(\xi_{\chi,h}\otimes\xi_1)(n_1n_2h_1,n_2^{h_1}h_1^{-1}h_2)\\ &\qquad\times\overline{(\xi_{\chi,h}\otimes\xi_2)(n_1h_1,n_2h_2)}\\ &= \frac{1}{\#N}\sum_{n_1,h_1,n_2,h_2} \tilde{\zeta}_{n_2}(h_1,h_1^{-1}h_2)\delta_{h,h_1}\chi(n_1n_2)\xi_1(n_2^{h_1}h_1^{-1}h_2)\\ &\qquad\times\delta_{h,h_1}\overline{\chi(n_1)\xi_2(n_2h_2)}\\ &= \frac{1}{\#N}\sum_{n_1,n_2,h_2} \tilde{\zeta}_{n_2}(h,h^{-1}h_2)\chi(n_1n_2)\xi_1(n_2^{h}h^{-1}h_2)\overline{\chi(n_1)\xi_2(n_2h_2)}.\end{aligned}$$

Since $\chi(n_1n_2)\overline{\chi(n_1)}=\chi(n_2)$, the dependency on $n_1$ here disappears and the above becomes the sum over $n_2$'s and $h_2$'s (without the coefficient $\frac{1}{\#N}$). Therefore, we conclude

$$(U_{\chi,h}\xi)(n'h')=\tilde{\zeta}_{n'}(h,h^{-1}h')\chi(n')\xi(n'^{h}h^{-1}h'),$$

which means $U_{\chi,h}=A_{\chi,h}$.

The product rule for $A$'s has been already known so that we know the group structure of $G(\mathcal{A})$. Recall that $H_0$ acts on $Hom(N,\mathbf{T})$ via

$$\chi\longrightarrow\chi^{h^{-1}}=\chi(\cdot^{h})=\chi(h^{-1}\cdot h).$$

Due to the renormalization (6.3) of $\tilde{\eta}_h$ $(h\in H_0)$ and the fundamental relation between $\tilde{\eta}$ and $\tilde{\zeta}$, for each $h_1,h_2\in H_0$

$$\tilde{\zeta}_{\cdot}(h_1,h_2):\ n\in N\longrightarrow\tilde{\zeta}_n(h_1,h_2)\in\mathbf{T}$$

is a character. Therefore, we have $\tilde{\zeta}_{\cdot}|_{H_0}\in Z^2(H_0,Hom(N,\mathbf{T}))$ (relative to the above action). By summing up the discussions so far, we get

THEOREM 6.1. *The intrinsic group $G(\mathcal{A})$ is the extension*

$$1\longrightarrow Hom(N,\mathbf{T})\longrightarrow G(\mathcal{A})\longrightarrow H_0\longrightarrow 1$$

*corresponding to the cocycle $\tilde{\zeta}_{\cdot}|_{H_0}$ (relative to the natural action $\chi\longrightarrow\chi^{h^{-1}}$ of $H_0$).*

We next deal with the intrinsic group $G(\hat{\mathcal{A}})$. It is plain to see as before that the unit vector $\xi_{n,\chi}\in\ell^2(G)$ $(n\in N_0,\chi\in Hom(H,\mathbf{T}))$ defined by

$$\xi_{n,\chi}(n'h')=\delta_{n,n'}\times\frac{1}{\sqrt{\#H}}\chi(h')$$

satisfies $q_{n,\chi}=\xi_{n,\chi}\otimes\xi^c_{n,\chi}$. Therefore, thanks to (6.1) (with $\xi_0=\xi_{n,\chi}$), the corresponding group-like element $U_{n,\chi}\in\hat{\mathcal{A}}$ is given by

$$(U_{n,\chi}\xi_1,\xi_2)=(V(\xi_1\otimes\xi_{n,\chi}),(\xi_2\otimes\xi_{n,\chi})).$$

We recall the definition of $V$, and the above inner product is equal to

$$\begin{aligned}&\sum_{n_1,h_1,n_2,h_2}\tilde{\eta}_{h_1}(n_1,n_2)(\xi_1\otimes\xi_{n,\chi})(n_1n_2h_1,n_2^{h_1}h_1^{-1}h_2)\times\overline{(\xi_2\otimes\xi_{n,\chi})(n_1h_1,n_2h_2)}\\ &=\frac{1}{\#H}\sum_{n_1,h_1,n_2,h_2}\tilde{\eta}_{h_1}(n_1,n_2)\xi_1(n_1n_2h_1)\delta_{n,n_2^{h_1}}\chi(h_1^{-1}h_2)\times\overline{\xi_2(n_1h_1)\delta_{n,n_2}\chi(h_2)}.\end{aligned}$$

Notice that the second Kronecker symbol says that we can assume $n_2 = n$ (and in this case $n_2^{h_1} = n$ since $n_2 = n \in N_0$). Therefore, the above quantity is equal to

$$\frac{1}{\#H} \sum_{n_1,h_1,h_2} \tilde{\eta}_{h_1}(n_1,n)\xi_1(n_1nh_1)\chi(h_1^{-1})\overline{\xi_2(n_1h_1)}$$
$$= \sum_{n_1,h_1} \tilde{\eta}_{h_1}(n_1,n)\chi(h_1^{-1})\xi_1(n_1nh_1)\overline{\xi_2(n_1h_1)}.$$

Thus, (by changing $\chi$ to $\overline{\chi}$) we get

$$U_{n,\chi} = \sum_{h\in H} \chi(h)B_{n,h}.$$

For $n_1, n_2 \in N_0$ and $\chi_1, \chi_2 \in Hom(H, \mathbf{T})$, we observe

$$U_{n_1,\chi_1}U_{n_2,\chi_2} = \sum_{h,h'} \chi_1(h)\chi_2(h')B_{n_1,h}B_{n_2,h'}$$
$$= \sum_{h,h'} \chi_1(h)\chi_2(h') \times \delta_{h,h'}\tilde{\eta}_h(n_1,n_2)B_{n_1n_2,k}$$
$$= \sum_{h} \chi_1(h)\chi_2(h)\tilde{\eta}_h(n_1,n_2)B_{n_1n_2,h}.$$

The renormalization (6.4) of $\tilde{\zeta}_n$ ($n \in N_0$) and the fundamental relation between $\tilde{\eta}$ and $\tilde{\zeta}$ say that for each $n_1, n_2 \in N_0$

$$\tilde{\eta}_\cdot(n_1,n_2) : \ h \in H \longrightarrow \tilde{\eta}_h(n_1,n_2) \in \mathbf{T}$$

is a character, and the above computation means $U_{n_1,\chi_1}U_{n_2,\chi_2} = U_{n_1n_2,\chi}$ with $\chi = \chi_1\chi_2\tilde{\eta}_\cdot(n_1,n_2) \in Hom(H,\mathbf{T})$ (the product of characters). Note that $N_0$ is not acting on $Hom(H,\mathbf{T})$) this time, and we have $\tilde{\eta}_\cdot|_{N_0} \in Z^2(N_0, Hom(H,\mathbf{T}))$.

THEOREM 6.2. *The intrinsic group $G(\hat{\mathcal{A}})$ is the extension*

$$1 \longrightarrow Hom(H,\mathbf{T}) \longrightarrow G(\hat{\mathcal{A}}) \longrightarrow N_0 \longrightarrow 1$$

*corresponding to the cocycle $\tilde{\eta}_\cdot|_{N_0}$ (relative to the trivial action of $N_0$).*

**Remarks.**
(i) When no cocycle is around, we have

$$\mathcal{A} = \ell^\infty(N) \rtimes H \text{ and } \hat{\mathcal{A}} = \mathbf{C}(N) \otimes \ell^\infty(H),$$

where the latter is the direct sum of $\#H$ copies of the ordinary group ring $\mathbf{C}(N)$. The two intrinsic groups here are the semi-direct product $G(\mathcal{A}) = Hom(N,\mathbf{T}) \rtimes H$ and the product group $G(\hat{\mathcal{A}}) = Hom(H,\mathbf{T}) \times N_0$ with $N_0 = \{n \in N;\ n^h = n \text{ for each } n \in H\}$. Note that this is the situation considered in [**12**] (see also [**42, 75**] for related topics).
(ii) The Kac algebra $\mathcal{A}$ ($= \ell^\infty(\Gamma)$) is commutative if and only if (1) $H$ is abelian, (2) $H$ acts trivially on $N$, and (3) $\zeta_n \in B^2(H,\mathbf{T})$ for each $n \in N$. In this case the group $\Gamma$ is determined by the extension

$$1 \longrightarrow Hom(H,\mathbf{T}) \longrightarrow \Gamma \longrightarrow N \longrightarrow 1$$

in Theorem 6.2. The dual Kac algebra $\hat{\mathcal{A}}$ $(= \ell^\infty(\Gamma'))$ is commutative (i.e., $\mathcal{A}$ is co-commutative) if and only if (1) $N$ is abelian, and (2) $\eta_h \in B^2(N, \mathbf{T})$ for each $h \in H$. In this case the group $\Gamma'$ is determined by the extension

$$1 \longrightarrow Hom(N, \mathbf{T}) \longrightarrow \Gamma' \longrightarrow H \longrightarrow 1$$

in Theorem 6.1. Note that $\mathcal{A}$ is always cocommutative in the case that $N$ is abelian and $H^2(N, \mathbf{T}) = 1$.

The above groups $\Gamma, \Gamma'$ are the intrinsic groups $G(\hat{\mathcal{A}}), G(\mathcal{A})$ respectively. In fact, we point out the following general fact:

PROPOSITION 6.3. *If $\mathcal{N} \subseteq \mathcal{M}$ is an irreducible inclusion of factors of depth 2 with the corresponding Kac algebra $\mathcal{A}$, then we have (i) $G(\mathcal{A})$ can be identified with the Galois group of $\mathcal{N} \subseteq \mathcal{M}$ and (ii) $G(\hat{\mathcal{A}})$ can be identified with the Weyl group of $\mathcal{N} \subseteq \mathcal{M}$.*

Here, the Galois and Weyl groups mean

$$\begin{aligned} \text{Galois } (\mathcal{M} \supseteq \mathcal{N}) &= \{\alpha \in Aut(\mathcal{M});\ \alpha|_{\mathcal{N}} = id_{\mathcal{N}}\}, \\ \text{Weyl } (\mathcal{M} \supseteq \mathcal{N}) &= \mathbf{N}(\mathcal{N})/\mathbf{U}(\mathcal{N}), \end{aligned}$$

where $\mathbf{N}(\mathcal{N})$ is the normalizer of $\mathcal{N}$ in $\mathcal{M}$ and $\mathbf{U}(\mathcal{N})$ is the unitary group of $\mathcal{N}$. It is well-known that

$$\text{Weyl } (\mathcal{M} \supseteq \mathcal{N}) \cong \text{Galois } (\mathcal{M}_1 \supseteq \mathcal{M})$$

with the basic extension $\mathcal{M}_1$ of $\mathcal{M} \supseteq \mathcal{N}$.

Note that (i) and (ii) are symmetric, and we just sketch (i). Let $\mathcal{N} = \rho(\mathcal{M})$ ($\rho \in End(\mathcal{M})$) with the canonical endomorphism $\gamma = \rho\bar{\rho}$. Let $\mathcal{H} = (\rho, \rho\bar{\rho}\rho)$ be the Hilbert space of intertwiners with a basis $\{S_i\}_{i=1,2,\cdots,n}$, and recall (see the beginning of Chapter 4) that the multiplicative unitary is $V = \sum_{i=1}^n \gamma(S_i)S_i^* F$ $(\in \mathcal{H}^2\mathcal{H}^{*2})$ with the flip $F$. Let $\{\theta_g\}$ be the Galois group. For an isometry $S \in (id, \gamma)$ we set $S_g = \theta_g(S)$. We remark $S_g \in (\theta_g, \gamma)$ because of

$$S_g\theta_g(m) = \theta_g(Sm) = \theta_g(\gamma(m)S) = \gamma(m)S_g.$$

Therefore, the range projection $e_g = S_g S_g^*$ is minimal and central in $\mathcal{M} \cap \gamma(\mathcal{M})' = \hat{\mathcal{A}}'$. Let $U_g$ be the corresponding group-like element in $G(\mathcal{A})$. Since $S_g U_g S_g^*$ $(\cong e_g \otimes U_g) = e_g V$ ((6.2) means $U_g = S_g^* V S_g$ in the present setting), we compute

$$S_g U_g S_g^* = \sum_{i=1}^n S_g S_g^* \gamma(S_i) S_i^* F = \sum_{i=1}^n S_g \theta_g(S_i) S_g^* S_i^* F = \sum_{i=1}^n S_g \theta_g(S_i) S_i^* S_g^*.$$

Therefore, we conclude $U_g = \sum_{i=1}^n \theta_g(S_i)S_i^*$, and $g \longrightarrow U_g$ is isomorphic.

CHAPTER 7

# Examples of finite-dimensional Kac algebras

In this chapter we present a variety of concrete examples of Kac algebras obtained from various cocycles via our construction.

## 1. $(\mathbf{Z}_n \times \mathbf{Z}_n) \rtimes \mathbf{Z}_2$ with $n \geq 3$

Let $\beta$ be an outer action on a factor $\mathcal{L}$ of $G = N \rtimes H$, and let $\xi \in Z^2(N, \mathbf{T})$ be a 2-cocycle. Then, one can find unitaries $\{v_n\}_{n \in N}$ in $\mathcal{L}$ satisfying

$$v_{n_1}\beta_{n_1}(v_{n_2}) = \xi(n_1, n_2)v_{n_1 n_2}$$

(see for example [**68**]). For each $n \in N$ we set $\alpha_n = Adv_n\beta_n \in Aut(\mathcal{L})$ so that $\alpha$ is an action of $N$. Notice that $(\alpha, \beta = \beta\mid_H)$ is an outer action of the matched pair $(N, H)$. Since

$$\beta_h \alpha_{n^h} \beta_h = Ad(\beta(v_{n^h}))\beta_{hn^h h^{-1}} = Ad(\beta(v_{n^h}))\beta_h = Ad(\beta(v_{n^h})v_h^*)\alpha_n,$$

we can set

$$u(h, n) = \beta_h(v_{n^h})v_n^*.$$

Recall that Lemma 2.4,(iii) shows

$$\eta_h(n_1, n_2) = \xi(n_1, n_2)\overline{\xi(n_1^h, n_2^h)} \quad \text{and} \quad \zeta = 1.$$

For example, let $G = (\mathbf{Z}_n \times \mathbf{Z}_n) \rtimes \mathbf{Z}_2$ $(n \geq 3)$. Here, $H = \mathbf{Z}_2 = \{e, \epsilon\}$ and $\epsilon$ acts on $N = \mathbf{Z}_n \times \mathbf{Z}_n$ as a flip. Let $\omega$ be an $n$-th root of 1, and we consider the following standard cocycle (in $Z^2(N, \mathbf{T})$):

$$\xi((a, b), (c, d)) = \omega^{ad}.$$

Since $\zeta = 1$, we have $\mathcal{A} = \ell^\infty(\mathbf{Z}_n \times \mathbf{Z}_n) \times \mathbf{Z}_2$. Notice the $n$ points $\{(k, k);\ k = 0, 1, \cdots, n-1\}$ in $\mathbf{Z}_n \times \mathbf{Z}_n$ are fixed by the $\mathbf{Z}_2$-action. All the other $\frac{n^2-n}{2}$ orbits (in $\mathbf{Z}_n \times \mathbf{Z}_n$) consists of two points. Therefore, we conclude that the Kac algebra $\mathcal{A}$ is the following direct sum:

$$\mathcal{A} = \underbrace{\mathbf{C} \oplus \mathbf{C} \oplus \cdots \oplus \mathbf{C}}_{2n} \oplus \underbrace{M_2(\mathbf{C}) \oplus M_2(\mathbf{C}) \oplus \cdots \oplus M_2(\mathbf{C})}_{\frac{n^2-n}{2}}.$$

On the other hand, the dual Kac algebra is $\hat{\mathcal{A}} = \mathbf{C}(\mathbf{Z}_n \times \mathbf{Z}_n) \oplus \mathbf{C}_{\eta_\epsilon}(\mathbf{Z}_n \times \mathbf{Z}_n)$. Since $\mathbf{Z}_n \times \mathbf{Z}_n$ is abelian, so is the group ring $\mathbf{C}(\mathbf{Z}_n \times \mathbf{Z}_n)$. Notice

$$\eta_\epsilon((a, b), (c, d)) = \xi((a, b), (c, d))\overline{\xi((b, a), (d, c))} = \omega^{ad-bc}.$$

Let $x = \sum_{a,b} c_{a,b}\lambda_{(a,b)} \in \mathbf{C}_{\eta_\epsilon}(\mathbf{Z}_n \times \mathbf{Z}_n)$ with coefficients $c_{a,b} \in \mathbf{C}$. It is in the center if and only if

$$\sum_{a,b} c_{a,b}\omega^{ad-bc}\lambda_{(a+c,b+d)} = \sum_{a,b} c_{a,b}\omega^{bc-ad}\lambda_{(a+c,b+d)} \tag{7.1}$$

for each $c, d$. It is easy to see that this is satisfied (for each $c, d$) if and only if the coefficients $c_{a,b}$ vanish unless $\omega^{2a} = \omega^{2b} = 1$. Let $p_0$ be the period of $\omega^2$. From the above the center of $\mathbf{C}_{\eta_\epsilon}(\mathbf{Z}_n \times \mathbf{Z}_n)$ is $(\frac{n}{p_0})^2$-dimensional, and we conclude that $\mathbf{C}_{\eta_\epsilon}(\mathbf{Z}_n \times \mathbf{Z}_n)$ is the direct sum of $(\frac{n}{p_0})^2$ copies of $M_{p_0}(\mathbf{C})$. (To determine the size of a full matrix algebra, one can compute the trace value of the corresponding central minimal projection.) In particular, if $n$ ($n \geq 3$) is odd and $\omega = \exp(\frac{2\pi i}{n})$, then we have

$$\hat{\mathcal{A}} = \underbrace{\mathbf{C} \oplus \mathbf{C} \oplus \cdots \oplus \mathbf{C}}_{n^2} \oplus M_n(\mathbf{C}).$$

Therefore, we have seen that with a 2-cocycle $\xi \in Z^2(\mathbf{Z}_n \times \mathbf{Z}_n, \mathbf{T})$ a cocommutative Kac algebra can be easily deformed into a non-trivial (i.e., non-commutative and non-cocommutative) Kac algebra. In the above last example, we have $H_0 = \{e\}$ because of $\eta_\epsilon \neq 1 \in H^2(N, \mathbf{T})$ so that

$$G(\mathcal{A}) = Hom(N, \mathbf{T}) \cong \mathbf{Z}_n \times \mathbf{Z}_n.$$

On the other hand, $\zeta = 1$ and we get $\eta_\epsilon((a, a)(c, c)) = 1$ for $a = b, c = d$. Therefore, we get $N_0 = \{(k, k);\ k = 0, 1, \cdots, n-1\} \cong \mathbf{Z}_n$ and $\eta_\cdot|_{N_0} = 1$ so that we conclude

$$C(\hat{\mathcal{A}}) = \mathbf{Z}_n \times \mathbf{Z}_2.$$

**Remarks.** Let $\mathcal{N} \subseteq \mathcal{M}$ be an inclusion with index strictly less than 4. Then the (unitary equivalence class of) the $\mathcal{M}$-$\mathcal{M}$ bimodule ${}_{\mathcal{M}}L^2(\mathcal{M}_1)_{\mathcal{M}}$ determines the inner conjugacy class of the subfactor $\mathcal{N}$ ([**33**]). This is not valid any longer when the index is 4. In fact, counter-examples occur for the inclusions obtained from the alternating groups $\mathfrak{A}_4 \supseteq \mathfrak{A}_3$ or $\mathbf{Z}_2 \times \mathbf{Z}_2 \supseteq \{e\}$. However, thanks to [**35**], subfactors obtained in this way are all conjugate (when factors are hyperfinite $II_1$). Therefore, the obvious question is whether the conjugacy class of $\mathcal{N}$ is determined by the $\mathcal{M}$-$\mathcal{M}$ bimodule ${}_{\mathcal{M}}L^2(\mathcal{M}_1)_{\mathcal{M}}$ (when factors are hyperfinite $II_1$). The answer to this question is negative. To see this, we set $\mathcal{N} = \mathcal{L}^{(\mathcal{N},\alpha)} = \rho_1(\mathcal{L})$, $\mathcal{L}^{(H,\beta)} = \rho_2(\mathcal{L})$ and $\rho = \bar{\rho}_1\rho_2$ as usual. Our inclusion $\mathcal{N} \subseteq \mathcal{M} = \mathcal{L} \rtimes_\beta H$ is conjugate to $\rho(\mathcal{L}) \subseteq \mathcal{L}$ so that the canonical endomorphism of the conjugate inclusion $\bar{\rho}(\mathcal{L}) \subseteq \mathcal{L}$ is

$$\bar{\rho}\rho = \bar{\rho}_2\rho_1\bar{\rho}_1\rho_2 = \oplus_{n \in N}\bar{\rho}_2\alpha_n\rho_2.$$

Note $\alpha_n = \beta_n$ as sectors from the definition of $\alpha$. This means that $\bar{\rho}\rho$ does not change when $\rho_1$ is replaced by a new $\rho_1'$ satisfying $\mathcal{L}^{(N,\beta)} = \rho_1'(\mathcal{L})$. Let $\rho' = \overline{\rho_1'}\rho_2$. The two inclusions $\bar{\rho}(\mathcal{L}) \subseteq \mathcal{L}$ and $\overline{\rho'}(\mathcal{L}) \subseteq \mathcal{L}$ admit the same canonical endomorphism, but they are not conjugate. Indeed, $\mathcal{L} \cap \rho\bar{\rho}(\mathcal{L})' = \hat{\mathcal{A}}$ while $\mathcal{L} \cap \rho'\overline{\rho'}(\mathcal{L})' = \mathbf{C}(\mathbf{Z}_n \times \mathbf{Z}_n) \otimes \ell^\infty(\mathbf{Z}_2)$ since no cocycle is involved in the latter inclusion. Since $\mathbf{C}(\mathbf{Z}_n \times \mathbf{Z}_n) \otimes \ell^\infty(\mathbf{Z}_2)$ is abelian, $\mathcal{L}$ is the crossed product of $\overline{\rho'}(\mathcal{L})$ relative to a group action. Notice that the reasoning so far works as long as $N$ is abelian and there exist a cocycle $\xi \in Z^2(N, \mathbf{T})$ such that

$$\eta_h(n_1, n_2) = \xi(n_1, n_2)\xi(n_1^h, n_2^h) \notin B^2(N, \mathbf{T}) \quad \text{for some } h \in H.$$

## 2. $(\mathbf{Z}_2 \times \mathbf{Z}_2) \rtimes \mathbf{Z}_2$

For $G = (\mathbf{Z}_2 \times \mathbf{Z}_2) \rtimes \mathbf{Z}_2$, the construction in the previous section is not so useful since the period of $(-1)^2$ is 1. In the appendix of [**1**], the following $\eta_\epsilon$ is considered instead:

$$\eta_\epsilon((a, b), (c, d)) = i^{2ad - ab - cd + (a+c)(b+d)} \quad \text{for } (a, b), (c, d) \in \mathbf{Z}_2 \times \mathbf{Z}_2, \tag{7.2}$$

where $a+c, b+d$ are computed in $mod$ 2.

| | (0,0) | (1,0) | (0,1) | (1,1) |
|---|---|---|---|---|
| (0,0) | 1 | 1 | 1 | 1 |
| (1,0) | 1 | 1 | $-i$ | $i$ |
| (0,1) | 1 | $i$ | 1 | $-i$ |
| (1,1) | 1 | $-i$ | $i$ | 1 |

Note (from the above table) that we have the $\mathbf{Z}_2$-equivariance (which simply means the following in the current case):

$$\eta_\epsilon((a,b),(c,d))\eta_\epsilon((b,a),(d,c)) = \eta_\epsilon((a,b),(c,d))(\epsilon \cdot \eta_\epsilon)((a,b),(c,d)) = 1.$$

We actually have $H((\mathbf{Z}_2 \times \mathbf{Z}_2, \mathbf{Z}_2), \mathbf{T}) = \mathbf{Z}_2$ and $(\eta, \zeta = 1)$ is a generator (see the argument for the proof of Proposition 7.5).

The cocycle $(\eta, \zeta = 1)$ gives us the 8-dimensional Kac-Paljutkin algebra ([**38**])

$$\mathcal{A} = \hat{\mathcal{A}} = \mathbf{C} \oplus \mathbf{C} \oplus \mathbf{C} \oplus \mathbf{C} \oplus M_2(\mathbf{C})$$

since $\eta_\epsilon \in Z^2(\mathbf{Z}_2 \times \mathbf{Z}_2, \mathbf{T})$ is cohomologous to the standard cocycle

$$((a,b),(c,d)) \in \mathbf{Z}_2 \times \mathbf{Z}_2 \longrightarrow (-1)^{ad}$$

and $\mathbf{C}_{\eta_\epsilon}(\mathbf{Z}_2)$ is a non-commutative 4-dimensional algebra. This is the only non-trivial Kac algebra of dimension 8. We have $\eta_\epsilon \neq 1 \in H^2(N, \mathbf{T})$ and $H_0 = \{e\}$ so that we see

$$G(\mathcal{A}) = Hom(N, \mathbf{T}) \cong \mathbf{Z}_2 \times \mathbf{Z}_2.$$

On the other hand, $\zeta = 1$ and $N_0 = \{(0,0),(1,1)\} \cong \mathbf{Z}_2$. Note $\eta_\epsilon((1,1)(1,1)) = 1$, which means $\eta.|_{N_0} = 1$. Therefore, we conclude

$$G(\hat{\mathcal{A}}) = N_0 \times Hom(H, \mathbf{T}) = \mathbf{Z}_2 \times \mathbf{Z}_2.$$

Let $\alpha, \beta$ are period 2 automorphisms of a factor $\mathcal{L}$, and we assume that the outer period of the composition $\alpha\beta$ is 4. We look at the inclusion

$$\mathcal{N} = \mathcal{L}^\alpha \subseteq \mathcal{M} = \mathcal{L} \rtimes_\beta \mathbf{Z}_2$$

of index 4. In Chapter 8 we will see that the Kac-Paljutkin algebra can be naturally obtained from this inclusion when the Connes obstruction of $\alpha\beta$ (one of the 4-th roots of 1) is exactly $-1$.

The same construction works for $(\mathbf{Z}_n \times \mathbf{Z}_n) \rtimes \mathbf{Z}_2$ as long as $n$ is even. Let $\eta_\epsilon$ is given by (7.2) again ($a+c, b+d$ are computed in $mod$ $n$) and $\zeta = 1$. We have the equivariance $\eta_\epsilon(\epsilon \cdot \eta_\epsilon) = 1$ because of

$$(2ad - ab - cd + (a+c)(b+d)) + (2bc - ba - dc + (b+d)(a+c)) = -4(ab+cd),$$

a multiple of 4. As in §1 we have

$$\mathcal{A} = \underbrace{\mathbf{C} \oplus \mathbf{C} \oplus \cdots \oplus \mathbf{C}}_{2n} \oplus \underbrace{M_2(\mathbf{C}) \oplus M_2(\mathbf{C}) \oplus \cdots \oplus M_2(\mathbf{C})}_{\frac{n^2-n}{2}}.$$

The dual Kac algebra is $\hat{\mathcal{A}} = \mathbf{C}(\mathbf{Z}_n \times \mathbf{Z}_n) \oplus \mathbf{C}_{\eta_\epsilon}(\mathbf{Z}_n \times \mathbf{Z}_n)$. As above $\eta_\epsilon \in Z^2(\mathbf{Z}_n \times \mathbf{Z}_n, \mathbf{T})$ is cohomologous to

$$((a,b),(c,d)) \in \mathbf{Z}_n \times \mathbf{Z}_n \longrightarrow (-1)^{ad}.$$

With this cocycle, the computation (7.1) of the center of $\mathbf{C}_{\eta_\epsilon}(\mathbf{Z}_n \times \mathbf{Z}_n)$ becomes

$$\sum_{a,b} c_{a,b}(-1)^{ad}\lambda_{(a+c,b+d)} = \sum_{a,b} c_{a,b}(-1)^{bc}\lambda_{(a+c,b+d)}.$$

Therefore, $x = \sum_{a,b} c_{a,b}\lambda_{(a,b)} \in \mathbf{C}_{\eta_\epsilon}(\mathbf{Z}_n \times \mathbf{Z}_n)$ is in the center if and only if $c_{a,b} = 0$ when either $a$ or $b$ is odd. Hence the dimension of the center is $(\frac{n}{2})^2$, and we get

$$\hat{\mathcal{A}} = \underbrace{\mathbf{C} \oplus \mathbf{C} \oplus \cdots \oplus \mathbf{C}}_{n^2} \oplus \underbrace{M_2(\mathbf{C}) \oplus M_2(\mathbf{C}) \oplus \cdots \oplus M_2(\mathbf{C})}_{(\frac{n}{2})^2}.$$

We note $H_0 = \{e\}$ and $N_0 = \{(k,k);\ k = 1, 2, \cdots, n-1\} \cong \mathbf{Z}_n$. We have $\eta_\cdot|_{N_0} = 1$ because of

$$\eta_\epsilon((a,a)(c,c)) = i^{2ac - a^2 - c^2 + (a+c)^2} = i^{4ac} = 1.$$

Therefore, the intrinsic groups are

$$G(\mathcal{A}) = Hom(N, \mathbf{T}) = \mathbf{Z}_n \times \mathbf{Z}_n \text{ and } G(\hat{\mathcal{A}}) = \mathbf{Z}_n \times \mathbf{Z}_2.$$

### 3. $(\mathbf{Z}_n \times \mathbf{Z}_n) \rtimes \mathbf{Z}_2$ with $n \geq 2$

Let $G = (\mathbf{Z}_n \times \mathbf{Z}_n) \rtimes \mathbf{Z}_2$ as usual, but we assume $n \geq 2$. For an $n$-th root $\omega$ of 1, we define $\eta_\epsilon \in Z^2(\mathbf{Z}_n \times \mathbf{Z}_n, \mathbf{T})$ by

$$\eta_\epsilon((a,b),(c,d)) = \omega^{ad}$$

(and $\eta_e = 1$). We compute

$$\frac{\eta_\epsilon((a,b),(c,d))\eta_\epsilon((a,b)^\epsilon,(c,d)^\epsilon)}{\eta_e((a,b),(c,d))} = \eta_\epsilon((a,b),(c,d))\eta_\epsilon((b,a)(d,c)) = \omega^{ad+bc},$$

showing that the $\mathbf{Z}_2$-equivariance is not available. However, we can define

$$\zeta_{(a,b)}(\epsilon,\epsilon) = \omega^{ab}$$

(and $\zeta_\cdot(e,e) = \zeta_\cdot(e,\epsilon) = \zeta_\cdot(\epsilon,e) = 1$). It is easy to see $\zeta \in Z^2(\mathbf{Z}_2, \mathbf{U}(\ell^\infty)(\mathbf{Z}_n \times \mathbf{Z}_n))$. We compute

$$\frac{\zeta_{(a,b)(c,d)}(\epsilon,\epsilon)}{\zeta_{(a,b)}(\epsilon,\epsilon)\zeta_{(c,d)}(\epsilon,\epsilon)} = \frac{\omega^{(a+c)(b+d)}}{\omega^{ab}\omega^{cd}} = \omega^{(a+c)(b+d)-ab-cd} = \omega^{ad+bc}.$$

Therefore, the fundamental relation between $\eta$ and $\zeta$ is satisfied, and from this data we get a non-trivial Kac algebra. The underlying algebra structure of $\mathcal{A} = \ell^\infty(\mathbf{Z}_n \times \mathbf{Z}_n) \rtimes_\zeta \mathbf{Z}_2$ is not changed by $\zeta$, i.e., the same as the one in §1. (In fact, firstly each fixed point in $N$ under the $\mathbf{Z}_2$-action gives us the group ring $\mathbf{C}(\mathbf{Z}_2) = \mathbf{C} \oplus \mathbf{C}$ because of $H^2(\mathbf{Z}_2, \mathbf{T}) = 1$. Secondly, each orbit consisting of two points gives us $M_2(\mathbf{C})$ anyway.) The computation (7.1) of the center of $\mathbf{C}_{\eta_\epsilon}(\mathbf{Z}_n \times \mathbf{Z}_n)$ becomes

$$\sum_{a,b} c_{a,b}\omega^{ad}\lambda_{(a+c,b+d)} = \sum_{a,b} c_{a,b}\omega^{bc}\lambda_{(a+c,b+d)} \quad \text{for each } b, c$$

in the present case. Therefore, what is relevant here is the period of $\omega$ (instead of $\omega^2$). Let $p_0$ be the period, and the dimension of the center is $(\frac{n}{p_0})^2$. The underlying algebra structure of the dual Kac algebra is

$$\hat{\mathcal{A}} = \underbrace{\mathbf{C} \oplus \mathbf{C} \oplus \cdots \oplus \mathbf{C}}_{n^2} \oplus \underbrace{M_{p_0}(\mathbf{C}) \oplus M_{p_0}(\mathbf{C}) \oplus \cdots M_{p_0}(\mathbf{C})}_{(\frac{n}{p_0})^2}.$$

When $n$ is even, our cocycle with $\omega = -1$ is cohomologous to the one at the end of §2. We remark $H^2(((\mathbf{Z}_n \times \mathbf{Z}_n), \mathbf{Z}_n), \mathbf{T}) = \mathbf{Z}_n$. In fact, we have already had $(\eta_\epsilon^\omega, \zeta)$ and $H^2(\mathbf{Z}_n \times \mathbf{Z}_n, \mathbf{T}) = \mathbf{Z}_n$ is known. What we have to show here is that there is no further freedom for changing $\zeta$. This means that (after assuming $\eta = 1$) we have to show $H^2(\mathbf{Z}_2, Hom(\mathbf{Z}_n \times \mathbf{Z}_n, \mathbf{T})) = 1$. To this end, let $\zeta'$ be a cocycle, and as usual we assume $\zeta'_\cdot(e,e) = \zeta'_\cdot(\epsilon,e) = \zeta'_\cdot(e,\epsilon) = 1$ and $n \in \mathbf{Z}_2 \times \mathbf{Z}_2 \longrightarrow \zeta'_n(\epsilon,\epsilon)$ is a character. Since $\zeta'_n(\epsilon,\epsilon) = \zeta'_{n^\epsilon}(\epsilon,\epsilon)$, it must be of the form $\zeta'_n(\epsilon,\epsilon) = \omega^{i+j}$ with $\omega^n = 1$. But this is a coboundary since $\xi(e,(i,j)) = 1$ and $\xi(\epsilon,(i,j)) = \omega^{-i}$ give us

$$\xi(e,(i,j))\overline{\xi(\epsilon,(i,j))\xi(\epsilon,(i,j)^\epsilon)} = \omega^i \times \omega^j.$$

## 4. $(\mathbf{Z}_n \times \mathbf{Z}_n) \rtimes \mathbf{Z}_2$ with $n$ even

So far we have been dealing with the flip action of $\mathbf{Z}_2$ on $\mathbf{Z}_n \times \mathbf{Z}_n$. This time we set $G = (\mathbf{Z}_n \times \mathbf{Z}_n) \rtimes \mathbf{Z}_2$ with $n = 2m$ $(\geq 4)$ even with the following action of $H = \mathbf{Z}_2 = \{e, \epsilon\}$:

$$(a,b)^\epsilon = (a,-b) \quad \text{for } (a,b) \in N = \mathbf{Z}_n \times \mathbf{Z}_n.$$

For an $n$-th root $\omega$ of 1, we set

$$\eta^\omega_\epsilon((a,b),(c,d)) = \omega^{ad}$$

(and $\eta^\omega_e = 1$). We have the $\mathbf{Z}_2$-equivariance

$$\eta^\omega_\epsilon((a,b),(c,d))\eta^\omega_\epsilon((a,b)^\epsilon(c,d)^\epsilon) = \eta^\omega_\epsilon((a,b),(c,d))\eta^\omega_\epsilon((a,-b)(c,-d)) = \omega^{ad-ad} = 1,$$

i.e., $[\eta^\omega, \zeta = 1] \in H^2((\mathbf{Z}_n \times \mathbf{Z}_n, \mathbf{Z}_2), \mathbf{T})$. We actually prove

PROPOSITION 7.1. *We have* $H^2((\mathbf{Z}_n \times \mathbf{Z}_n, \mathbf{Z}_2), \mathbf{T}) = \mathbf{Z}_n \times \mathbf{Z}_2 \times \mathbf{Z}_2$.

PROOF. As before, we have to worry about is the ambiguity for $\zeta$ and hence we have to show $H^2(\mathbf{Z}_2, Hom(\mathbf{Z}_n \times \mathbf{Z}_n)) = \mathbf{Z}_2 \times \mathbf{Z}_2$. The action is $(i,j)^\epsilon = (i,-j)$ so that the requirement $\zeta'_n(\epsilon,\epsilon) = \zeta'_{n^\epsilon}(\epsilon,\epsilon)$ in the argument at the last part of §3 becomes $\zeta'_n(\epsilon,\epsilon) = \omega_1^i\omega_2^j$ with $\omega_1^n = 1$ and $\omega_2^2 = 1$. We now try to look for a character $\xi(\epsilon,\cdot)$ (and $\xi(e,\cdot) = 1$) satisfying

$$\xi(e,(i,j))\overline{\xi(\epsilon,(i,j))\xi(\epsilon,(i,j)^\epsilon)} = \omega_1^i\omega_2^j.$$

By setting $\xi(\epsilon,(i.j)) = \omega_3^i\omega_4^j$ $(\omega_3^n = \omega_4^n = 1)$, this condition means $\omega_1^i\omega_2^j = \omega_3^{-2i}$. Therefore, if $\omega_1^{\frac{n}{2}} = 1$ and $\omega_2 = 1$, then $\zeta'$ is a coboundary. Moreover, it is also clear from the computation so far that we have $H^2(\mathbf{Z}_2, Hom(\mathbf{Z}_n \times \mathbf{Z}_n, \mathbf{T})) = \mathbf{Z}_2 \times \mathbf{Z}_2$, and we have the following representatives:

$$\zeta_{(i,j)}(\epsilon,\epsilon) = 1,\ \omega_0^i,\ (-1)^j,\ \omega_0^i(-1)^j$$

with $\omega_0 = e^{\frac{2\pi i}{n}}$. □

Let $(\eta^\omega, \zeta)$ be our cocycle. Recall that the action of $\epsilon$ is $(i,j) \longrightarrow (i,-j)$. Hence, the following $2n$ points in $N = \mathbf{Z}_n \times \mathbf{Z}_n$ are fixed under the $\mathbf{Z}_2$-action:

$$\{(k,0), (k, m = \tfrac{n}{2});\ k = 0,1,\cdots,n-1\}.$$

Therefore, the Kac algebra $\mathcal{A}$ contains $4n$ copies of $\mathbf{C}$. The underlying algebra structure of $\mathcal{A}$ is not effected by $\zeta$, and we have

$$\mathcal{A} = \underbrace{\mathbf{C} \oplus \mathbf{C} \oplus \cdots \oplus \mathbf{C}}_{4n} \oplus \underbrace{M_2(\mathbf{C}) \oplus M_2(\mathbf{C}) \oplus \cdots \oplus M_2(\mathbf{C})}_{\frac{n^2-2n}{2}}.$$

(Compare this with [**63**].) Also (as in the §3) the underlying algebra structure of the dual Kac algebra $\hat{\mathcal{A}}$ is

$$\begin{aligned}\hat{\mathcal{A}} &= \mathbf{C}(\mathbf{Z}_n \times \mathbf{Z}_n) \oplus \mathbf{C}_{\eta_\epsilon}(\mathbf{Z}_n \times \mathbf{Z}_n) \\ &= \underbrace{\mathbf{C} \oplus \mathbf{C} \oplus \cdots \oplus \mathbf{C}}_{n^2} \oplus \underbrace{M_{p_0}(\mathbf{C}) \oplus M_{p_0}(\mathbf{C}) \oplus \cdots M_{p_0}(\mathbf{C})}_{(\frac{n}{p_0})^2},\end{aligned}$$

where $p_0$ is the period of $\omega$.

When $\omega = 1$, $\hat{\mathcal{A}}$ is abelian and hence $\mathcal{A}$ is cocommutative. For example, when $\zeta = 1$, it is easy to see that $\mathcal{A}$ is just the group ring of the original group $G = (\mathbf{Z}_n \times \mathbf{Z}_n) \rtimes \mathbf{Z}_2$. (We have $H_0 = H$ naturally acting on $Hom(N, \mathbf{T}) = \mathbf{Z}_n \times \mathbf{Z}_n$, and hence $G(\mathcal{A}) = Hom(N, \mathbf{T}) \rtimes H \cong G$. But, $\zeta \neq 1$ gives us a different group.) In what follows, we assume $\omega \neq 1$. Therefore, we have $H_0 = \{e\}$ and

$$G(\mathcal{A}) = Hom(N, \mathbf{T}) = \mathbf{Z}_n \times \mathbf{Z}_n$$

always. So far the effect of the cocycle $\zeta$ is not so relevant. However, the intrinsic group $G(\hat{\mathcal{A}})$ very much depends on $\zeta$ as will be seen below.

We note

$$\begin{aligned}N_0 &= \{(i,j) \in N;\ (i,j)^\epsilon = (i,-j) = (i,j) \text{ and } \zeta_{(i,j)}(\cdot,\cdot) \in B^2(\mathbf{Z}_2, \mathbf{T})\} \\ &= \{(k,0), (k,m) : k = 0, 1, \cdots, n-1\},\end{aligned}$$

which is isomorphic to $\mathbf{Z}_n \times \mathbf{Z}_2$

**(1).** The case $\zeta = 1$

Remark that $\eta_\cdot|_{N_0}$ is given by

$$\begin{array}{ll}\eta_\epsilon((k,0),(k',0)) = 1 & \eta_\epsilon((k,m),(k',0)) = 1 \\ \eta_\epsilon((k,0),(k',m)) = \omega^{km} & \eta_\epsilon((k,m),(k',m)) = \omega^{km}.\end{array}$$

When $\omega^m = 1$, $\eta_\cdot|_{N_0}$ is trivial and from

$$1 \longrightarrow Hom(H, \mathbf{T}) = \mathbf{Z}_2 \longrightarrow G(\hat{\mathcal{A}}) \longrightarrow N_0 \longrightarrow 1$$

we get $G(\hat{\mathcal{A}}) = N_0 \times \mathbf{Z}_2 = \mathbf{Z}_n \times \mathbf{Z}_2 \times \mathbf{Z}_2$.

We next deal with the case $\omega^m = -1$. Then $\eta_\cdot|_{N_0}$ is

$$\begin{aligned}&\eta_\epsilon((k,0),(k',0)) = \eta_\epsilon((k,m),(k',0)) = 1, \\ &\eta_\epsilon((k,0),(k',m)) = \eta_\epsilon((k,m),(k',m)) = (-1)^k.\end{aligned}$$

We set

$$a = (0,(1,0)),\ b = (0,(0,m)), c = (1,(0,0)) \in Hom(H,\mathbf{T}) \times N_0 = \mathbf{Z}_2 \times (\mathbf{Z}_n \times \mathbf{Z}_2).$$

We obviously have

$$a^n = b^2 = c^2 = e, ac = ca, bc = cb.$$

But, due to the presence of the cocycle we have

$$\begin{aligned}ab &= (0,(1,0))(0,(0,m)) = (1,(1,m)), \\ abc &= (0,(1,m)) = ba.\end{aligned}$$

Hence, $a(bc)a^{-1} = b$ and $aba^{-1} = aba^{-1}c^2 = abca^{-1}c = baa^{-1}c = bc$. The intrinsic group $G(\hat{\mathcal{A}})$ generated by $a, b, c$ is isomorphic to $(\mathbf{Z}_2 \times \mathbf{Z}_2) \rtimes \mathbf{Z}_n$ with the action

$$(0,0)^a = (0,0), (1,1)^a = (1,1), (1,0)^a = (0,1),$$

where $a$ $(a^n = e)$ is the generator of $\mathbf{Z}_n$. Note that this action makes sense since $n$ is even, and the group is the same as $((\mathbf{Z}_2 \times \mathbf{Z}_2) \rtimes \mathbf{Z}_2) \times \mathbf{Z}_{\frac{n}{2}}$.

**(2).** The case $\zeta_{(i,j)}(\epsilon, \epsilon) = (-1)^j$
At first we have to renormalize $(\eta^\omega, \zeta)$ so that $\zeta_n = 1$, $n \in N_0$. If $m = \frac{n}{2}$ is even, $(-1)^m = 1$ and hence $\zeta_n = 1$, $n \in N_0$. Therefore, renormalization is not necessary and the computation of $G(\hat{\mathcal{A}})$ is the same as in **(1)**. So let us assume that $m$ is odd. We set

$$\xi(\epsilon, (i, m)) = \sqrt{-1} \ (i = 0, 1, \cdots, n-1)$$

(and other $\xi(\cdot, \cdot)$'s are 1) so that

$$\zeta_{(i,m)} \times (\partial^H \xi)_{(i,m)} = 1.$$

Recall

$$(\partial^N \xi)_\epsilon(n_1, n_2) = \overline{\xi(\epsilon, n_1 n_2)}\xi(\epsilon, n_1)\xi(\epsilon, n_2).$$

It is easy to see that $\eta_\epsilon$ (on $N_0$) is changed to

$$\eta_\epsilon((k, m)(k', m)) = -\omega^{km}.$$

(Other values are not changed.)

Thus, when $\omega^m = 1$, $\eta_\cdot|_{N_0}$ is given by

$$\eta_\epsilon((k, m)(k', m)) = -1 \text{ and other } \eta_\epsilon \text{ is } 1.$$

Thus, the group $G(\hat{\mathcal{A}})$ is generated by commuting $a, b, c$ satisfying

$$a^n = c^2 = e, b^2 = c.$$

(note $(0, (0, m))(0, (0, m)) = (1, (0, 0))$ due to the cocycle) Notice that $b, c$ generate the group $\mathbf{Z}_4$ and $G(\hat{\mathcal{A}}) = \mathbf{Z}_4 \times \mathbf{Z}_n$.

We next assume $\omega^m = -1$. In this case, the (renormalized) $\eta_\epsilon$ is

$$\begin{aligned}
\eta_\epsilon((k, 0), (k', 0)) &= \eta_\epsilon((k, m), (k', 0)) = 1, \\
\eta_\epsilon((k, 0), (k', m)) &= (-1)^k, \\
\eta_\epsilon((k, m), (k', m)) &= -(-1)^k.
\end{aligned}$$

Thus, the generators $a, b, c$ satisfy

$$a^n = c^2 = e, b^2 = c, ca = ac, cb = bc, abc = ba.$$

Thus, $a$ acts on $\mathbf{Z}_4$ generated by $b, c$ via $a1a^{-1} = 3$, and the intrinsic group is $\mathbf{Z}_4 \rtimes \mathbf{Z}_n = (\mathbf{Z}_4 \rtimes \mathbf{Z}_2) \times \mathbf{Z}_{\frac{n}{2}}$.

**(3).** The case $\zeta_{(i,j)}(\epsilon, \epsilon) = \omega_0^i$
We set

$$\xi(\epsilon, (i, *)) = \omega_1^i, \ * = 0, m$$

with $\omega_1 = e^{\frac{\pi i}{n}}$. (Other $\xi(\cdot, \cdot)$'s are 1.) Since $\omega_1^2 = \omega_0$, we have

$$(\partial^H \xi)_{(i,*)}(\epsilon, \epsilon) = \xi(e, (i, *))\overline{\xi(\epsilon, (i, *))\xi(\epsilon, (i, *))} = \omega_1^{-2i} = \omega_0^{-i}$$

for $(i, *) \in N_0$. Therefore, $(\partial^H \xi)_{(i,*)} \times \zeta_{(i,*)} = 1$ for $* = 0, m$. We compute

$$(\partial^N \xi)_\epsilon((i, *), (j, *)) = \overline{\omega_1^{i+j}}\omega_1^i \omega_1^j$$

for $* = 0, m$. Note that $i + j$ here should be computed in *mod* $n$ so that we get the value $-1$ when $i + j \geq n$. We set

$$f(i, j) = \begin{cases} 1 & \text{if } i + j < n, \\ -1 & \text{if } i + j \geq n \end{cases}$$

so that the product $\eta_\epsilon((i, *), (j, *)) \times f(i, j)$ is a renormalized cocycle.

At first, when $\omega^m = 1$, the original $\eta_\epsilon$ is trivial on $N_0 = \mathbf{Z}_n \times \mathbf{Z}_2$. Hence, by the cocycle $f$, we easily get $G(\hat{\mathcal{A}}) = \mathbf{Z}_{2n} \times \mathbf{Z}_2$. When $\omega^m = -1$, we have

$$\eta_\epsilon((k,0),(k',0)) = \eta_\epsilon((k,m)(k',0)) = f(k,k'),$$
$$\eta_\epsilon((k,0),(k',m)) = \eta_\epsilon((k,m)(k',m)) = f(k,k')(-1)^k.$$

The generators $a, b, c$ satisfy

$$a^n = c, c^2 = b^2 = e, ca = ac, cb = bc, abc = ba.$$

In particular, we compute $bab = abcb = ab^2c = ac$. Note that $a, c$ generate $\mathbf{Z}_{2n}$ and that $b \in \mathbf{Z}_2 = \{e, b\}$ acts on $\mathbf{Z}_{2n}$ by $b1b = n+1$, i.e.,

$$bib^{-1} = \begin{cases} i & \text{when } i \in \mathbf{Z}_{2n} \text{ is even,} \\ i+n \ (mod\ 2n) & \text{when } i \in \mathbf{Z}_{2n} \text{ is odd.} \end{cases}$$

The intrinsic group $G(\hat{\mathcal{A}})$ is thus the semi-direct product $\mathbf{Z}_{2n} \rtimes \mathbf{Z}_2$ relative to this action.

**(4).** The case $\zeta_{(i,j)}(\epsilon,\epsilon) = \omega_0^i(-1)^j$
We assume that $m$ is odd. (Otherwise, $(-1)^m = 1$ and the cocycle is the same as the one in **(3)** on $N_0$.) In this case, the effect of the previous two cases appear simultaneously and hence the renormalized cocycle is given by

$$\eta_\epsilon((k,0),(k',0)) = f(k,k') \qquad \eta_\epsilon((k,m),(k',0)) = f(k,k')$$
$$\eta_\epsilon((k,0),(k',m)) = f(k,k')(\omega^m)^k \quad \eta_\epsilon((k,m),(k',m)) = -f(k,k')(\omega^m)^k.$$

At first we assume $\omega^m = 1$. Then, $(\omega^m)^k = 1$ in the above formula, and the commuting generators $a, b, c$ satisfy

$$a^n = c, c^2 = e, b^2 = c.$$

Note that $b^3 = bc = (1,(0,m))$ has order 4:

$$(bc)^2 = b^6 = c = (1,(0,0)),\ (bc)^3 = b^9 = b = (0,(0,m)),\ (bc)^4 = c^2 = e$$

On the other hand, $(ab)^n = a^nb^n = cb^{2m} = c^{m+1} = e$ since $m$ is odd, and

$$ab = (0,(1,m)), (ab)^2 = (1,(2,0)), (ab)^3 = (1,(3,m)), (ab)^4 = (0,(4,0)), \cdots.$$

Everything is commuting here, and let us assume $(ab)^i(bc)^j = e$. Since $(ab)^i(bc)^j$ is of the form $(*,(i,*))$, we have $i = 0\ (mod\ n)$ and hence $j = 0\ (mod\ 4)$. We thus conclude $G(\hat{\mathcal{A}}) = \mathbf{Z}_n \times \mathbf{Z}_4$.

Secondly, we assume $\omega^n = -1$ so that the factor $(-1)^k$ appears in the above formula. Therefore, the group $G(\hat{\mathcal{A}})$ is given by generators $a, b, c$ satisfying

$$a^n = c, c^2 = 1, b^2 = c, ca = ac, cb = bc, abc = ba.$$

## 5. $D_{2n} \rtimes \mathbf{Z}_2$ with $n$ odd

Let $N = D_{2n} = \mathbf{Z}_n \rtimes \mathbf{Z}_2$ be the dihedral group of order $2n$, where $(0,1) \in \mathbf{Z}_2$ acts on $\mathbf{Z}_n$ via $(k,0) \longrightarrow (n-k,0)$. Let $a = (0,1), b = (n-1,1)$, and notice

$$a^2 = b^2 = (ab)^n = e.$$

We let $H = \mathbf{Z}_2 = \{e, \epsilon\}$ act on $N$ by exchanging the two generators $a$ and $b$, and we set $G = N \rtimes H$. Here, we assume that $n$ is odd, and see that the Kac algebra structure (but not underlying algebra structure) can be deformed by using $\zeta$.

Since $n$ is odd, $Hom(N, \mathbf{T}) = \mathbf{Z}_2$ and the non-trivial character $\chi$ on $N = D_{2n}$ is given by

$$\chi(k,0) = 1, \ \chi(k,1) = -1$$

(i.e., $\chi(\cdot)$ is the parity of the "second variable"). Let $\eta_e = \eta_\epsilon = 1$ and we define $\zeta$ by

$$\zeta_n(\alpha,\beta) = \begin{cases} 1 & \text{when } (\alpha,\beta) = (\epsilon,\epsilon), \ n = (k,0) \in D_{2n} = \mathbf{Z}_n \rtimes \mathbf{Z}_2, \\ -1 & \text{when } (\alpha,\beta) = (\epsilon,\epsilon), \ n = (k,1) \in D_{2n} = \mathbf{Z}_n \rtimes \mathbf{Z}_2, \\ 1 & \text{when } (\alpha,\beta) \neq (\epsilon,\epsilon). \end{cases}$$

It is plain to see $\zeta \in Z^2(\mathbf{Z}_2, \mathbf{U}(\ell^\infty(D_{2n})))$. Also note $n \longrightarrow \zeta_n(\epsilon,\epsilon)$ is exactly $\chi$. From this we observe the character property $\zeta_n\zeta_{n'} = \zeta_{nn'}$, and hence $(\eta = 1, \zeta)$ gives us a Kac algebra.

Since $D_{2n}$ (with $n$ odd) possesses two 1-dimensional characters and $\frac{n-1}{2}$ 2-dimensional irreducible representations, we have

$$\begin{aligned} \hat{\mathcal{A}} &= \mathbf{C}(D_{2n}) \oplus \mathbf{C}(D_{2n}) \\ &= \mathbf{C} \oplus \mathbf{C} \oplus \mathbf{C} \oplus \mathbf{C} \oplus \underbrace{M_2(\mathbf{C}) \oplus M_2(\mathbf{C}) \oplus \cdots \oplus M_2(\mathbf{C})}_{n-1} \end{aligned}$$

with or without $\zeta$. Notice that $D_{2n}$ has two fixed points $\{e = (0,0), (\frac{n-1}{2}, 1)\}$ under the $\mathbf{Z}_2$-action. It is easy to see that with or without the cocycle $\zeta$ the Kac algebra $\mathcal{A}$ has the same underlying algebra and it is the same as the above $\hat{\mathcal{A}}$. However, the introduction of the cocycle $\zeta$ does indeed change the Kac algebra structure (see Lemma 7.2 below). We also point out

$$H^2((D_{2n}, H), \mathbf{T}) = H^2(H, Hom(D_{2n})) = H^2(\mathbf{Z}_2, \mathbf{Z}_2) = \mathbf{Z}_2$$

thanks to $H^2(D_{2n}, \mathbf{T}) = 1$.

When $\zeta = 1$, we have $N_0 = \{(0,0), (\frac{n-1}{2}, 1)\} \cong \mathbf{Z}_2$ and $H = H_0 = \{e, \epsilon\} = \mathbf{Z}_2$. In the present case, $H_0$ acts trivially to $Hom(N, \mathbf{T}) \cong \mathbf{Z}_2$. Also, $Hom(H, \mathbf{T}) \cong \mathbf{Z}_2$. Therefore, we conclude (without cocycles)

$$G(\mathcal{A}) = G(\hat{\mathcal{A}}) = \mathbf{Z}_2 \times \mathbf{Z}_2.$$

LEMMA 7.2. *When we have the above cocycle $\zeta$, we get*

$$G(\mathcal{A}) = G(\hat{\mathcal{A}}) = \mathbf{Z}_4.$$

PROOF. Note $N_0 = \{(0,0), (\frac{n-1}{2}, 1)\} \cong \mathbf{Z}_2$ because $H^2(H, \mathbf{T}) = H^2(\mathbf{Z}_2, \mathbf{T})$ is trivial. Also, since $\eta = 1$ in the present case, we have $H_0 = H$. Since $\eta = 1$, we need not renormalize $\zeta$ and $\zeta_\cdot|_{H_0} \in Z^2(H_0, Hom(N, \mathbf{T}))$ is indeed the restriction. The action of $H_0$ on $Hom(N, \mathbf{T}) \cong \mathbf{Z}_2$ is trivial, and $\zeta_\cdot|_{H_0}$ represents the non-trivial element in $H^2(H_0, Hom(N, \mathbf{T})) = H^2(\mathbf{Z}_2, \mathbf{Z}_2) \cong \mathbf{Z}_2$. Therefore, the corresponding extension

$$1 \longrightarrow Hom(N, \mathbf{T}) \longrightarrow G(\mathcal{A}) \to H_0 \longrightarrow 1$$

is non-trivial, and we conclude $G(\mathcal{A}) = \mathbf{Z}_4$.

Note that $\zeta_{(\frac{n-1}{2}, 1)}(\cdot, \cdot)$ is a coboundary ($H^2(\mathbf{Z}_2, \mathbf{T}) = 1$), but not identically 1. Thus, we have to renormalize $(\eta = 1, \zeta)$. For example, we can use $\xi(\cdot, \cdot)$ satisfying

$$\xi\left(\epsilon, \left(\frac{n-1}{2}, 1\right)\right) = i, \ \xi\left(e, \left(\frac{n-1}{2}, 1\right)\right) = 1.$$

Thus, the renormalized $\eta$ (i.e., $1 \times \partial^H \xi$) satisfies

$$\eta_\epsilon\left(\left(\frac{n-1}{2},1\right),\left(\frac{n-1}{2},1\right)\right) = -1.$$

Consequently, $\eta.|_{N_0} \in Z^2(N_0, Hom(H,\mathbf{T}))$ gives rise to the non-trivial element in $H^2(N_0, Hom(H,\mathbf{T})) \cong H^2(\mathbf{Z}_2,\mathbf{Z}_2) \cong \mathbf{Z}_2$. Hence, the corresponding extension

$$1 \longrightarrow Hom(H,\mathbf{T}) \longrightarrow G(\hat{\mathcal{A}}) \longrightarrow H_0 \longrightarrow 1$$

is once again non-trivial, and we have $G(\hat{\mathcal{A}}) = \mathbf{Z}_4$. □

We point out that (when $n = 3$) in this way we get the two 12-dimensional non-trivial Kac algebras with the identical underlying algebra

$$\mathcal{A} = \hat{\mathcal{A}} = \mathbf{C} \oplus \mathbf{C} \oplus \mathbf{C} \oplus \mathbf{C} \oplus M_2(\mathbf{C}) \oplus M_2(\mathbf{C})$$

(but with different intrinsic groups). It turns out that they are the only 12-dimensional non-trivial ones (as will be shown in Theorem 10.10 and Theorem 10.7,(iv)). Some related results were also obtained in [**14**].

## 6. $D_{2n} \rtimes \mathbf{Z}_2$ with $n$ even

Let $G = D_{2n} \rtimes \mathbf{Z}_2$ as in the previous part, but we now assume that $n$ is even. At first, notice that $D_{2n}$ has the two points $\{e,(m,0)\}$ fixed by the $\mathbf{Z}_2$-action. (We have $\frac{2n-2}{2}$ remaining orbits, and all of them consists of two points.) Recall that $D_{2n}$ has four characters and $(\frac{n}{2}-1)$ 2-dimensional irreducible representations. (The four characters assign values $(1,1),(1,-1),(-1,1),(-1,-1)$ to the two generators $a,b$) Therefore, without introducing cocycles we have the following underlying algebra structure:

$$\begin{aligned}
\mathcal{A} &= \mathbf{C}(D_{2n}) \rtimes \mathbf{Z}_2 \\
&= \mathbf{C} \oplus \mathbf{C} \oplus \mathbf{C} \oplus \mathbf{C} \oplus \underbrace{M_2(\mathbf{C}) \oplus M_2(\mathbf{C}) \oplus \cdots \oplus M_2(\mathbf{C})}_{n-1}, \\
\hat{\mathcal{A}} &= \mathbf{C}(D_{2n}) \oplus \mathbf{C}(D_{2n}) \\
&= \underbrace{\mathbf{C} \oplus \mathbf{C} \oplus \cdots \oplus \mathbf{C}}_{8} \oplus \underbrace{M_2(\mathbf{C}) \oplus M_2(\mathbf{C}) \oplus \cdots \oplus M_2(\mathbf{C})}_{n-2}.
\end{aligned}$$

In this case (i.e., $n$ is even) we will see that one can deform the relevant structure by introducing $\eta$ instead. We will indeed see that the underlying algebra structure of $\hat{\mathcal{A}}$ is changed (we will get $\hat{\mathcal{A}} = \mathcal{A}$).

Let $f$ be a $\{\pm 1\}$-valued function on $\mathbf{Z}$ of period $2n$ satisfying

$$f(n+k) = -f(k), \tag{7.3}$$

$$f(k)f(-k-1) = -1. \tag{7.4}$$

(All the values are determined by $f(0), f(1), \cdots, f(\frac{n}{2}-1)$.) We take $\omega \in \mathbf{T}$ satisfying $\omega^n = -1$, and consider $\eta_\epsilon((k,j)(m,j'))$ $(j,j' = 0,1)$ given by the following table:

| | $(m,0)$ | $(m,1)$ |
|---|---|---|
| $(k,0)$ | $1$ | $f(m)f(k+m)\omega^k$ |
| $(k,1)$ | $f(k)f(k-m)\omega^m$ | $f(k)f(m)\omega^{m-k}$ |

(We of course set $\eta_e = 1$.)

LEMMA 7.3. *The above $\eta_\epsilon$ is a normalized 2-cocycle on $D_{2n}$. In fact, it corresponds to the following projective representation of $D_{2n}$:*

$$\begin{aligned}\pi((k,0)) &= \omega^{-k}\begin{pmatrix}\omega^{-k} & 0\\ 0 & \omega^{k}\end{pmatrix},\\ \pi((k,1)) &= f(k)\begin{pmatrix}0 & \omega^{-k}\\ \omega^{k} & 0\end{pmatrix}.\end{aligned}$$

PROOF. The normalization property ($\eta_\epsilon(n,e) = \eta_\epsilon(e,n) = \eta_\epsilon(n,n^{-1}) = 1$) is guaranteed by the fact that the values of $f(i)$'s are just $\pm 1$. We compute

$$\begin{aligned}\pi((k,0))\pi((m,1)) &= f(m)\omega^{-k}\begin{pmatrix}\omega^{-k} & 0\\ 0 & \omega^{k}\end{pmatrix}\begin{pmatrix}0 & \omega^{-m}\\ \omega^{m} & 0\end{pmatrix}\\ &= f(m)\omega^{-k}\begin{pmatrix}0 & \omega^{-k-m}\\ \omega^{k+m} & 0\end{pmatrix},\\ \pi((k,0)(m,1)) &= \pi((k+m,1)) = f(k+m)\begin{pmatrix}0 & \omega^{-k-m}\\ \omega^{k+m} & 0\end{pmatrix},\\ \pi((k,1))\pi((m,0)) &= f(k)\omega^{-m}\begin{pmatrix}0 & \omega^{-k}\\ \omega^{k} & 0\end{pmatrix}\begin{pmatrix}\omega^{-m} & 0\\ 0 & \omega^{m}\end{pmatrix}\\ &= f(k)\omega^{-m}\begin{pmatrix}0 & \omega^{-k+m}\\ \omega^{k-m} & 0\end{pmatrix},\\ \pi((k,1)(m,0)) &= \pi((k+n-m,1))\\ &= f(k+n-m)\begin{pmatrix}0 & \omega^{-k-n+m}\\ \omega^{k+n-m} & 0\end{pmatrix}\\ &= f(k-m)\begin{pmatrix}0 & \omega^{-k+m}\\ \omega^{k-m} & 0\end{pmatrix},\\ \pi((k,1))\pi((m,1)) &= f(k)f(m)\begin{pmatrix}0 & \omega^{-k}\\ \omega^{k} & 0\end{pmatrix}\begin{pmatrix}0 & \omega^{-m}\\ \omega^{m} & 0\end{pmatrix}\\ &= f(k)f(m)\begin{pmatrix}\omega^{m-k} & 0\\ 0 & \omega^{k-m}\end{pmatrix},\\ \pi((k,1)(m,1)) &= \pi((k+n-m,0)) = \omega^{-k-n+m}\begin{pmatrix}\omega^{-k-n+m} & 0\\ 0 & \omega^{k+n-m}\end{pmatrix}\\ &= \omega^{m-k}\begin{pmatrix}\omega^{m-k} & 0\\ 0 & \omega^{k-m}\end{pmatrix}.\end{aligned}$$

In some of the above calculations, (7.3) and $\omega^n = -1$ were used. Notice that the respective ratios are $\eta_\epsilon((k,0),(m,1))$, $\eta_\epsilon((k,1),(m,0))$, and $\eta_\epsilon((k,1),(m,1))$, and hence $\eta_\epsilon$ is indeed a cocycle. □

Notice $(k,0)^\epsilon = (n-k,0)$ and $(k,1)^\epsilon = (n-k-1,1)$ ($ab = (1,0)$ and $(ab)^\epsilon = ba = (n-1,0)$). We have

LEMMA 7.4. *We have the $\mathbf{Z}_2$-equivariance:*

$$\eta_\epsilon(n_1,n_2)\eta_\epsilon(n_1^\epsilon,n_2^\epsilon) = 1 \quad \textit{for each } n_1,n_2 \in D_{2n}.$$

PROOF. By making use of (7.3),(7.4), and $\omega^n = -1$ we compute

$$\begin{aligned}
&\eta_\epsilon((k,0),(m,1))\eta_\epsilon((k,0)^\epsilon(m,1)^\epsilon)\\
&\quad = \eta_\epsilon((k,0),(m,1))\eta_\epsilon((n-k,0),(n-m-1,1))\\
&\quad = f(m)f(k+m)\omega^k \times f(n-m-1)f(2n-m-k-1)\omega^{n-k}\\
&\quad = f(m)f(-m-1)f(k+m)f(-m-k-1) = 1,\\
&\eta_\epsilon((k,1),(m,0))\eta_\epsilon((k,1)^\epsilon,(m,0)^\epsilon)\\
&\quad = \eta_\epsilon((k,1),(m,0))\eta_\epsilon((n-k-1,1),(n-m,0))\\
&\quad = f(k)f(k-m)\omega^m \times f(n-k-1)f(m-k-1)\omega^{n-m}\\
&\quad = -f(k)f(n-k-1)f(k-m)f(m-k-1)\\
&\quad = f(k)f(-k-1)f(k-m)f(m-k-1) \;=\; 1;\\
&\eta_\epsilon((k,1),(m,1))\eta_\epsilon((k,1)^\epsilon,(m,1)^\epsilon)\\
&\quad = \eta_\epsilon((k,1),(m,1))\eta_\epsilon((n-k-1,1),(n-m-1,1))\\
&\quad = f(k)f(m)\omega^{m-k} \times f(n-k-1)f(n-m-1)\omega^{k-m}\\
&\quad = f(k)f(-k-1)f(m)f(-m-1) \;=\; 1.
\end{aligned}$$

□

We now compute the center of the twisted group ring $\mathbf{C}_{\eta_\epsilon}(D_{2n})$. Let

$$x = \sum_{i=0}^{n-1} a_i \rho_\epsilon(i,0) + \sum_{j=0}^{n-1} b_j \rho_\epsilon(j,1) \in \mathbf{C}_{\eta_\epsilon}(D_{2n}),$$

and we require that $x$ commutes with $\rho_\epsilon(1,0)$. We compute

$$\rho_\epsilon(1,0)x \;=\; \sum_{i=0}^{n-1} a_i\rho_\epsilon(i+1,0) + \sum_{i=0}^{n-1} b_j f(j)f(j+1)\omega\rho_\epsilon(j+1,1), \tag{7.5}$$

$$x\rho_\epsilon(1,0) \;=\; \sum_{i=0}^{n-1} a_i\rho_\epsilon(i+1,0) + \sum_{j=0}^{n-1} b_j f(j)f(j-1)\omega\rho_\epsilon(j+n-1,1). \tag{7.6}$$

By comparing coefficients of $\rho_\epsilon(n-1,1), \rho_\epsilon(0,1)$, $(j = n-2, n-1$ in (7.5) and $j = 0,1$ in (7.6)), we get

$$\begin{aligned}
b_{n-2}f(n-2)f(n-1) &= b_0 f(0)f(-1) = b_0 f(n)f(n-1),\\
b_{n-1}f(n-1)f(n) &= b_1 f(1)f(0) = b_1 f(n+1)f(n)
\end{aligned}$$

(by (7.3)) and hence

$$b_{n-2}f(n-2) = b_0 f(n), \;\; b_{n-1}f(n-1) = b_1 f(n+1).$$

Also, by comparing coefficients of $\rho_\epsilon(1,1)\rho_\epsilon(2,1), \cdots, \rho_\epsilon(n-2,1)$ $(j = 0,1,\cdots,n-3$ in (7.5) and $j = 2,3,\cdots,n-1$ in (7.6)), we get

$$b_j f(j)f(j+1) = b_{j+2}f(j+2)f(j+1), \text{ i.e., } b_j f(j) = b_{j+2}f(j+2)$$

for $j = 0,1,\cdots,n-3$. We can thus compute

$$b_0 f(0) = b_2 f(2) = \cdots = b_{n-2}f(n-2) = b_0 f(n).$$

Since $f(n) = -f(0)$, we get $b_0 = 0$ and hence $b_0 = b_2 = \cdots = b_{n-1} = 0$. Similarly we get $b_1 f(1) = b_3 f(3) = \cdots = b_{n-1}f(n-1) = b_1 f(n+1)$ so that $b_1 = b_3 = \cdots = b_{n-1} = 0$, and hence all the $b_i$'s are 0.

We next require that $x$ (with already $b_i = 0$ as shown so far) commutes with $\rho_\epsilon(0,1)$. We have

$$\begin{aligned}\rho_\epsilon(0,1)x &= \sum_{i=0}^{n-1} a_i f(0)f(-i)\omega^i \rho_\epsilon(n-i,1),\\ x\rho_\epsilon(0,1) &= \sum_{i=0}^{n-1} a_i f(0)f(i)\omega^i \rho_\epsilon(i,1).\end{aligned}$$

We have no condition on $a_0$ whatsoever, but notice that

$$a_{\frac{n}{2}} f(0)f(-\tfrac{n}{2})\omega^{\frac{n}{2}} = a_{\frac{n}{2}} f(0)f(\tfrac{n}{2})\omega^{\frac{n}{2}}$$

by comparing coefficients of the middle $\rho_\epsilon(\frac{n}{2},1)$. Therefore, we must have $a_{\frac{n}{2}} = 0$ due to $f(-\frac{n}{2}) = -f(\frac{n}{2})$ (a consequence of (7.3)). The other $a_i$'s are related by

$$a_i f(0)f(-i)\omega^i = a_{n-i} f(0)f(n-i)\omega^{n-i} = a_{n-i} f(0)f(-i)\omega^{-i}$$

for $i = 1, 2, \cdots, \frac{n}{2} - 1$, which means $a_{n-i} = \omega^{2i} a_i$. Hence, the center $\mathcal{Z}(\mathbf{C}_{\eta_\epsilon}(D_{2n}))$ is spanned by

$$\rho_\epsilon(0,0) \text{ and } \omega^{-i}\rho_\epsilon(i,0) + \omega^i \rho_\epsilon(n-i,0), \ i = 1, 2, \cdots, \frac{n}{2} - 1,$$

and we have $\dim \mathcal{Z}(\mathbf{C}_{\eta_\epsilon}(D_{2n})) = \frac{n}{2}$.

When we have no cocycle, from the first computation we get $b_0 = b_2 = \cdots = b_{n-2}$ and $b_1 = b_3 = \cdots = b_{n-1}$, and these two values are arbitrary. From the second computation, we have $a_i = a_{n-i}$ and no condition on $a_0$ and $a_{\frac{n}{2}}$. Note that $\sum_{i=0,2,\cdots,n-2} \rho_\epsilon(i,1)$ and $\sum_{i=1,3,\cdots,n-1} \rho_\epsilon(i,1)$ commute with $\rho_\epsilon(0,1)$. Therefore, the commutativity with $\rho_\epsilon(0,1)$ does not give any restriction on $b_0, b_1$. Consequently, in this case the dimension of the center is $(\frac{n}{2} - 1) + 4 = \frac{n}{2} + 3$.

Now we go back to the situation with the cocycle $\eta$. For example, when $n = 4$ the two central minimal projections are

$$\frac{1}{2}\left(1 \pm \frac{\omega^{-1}\rho_\epsilon(1,0) + \omega\rho_\epsilon(3,0)}{\sqrt{2}}\right).$$

Thus, $\mathbf{C}_{\eta_\epsilon}(D_8) = M_2(\mathbf{C}) \oplus M_2(\mathbf{C})$, the above two central projections having the same trace value. For a general (even) $n$, $\mathbf{C}_{\eta_\epsilon}(D_{2n})$ is the direct sum of $\frac{n}{2}$ copies of $M_2(\mathbf{C})$. In fact, a 1-dimensional summand $\mathbf{C}$ does not appear in $\mathbf{C}_{\eta_\epsilon}(D_{2n})$ since $\eta_\epsilon$ comes from a projective representation, and $\dim M_2(\mathbf{C}) \times \frac{n}{2} = 2n$ is already the dimension of $\mathbf{C}_{\eta_\epsilon}(D_{2n})$ (matrix algebras of larger sizes cannot be involved here). This also follows from the fact that any irreducible projective representation of $D_{2n}$ is of degree 2. Therefore, with the $\mathbf{Z}_2$-equivariant cocycle $\eta_\epsilon$ (i.e., with $(\eta, \zeta = 1)$), the dual Kac algebra $\hat{\mathcal{A}}$ is deformed to

$$\begin{aligned}\hat{\mathcal{A}} &= \mathbf{C}(D_{2n}) \oplus \mathbf{C}_{\eta_\epsilon}(D_{2n})\\ &= \mathbf{C} \oplus \mathbf{C} \oplus \mathbf{C} \oplus \mathbf{C} \oplus \underbrace{M_2(\mathbf{C}) \oplus M_2(\mathbf{C}) \oplus \cdots \oplus M_2(\mathbf{C})}_{n-1} \ (\cong \mathcal{A})\end{aligned}$$

while the underlying algebra structure of $\mathcal{A}$ of course remains the same.

PROPOSITION 7.5. *We have $H^2((D_{2n}, \mathbf{Z}_2), \mathbf{T}) = \mathbf{Z}_2$ (regardless of the parity of $n$).*

Note that we obtained this result in §5 when $n$ is odd, and we are now working on the even case. The cocycle $(\eta, \zeta = 1)$ (arising from the $\mathbf{Z}_2$-equivariant $\eta_\epsilon$) represents a non-trivial element in $H^2((D_{2n}, \mathbf{Z}_2), \mathbf{T})$ (giving rise to a different Kac algebra). Since $H^2(D_{2n}, \mathbf{T}) = \mathbf{Z}_2$ and $\eta_\epsilon \neq 1 \in H^2(D_{2n}, \mathbf{T})$, the proposition follows from $H^2(\mathbf{Z}_2, Hom(D_{2n}, \mathbf{T})) = 1$ (see the last part of §3).

We now describe intrinsic groups.

LEMMA 7.6. *(i) With the trivial cocycle* $(\eta = 1, \zeta = 1)$ *we have*

$$G(\mathcal{A}) = D_8 \ (\textit{the dihedral group of order } 8),$$
$$G(\hat{\mathcal{A}}) = \mathbf{Z}_2 \times \mathbf{Z}_2.$$

*(ii) With the equivariant cocycle* $(\eta, \zeta = 1)$ *we have*

$$G(\mathcal{A}) = G(\hat{\mathcal{A}}) = \mathbf{Z}_2 \times \mathbf{Z}_2.$$

PROOF. (i) We have $N_0 = \{(0,0), (\frac{n}{2}, 0)\} \cong \mathbf{Z}_2$ and $H_0 = H = \mathbf{Z}_2$. Notice that the action of $H$ on $Hom(N, \mathbf{T}) \cong \mathbf{Z}_2 \times \mathbf{Z}_2$ looks like:

$$(0,0)^\epsilon = (0,0), (1,0)^\epsilon = (1,0), (0,1)^\epsilon = (1,1), (1,1)^\epsilon = (0,1).$$

(Recall the description of four characters in the proof of Lemma 7.5.) At first $G(\hat{\mathcal{A}}) = Hom(H, \mathbf{T}) \times N_0 = \mathbf{Z}_2 \times \mathbf{Z}_2$. The other intrinsic group is $G(\mathcal{A}) = Hom(N, \mathbf{T}) \rtimes H = D_8$. In fact, we set $\alpha = (1,1) \in Hom(N, \mathbf{T}) = \mathbf{Z}_2 \times \mathbf{Z}_2$ and $\beta = \epsilon$. Then, both of $\alpha, \beta$ are of period 2, and

$$(\alpha\beta)^2 = (1,1)\epsilon(1,1)\epsilon = (1,1)(1,1)^\epsilon = (1,0) \quad (\text{in } (\mathbf{Z}_2 \times \mathbf{Z}_2) \rtimes \mathbf{Z}_2),$$

i.e., $\alpha\beta$ is of period 4.

(ii) Here, we have $N_0 = \{(0,0), (\frac{n}{2}, 0)\}$ since $\zeta = 1$. Also, since $\eta_\epsilon$ is not a coboundary, we have $H_0 = \{e\}$. Notice $\zeta = 1$ and $\eta_\cdot|_{N_0} \in Z^2(N_0, Hom(H, \mathbf{T}))$ is identically 1 (since $\eta_\epsilon((\frac{n}{2}, 0), (\frac{n}{2}, 0)) = 1$). The arguments so far show that the both extensions

$$1 \longrightarrow Hom(N, \mathbf{T}) = \mathbf{Z}_2 \times \mathbf{Z}_2 \longrightarrow G(\mathcal{A}) \longrightarrow H_0 = \{e\} \longrightarrow 1$$
$$1 \longrightarrow Hom(H, \mathbf{T}) = \mathbf{Z}_2 \longrightarrow G(\hat{\mathcal{A}}) \longrightarrow N_0 = \mathbf{Z}_2 \longrightarrow 1$$

are trivial, and hence we get the conclusion. □

CHAPTER 8

# Inclusions with the Coxeter-Dynkin graph $D_6^{(1)}$ and the Kac-Paljutkin algebra

Let $\alpha, \beta$ be period 2 automorphisms on a factor $\mathcal{L}$, and we assume that the outer period of $\alpha\beta$ is 4, i.e., $(\alpha\beta)^4 = Ad(u)$ with a unitary $u \in \mathcal{L}$. Then, we have $\alpha\beta(u) = \omega u$ with $\omega^4 = 1$ (i.e., $\omega = \pm 1, \pm i$, the Connes obstruction [**4**]). The inclusion $\mathcal{L} \rtimes_\alpha \mathbf{Z}_2 \supseteq \mathcal{L}^\beta$ (of index 4) gives us the Coxeter-Dynkin graph $D_6^{(1)}$, and in the hyperfinite $II_1$ case such inclusions are classified by the Connes obstruction $\omega$ (see [**31**]). In this chapter (as Theorem 8.3) we show that from the inclusion corresponding to $\omega = -1$ one can construct the Kac-Paljutkin algebra ([**38**], see also §2 in Chapter 7) by considering a natural subfactor (of index 2) in $\mathcal{L}^\beta$.

## 1. The group of one-dimensional sectors

Notice that for an arbitrary $x \in \mathcal{L}$ one computes

$$\begin{aligned} u\alpha(u)x\alpha(u^*)u^* &= u\alpha\left(u\alpha(x)u^*\right)u^* = u\alpha\left((\alpha\beta)^4 \cdot \alpha(x)\right)u^* \\ &= u(\beta\alpha)^4(x)u^* = (\alpha\beta)^4 \cdot (\beta\alpha)^4(x) = x, \end{aligned}$$

showing that $u\alpha(u) = e^{i\theta}$ is a scalar. By changing $u$ to $e^{-\frac{i\theta}{2}}u$, we can assume $u\alpha(u) = 1$ (so that $\beta(u) = \alpha\alpha\beta(u) = \alpha(\omega u) = \omega\alpha(u)$). Hence, in what follows we will assume

$$u\alpha(u) = 1 \text{ and } u\beta(u) = \omega 1. \tag{8.1}$$

Let $\rho_1, \rho_2$ be endomorphisms of $\mathcal{L}$ with $\mathcal{L}^\beta = \rho_1(\mathcal{L})$ and $\mathcal{L}^\alpha = \rho_2(\mathcal{L})$ respectively, and we set

$$\rho = \bar{\rho}_2\rho_1 \in Sect(\mathcal{L}).$$

Hence, $\mathcal{L} \supseteq \rho(\mathcal{L})$ is conjugate to $\mathcal{L} \rtimes_\alpha \mathbf{Z}_2 \supseteq \mathcal{L}^\beta$, and we compute

$$\rho\bar{\rho} = \bar{\rho}_2\rho_1\bar{\rho}_1\rho_2 = \bar{\rho}_2(1 \oplus \beta)\rho_2 = 1 \oplus a \oplus \bar{\rho}_2\beta\rho_2.$$

(Note that $a$ is the "dual automorphism" of $\alpha$.) Notice $d(\bar{\rho}_2\beta\rho_1) = \sqrt{2} \times 1 \times \sqrt{2} = 2$ and $\bar{\rho}_2\beta\rho_1$ is irreducible. In fact, we compute

$$\dim(\bar{\rho}_2\beta\rho_2, \bar{\rho}_2\beta\rho_2) = \dim(\bar{\rho}_2\beta, \bar{\rho}_2\beta\rho_2\bar{\rho}_2) = \dim(\rho_2\bar{\rho}_2\beta, \beta\rho_2\bar{\rho}_2) = 1$$

because of $\rho_2\bar{\rho}_2 = 1 \oplus \alpha$ and $\alpha\beta \neq \beta\alpha$ (in $Out(\mathcal{L})$. We set

$$\epsilon = \bar{\rho}_2\beta\rho_2.$$

Figure 1

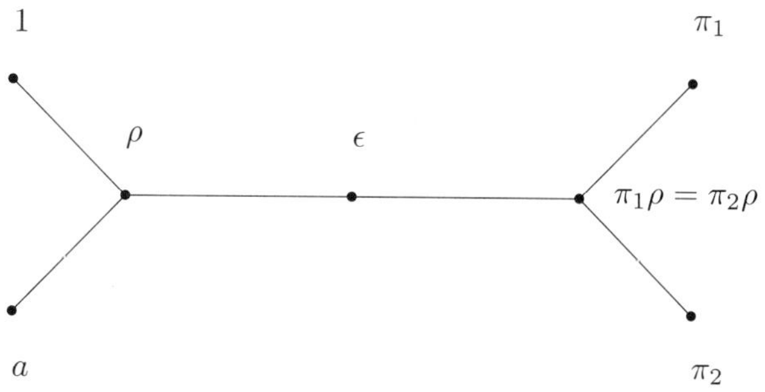


Notice

$$\begin{cases} \epsilon\rho\bar{\rho} = 1 \oplus a \oplus 2\epsilon \oplus \pi_1 \oplus \pi_2 \\ \epsilon(1 \oplus a) = 2\epsilon \end{cases}$$

because of $\epsilon a = \epsilon$ ($\epsilon a \prec (\rho\bar{\rho})^2, d(\epsilon a) = 2$, and $\epsilon a$ is irreducible). Since $\epsilon$ is self-dual and $\rho\bar{\rho} = 1 \oplus a \oplus \epsilon$, we conclude

$$\epsilon^2 = 1 \oplus a \oplus \pi_1 \oplus \pi_2.$$

On the other hand, we also compute

$$\epsilon^2 = \bar{\rho}_2\beta\rho_2\bar{\rho}_2\beta\rho_2 = \bar{\rho}_2\beta(1 \oplus \alpha)\beta\rho_2 = \bar{\rho}_2(1 \oplus \beta\alpha\beta)\rho_2 = 1 \oplus a \oplus \bar{\rho}_2\beta\alpha\beta\rho_2.$$

Comparing the two computations so far, we get

$$\pi_1 \oplus \pi_2 = \bar{\rho}_2\beta\alpha\beta\rho_2.$$

Recall $\bar{\rho}_2(\mathcal{L} \rtimes_\alpha \mathbf{Z}_2) = \mathcal{L}$, and we denote $\bar{\rho}_2$ considered as an $(\mathcal{L} \rtimes_\alpha \mathbf{Z}_2)$-$\mathcal{L}$ sector by $\bar{\varrho}_2$. (A readable account on "$\mathcal{M}$-$\mathcal{N}$ sectors" can be found in [**26**].) Notice that the statistical dimension of $\bar{\varrho}_2 \in Sect(\mathcal{L} \rtimes_\alpha \mathbf{Z}_2, \mathcal{L})$ is 1 and the conjugate sector is $\bar{\varrho}_2^{-1} \in Sect(\mathcal{L}, \mathcal{L} \rtimes_\alpha \mathbf{Z}_2)$. Notice $\mathcal{L}^\beta = \rho'(\mathcal{L} \rtimes_\alpha \mathbf{Z}_2)$ with

$$\rho' = \bar{\varrho}_2^{-1} \cdot \rho \cdot \bar{\varrho}_2 \in Sect(\mathcal{L} \rtimes_\alpha \mathbf{Z}_2).$$

We set

$$\begin{cases} \hat{\alpha} = \bar{\varrho}_2^{-1} \cdot a \cdot \bar{\varrho}_2 \\ \Pi_i = \bar{\varrho}_2^{-1} \cdot \pi_i \cdot \bar{\rho}_2 \quad (i = 1, 2). \end{cases}$$

They are automorphisms of $\mathcal{L} \rtimes_\alpha \mathbf{Z}_2$, and $\hat{\alpha}$ is indeed the dual action of $\alpha$ (since this automorphism is in the Galois group).

Figure 2

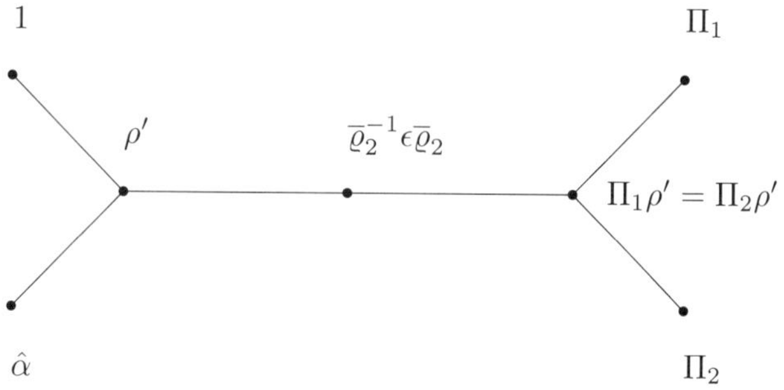

The previous computation obviously means

$$\Pi_1 \oplus \Pi_2 = \bar{\varrho}_2^{-1} \cdot \bar{\rho}_2\beta\alpha\beta\rho_2 \cdot \bar{\varrho}_2. \tag{8.2}$$

The product of the first two sectors in the right side of (8.2) is the inclusion map

$$\bar{\varrho}_2^{-1}\bar{\varrho}_2 = \iota_{\mathcal{L}\hookrightarrow\mathcal{L}\rtimes_\alpha \mathbf{Z}_2} \in Sect(\mathcal{L}, \mathcal{L} \rtimes_\alpha \mathbf{Z}_2)$$

Thus, if $\Pi$ is an automorphism of $\mathcal{L} \rtimes_\alpha \mathbf{Z}_2$ extending $\beta\alpha\beta \in Aut(\mathcal{L})$, then the product of the first five sectors in (8.2) is

$$\bar{\varrho}_2^{-1}\bar{\varrho}_2\beta\alpha\beta = \iota_{\mathcal{L}\hookrightarrow\mathcal{L}\rtimes_\alpha \mathbf{Z}_2}\beta\alpha\beta = \Pi \cdot \iota_{\mathcal{L}\hookrightarrow\mathcal{L}\rtimes_\alpha \mathbf{Z}_2} \in Sect(\mathcal{L}, \mathcal{L} \rtimes_\alpha \mathbf{Z}_2). \tag{8.3}$$

This computation means that to know what $\Pi_i$ are we have to write down an extension $\Pi \in Aut(\mathcal{L} \rtimes_\alpha \mathbf{Z}_2)$ explicitly. Let $x = x_0 + x_1\lambda$ be an element in $\mathcal{L} \rtimes_\alpha \mathbf{Z}_2$ (with the self-adjoint unitary $\lambda$), and we set

$$\Pi(x_0 + x_1\lambda) = \beta\alpha\beta(x_0) + \beta\alpha\beta(x_1)u^*\lambda. \tag{8.4}$$

At first note that $u^*\lambda$ is a self-adjoint unitary due to (8.1). Also we have the right covariance

$$u^*\lambda\left(\beta\alpha\beta(x)\right)\lambda^*u = u^*(\alpha\beta)^2(x)u = (\beta\alpha)^4(\alpha\beta)^2(x) = (\beta\alpha)^2(x) = \beta\alpha\beta(\alpha(x)).$$

Therefore, the above $\Pi$ is indeed an automorphism, and we have $\Pi(\mathcal{L}) = \mathcal{L}$ and $\Pi\mid_{\mathcal{L}} = \beta\alpha\beta$ from the construction. By (8.2) and (8.3) we have

$$\Pi_1 \oplus \Pi_2 = \Pi \cdot \iota_{\mathcal{L}\hookrightarrow\mathcal{L}\rtimes_\alpha \mathbf{Z}_2} \cdot \rho_2\bar{\varrho}_2.$$

Note $\rho_2\bar{\varrho}_2 = Ad(J_{\rho_2(\mathcal{L})}J_{\mathcal{L}}) \in Sect(\mathcal{L} \rtimes_\alpha \mathbf{Z}_2, \mathcal{L})$ and

$$\iota_{\mathcal{L}\hookrightarrow\mathcal{L}\rtimes_\alpha \mathbf{Z}_2} \cdot \rho_2\bar{\varrho}_2 = Ad(J_{\rho_2(\mathcal{L})}J_{\mathcal{L}}) = Ad(J_{\mathcal{L}}J_{\mathcal{L}\rtimes_\alpha \mathbf{Z}_2}) \in Sect(\mathcal{L} \rtimes_\alpha \mathbf{Z}_2)$$

is the sum of the two automorphism $id, \hat{\alpha}$ because of $\mathcal{L} = (\mathcal{L} \rtimes_\alpha \mathbf{Z}_2)^{\hat{\alpha}}$. Therefore, we conclude

$$\Pi_1 \oplus \Pi_2 = \Pi(id \oplus \hat{\alpha}) = \Pi \oplus \Pi\hat{\alpha},$$

and the four descendant sectors of statistical dimension 1 of $\rho'$ are

$$\langle [id], [\hat{\alpha}], [\Pi], [\Pi\hat{\alpha}] \rangle,$$

which form a finite group of order 4 (i.e., either $\mathbf{Z}_4$ or $\mathbf{Z}_2 \times \mathbf{Z}_2$).

For $x_0 + x_1\lambda$, by repeated use of (8.4) we get

$$\begin{aligned} \Pi^2(x_0 + x_1\lambda) &= \Pi(\beta\alpha\beta(x_0) + \beta\alpha\beta(x_1)u^*\lambda) \\ &= \beta\alpha\beta\beta\alpha\beta(x_0) + \beta\alpha\beta\beta\alpha\beta(x_1)\beta\alpha\beta(u^*)u^*\lambda \\ &= x_0 + x_1\beta\alpha\beta(u^*)u^*\lambda. \end{aligned}$$

Notice

$$u\beta\alpha\beta(u) = u\beta(\omega u) = \omega u\beta(u) = \omega^2 1$$

thanks to (8.1) together with $\alpha\beta(u) = \omega u$. Therefore, we conclude

$$(\Pi)^2(x_0 + x_1\lambda) = x_0 + (\bar{\omega})^2 x_1\lambda.$$

Note that when $\omega = \pm i$ we have

$$\Pi^2(x_0 + x_1\lambda) = x_0 - x_1\lambda, \text{ i.e., } (\Pi)^2 = \hat{\alpha}.$$

The arguments so far show

PROPOSITION 8.1. *Let $\alpha, \beta$ be period 2 automorphisms on a factor $\mathcal{L}$. Assume that $\alpha\beta$ has outer period 4 with Connes obstruction $\omega$. Then, the four descendant sectors of statistical dimension 1 associated with the inclusion $\mathcal{L} \rtimes_\alpha \mathbf{Z}_2 \supseteq \mathcal{L}^\beta$ of index 4 are $[id], [\hat{\alpha}], [\Pi], [\Pi\hat{\alpha}]$, where $\Pi$ is given by (8.4) and $\hat{\alpha}$ is the dual action of $\alpha$. Furthermore, we have*

$$\langle [id], [\hat{\alpha}], [\Pi], [\Pi\hat{\alpha}] \rangle \cong \begin{cases} \mathbf{Z}_4 & \text{when } \omega = \pm i, \\ \mathbf{Z}_2 \times \mathbf{Z}_2 & \text{when } \omega = \pm 1. \end{cases}$$

Note that this was also proved in [**20, 41**] (see also [**74**]).

## 2. Kac-Paljutkin algebra

We now assume $\omega = \pm 1$, and set

$$\mathcal{M} = \mathcal{L} \rtimes_\alpha \mathbf{Z}_2 \supseteq \mathcal{N} = \mathcal{L}^\beta \ (= \rho'(\mathcal{M})).$$

The sector $[\Pi_1]$ $(\prec (\rho'\overline{\rho'})^2)$ of statistical dimension 1 is self-conjugate due to Proposition 8.1. By the standard procedure, after inner perturbation we may and do assume

$$(\Pi_1)^2 = id \quad \text{and} \quad \Pi_1 \in Aut(\mathcal{M}, \mathcal{N}). \tag{8.5}$$

With this automorphism of period 2, we get the new inclusion

$$\mathcal{M} \supseteq \mathcal{P} = \mathcal{N}^{\Pi_1}$$

of index 8 and we set $\mathcal{P} = \tilde{\rho}(\mathcal{M})$ with $\tilde{\rho} \in End(\mathcal{M})$.

LEMMA 8.2. *The restriction $\Pi_1 |_\mathcal{N}$ is an outer automorphism of $\mathcal{N}$ (so that $\mathcal{P}$ is a factor). Moreover, we have*

$$\tilde{\rho}\bar{\tilde{\rho}} = id \oplus \hat{\alpha} \oplus \Pi_1 \oplus \Pi_2 \oplus 2\bar{\varrho}_2^{-1}\epsilon\bar{\varrho}_2.$$

*with $\epsilon$ in Figure 1.*

PROOF. If we had $\Pi_1(n) = unu^*$ $(n \in \mathcal{N})$ with some unitary $u \in \mathcal{N}$, then we would get $\Pi_1\rho'(x) = u\rho'(x)u^*$ $(x \in \mathcal{M})$ and $\Pi_1\rho' = \rho'$ as a sector, a contradiction. We set $\Pi'_1 = \rho'^{-1}\Pi_1\rho' \in Aut(\mathcal{M})$. Then, by the standard argument based on the Frobenius reciprocity, we know $\Pi'_1 \prec (\overline{\rho'}\rho')^2$. Notice

$$\mathcal{P} = \mathcal{N}^{\Pi_1} = (\rho'(\mathcal{M}))^{\Pi_1} = \rho'(\mathcal{M}^{\Pi'_1})$$

from the definition of $\Pi'_1$. Let $\eta$ be an endomorphism of $\mathcal{M}$ such that $\mathcal{M}^{\Pi'_1} = \eta(\mathcal{M})$ (so that $\eta\bar{\eta} = 1 \oplus \Pi'_1$). We can obviously assume $\tilde{\rho} = \rho'\eta$ and compute

$$\tilde{\rho}\bar{\tilde{\rho}} = \rho'\eta\bar{\eta}\overline{\rho'} = \rho'(1 \oplus \Pi'_1)\overline{\rho'} = \rho'\overline{\rho'} \oplus \rho'\Pi'_1\overline{\rho'} = 1 \oplus \hat{\alpha} \oplus \bar{\varrho}_2^{-1}\epsilon\bar{\varrho}_2 \oplus \rho'\Pi'_1\bar{\rho}'.$$

Since we observe

$$\rho'\Pi'_1\overline{\rho'} = \overline{\left(\overline{\Pi'_1\rho'}\right)}\overline{\rho'} = \Pi_1 \oplus \Pi_2 \oplus \bar{\varrho}_2^{-1}\epsilon\bar{\varrho}_2$$

from the figure below (Figure 3), we get the desired conclusion. □

Figure 3

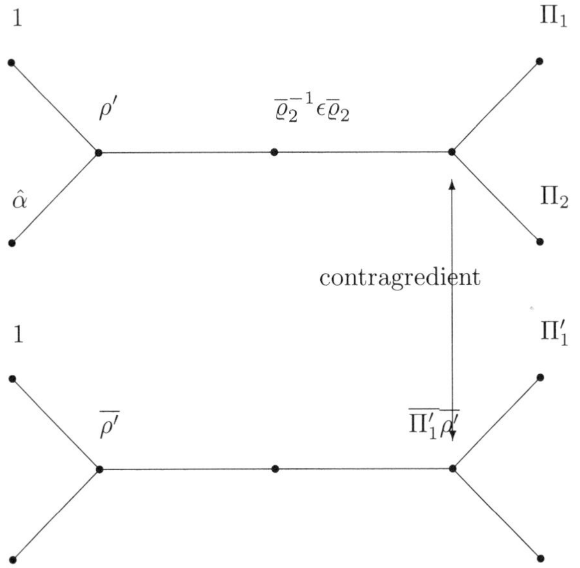


The lemma shows that the inclusion $\mathcal{M} \supseteq \mathcal{P}$ is of depth 2. When $\omega = 1$, nothing interesting happens here. In fact, it is straight-forward to see that the subfactor $\mathcal{P}$ is the fixed-point subalgebra under a $D_8$-action (the dihedral group of order 8). In the rest, we show

THEOREM 8.3. *When $\omega = -1$, the inclusion $\mathcal{M} \supseteq \mathcal{P}$ of depth 2 corresponds to the Kac-Paljutkin 8-dimensional algebra.*

As was explained above we have to find inner perturbation of $\Pi_1$ (satisfying (8.5)) so that we should start from a model as handy as possible. For this purpose we make use of characteristic invariants appeared in Jones' thesis [**35**]. Let $G = \mathbf{Z}_8 \rtimes \mathbf{Z}_2$ be the dihedral group of order 16, and we set $C = \{(0,0), (4,0)\}$. Obviously $C$ is a normal subgroup in $G$ (actually central), and we consider a characteristic invariant $(\mu, \lambda)$. Since $C$ is central, we set $\mu = 1 : C \times C \to \mathbf{T}$ and consider $\lambda : G \times C \to \mathbf{T}$ satisfying

$$\lambda(g, c_1c_2) = \lambda(g, c_1)\lambda(g, c_2),$$
$$\lambda(g_1g_2, c) = \lambda(g_1, c)\lambda(g_2, c),$$
$$\lambda(g, c) = 1 \text{ when } g \in C.$$

Then, $(C, (\mu, \lambda))$ becomes an admissible range of the Jones invariant of a $G$-action, i.e., we have an action $\sigma : G \to Aut(\mathcal{L})$ such that $\sigma_c = Ad(u_c)$ and $\sigma_g(u_c) = \lambda(g, c)u_c, u_{cc'} = u_cu_{c'}$ (since $C$ is central and $\mu = 1$). We set

$$\lambda((i, 0), (4, 0)) = \lambda((i, 1), (4, 0)) = (-1)^i, \ \lambda(\cdot, (0, 0)) = 1.$$

Then the above three conditions are satisfied, and we have the corresponding action $\sigma$ as above. We set

$$\alpha = \sigma_{(0,1)}, \ \beta = \sigma_{(1,1)}.$$

Note that $\alpha, \beta$ are automorphisms of period 2 and

$$(\alpha\beta)^4 = \sigma_{(4,0)} = Ad(u) \in Int(\mathcal{L}).$$

Because of $\lambda((0,1),(4,0)) = 1, \lambda((1,1),(4,0)) = -1$, we have

$$u^2 = 1,\ \alpha(u) = u,\ \beta(u) = -u.$$

(In particular, $\alpha\beta(u) = -1$ and the Connes obstruction is $-1$.) Since $u$ is a self-adjoint unitary, it is of the form

$$u = e_+ - e_- \text{ with projections } e_\pm = \frac{1}{2}(1 \pm u).$$

Notice

$$\alpha(e_+) = e_+, \alpha(e_-) = e_-, \beta(e_+) = e_-, \beta(e_-) = e_+.$$

We set

$$u^x = e_+ + e^{ix\pi}e_-,$$

and it is plain to see

$$u^{x+y} = u^x u^y, u^1 = u, u^0 = u^2 = 1, \alpha(u^x) = u^x, \beta(u^x) = e^{ix\pi}u^{-x}.$$

LEMMA 8.4. *We set* $\theta = Ad(u^{-\frac{1}{2}}) \cdot (\alpha\beta)^2 \in Aut(\mathcal{L})$. *Then, we have*

$$\langle id, \theta, \beta, \theta\beta\rangle \cong \mathbf{Z}_2 \times \mathbf{Z}_2.$$

PROOF. At first we have

$$\theta^2 = Ad\left(u^{-\frac{1}{2}}(\alpha\beta)^2(u^{-\frac{1}{2}})\right)(\alpha\beta)^4 = Ad(u^{-\frac{1}{2}}u^{-\frac{1}{2}})(\alpha\beta)^4 = Ad(u^{-1}u) = id$$

since $(\alpha\beta)(u^{-\frac{1}{2}}) = -iu^{\frac{1}{2}}$ and $(\alpha\beta)^2(u^{-\frac{1}{2}}) = u^{-\frac{1}{2}}$. We compute

$$\theta\beta = Ad(u^{-\frac{1}{2}})\alpha\beta\alpha,\ \beta\theta = Ad(\beta(u^{-\frac{1}{2}}))\beta\alpha\beta\alpha\beta,$$

but they are the same because of $\beta(u^{-\frac{1}{2}}) = -iu^{\frac{1}{2}}$ and $(\alpha\beta)^4 = Ad(u)$. From the first we notice

$$(\theta\beta)^2 = Ad\left(u^{-\frac{1}{2}}\alpha\beta\alpha(u^{-\frac{1}{2}})\right)\alpha\beta\alpha\alpha\beta\alpha = id$$

due to $\alpha\beta\alpha(u^{-\frac{1}{2}}) = -iu^{\frac{1}{2}}$. □

We consider the following inner perturbation of $\Pi$ (see (8.4)):

$$\Pi' = Ad(u^{-\frac{1}{2}}\lambda) \cdot \Pi.$$

Then, we compute

$$\begin{aligned}\Pi'(x_0 + x_1\lambda) &= u^{-\frac{1}{2}}\lambda\left(\beta\alpha\beta(x_0) + \beta\alpha\beta(x_1)u^*\lambda\right)\lambda^*(u^{-\frac{1}{2}})^* \\ &= u^{-\frac{1}{2}}\left((\alpha\beta)^2(x_0) + (\alpha\beta)^2(x_1)\alpha(u^*)\lambda\right)(u^{-\frac{1}{2}})^* \\ &= \theta(x_0) + \theta(x_1)u^*\lambda\end{aligned}$$

by using $\lambda u^x\lambda^* = \alpha(u^x) = u^x$, and hence

$$\begin{aligned}(\Pi')^2(x_0 + x_1\lambda) &= \Pi'(\theta(x_0) + \theta(x_1)u^*\lambda) = \theta^2(x_0) + \theta^2(x_1)\theta(u^*)u^*\lambda \\ &= x_0 + x_1\theta(u^*)u^*\lambda = x_0 + x_1\lambda \quad (\text{i.e., } (\Pi')^2 = id)\end{aligned}$$

due to $\theta(u) = Ad(u^{-\frac{1}{2}})(\alpha\beta)^2(u) = Ad(u^{-\frac{1}{2}})u = u$ and $u^2 = 1$. Notice $\Pi'(\mathcal{L}) = \mathcal{L}$ and $\Pi'|_{\mathcal{L}} = \theta$. Furthermore, since $\theta\beta = \beta\theta$, $\Pi'$ leaves $\mathcal{N} = \mathcal{L}^\beta$ invariant. Therefore, $\Pi' \in Aut(\mathcal{M} = \mathcal{L} \rtimes_\alpha \mathbf{Z}_2)$ serves as a right representative of $[\Pi_1]$ (recall (8.5)). We obviously have

$$\mathcal{M} = \mathcal{L} \rtimes_\alpha \mathbf{Z}_2 \supseteq \mathcal{P} = \mathcal{N}^{\Pi'} = (\mathcal{L}^\beta)^{\Pi'} = \mathcal{L}^{(\langle\beta,\theta\rangle, \mathbf{Z}_2\times\mathbf{Z}_2)}.$$

Therefore, the Kac algebra associated to the inclusion $\mathcal{M} \supseteq \mathcal{P}$ of depth 2 is the one corresponding to the outer action $(\beta', \alpha)$ of the matched pair $((\mathbf{Z}_2 \rtimes \mathbf{Z}_2), \mathbf{Z}_2)$ (see the computations below):

$$\beta'_{(1,0)} = \beta,\ \beta'_{(0,1)} = \theta\beta,\ \beta'_{(1,1)} = \theta \quad \text{and} \quad \alpha_\epsilon = \alpha \quad (\epsilon \in \mathbf{Z}_2).$$

To complete the proof of Theorem 8.3, we have to show that the associated invariant $(\eta, \zeta)$ is the one corresponding to the Kac-Paljutkin algebra (see §2 in the previous chapter). We compute

$$\begin{aligned}
\alpha\theta\alpha^{-1} &= Ad(\alpha(u^{-\frac{1}{2}}))\alpha(\alpha\beta)^2\alpha \\
&= Ad(u^{-\frac{1}{2}})(\beta\alpha)^2 \\
&= Ad(u^{-\frac{1}{2}})(\alpha\beta)^6 \\
&\qquad (\text{since } (\alpha\beta)^8 = id) \\
&= Ad(u^{-\frac{1}{2}})(\alpha\beta)^4(\alpha\beta)^2 \ = \ Ad(u^{-\frac{1}{2}}u)(\alpha\beta)^2 \\
&= Ad(u^{\frac{1}{2}})Ad(u^{\frac{1}{2}})\theta \ = \ Ad(u)\theta, \\
\alpha\beta\alpha^{-1} &= (\alpha\beta)^2\beta \ = \ Ad(u^{\frac{1}{2}})\theta\beta, \\
\alpha(\theta\beta)\alpha^{-1} &= Ad(\alpha(u^{-\frac{1}{2}}))\alpha(\alpha\beta)^2\beta\alpha \\
&= Ad(u^{-\frac{1}{2}})\beta.
\end{aligned}$$

Therefore, $H = \mathbf{Z}_2$ acts on $N = \mathbf{Z}_2 \times \mathbf{Z}_2$ as a flip, and for $\epsilon = ((0,0),1) \in H = \mathbf{Z}_2$ $(\subseteq N \rtimes H)$ we can choose

$$\begin{aligned}
u(\epsilon, (1,0)) &= u^{-\frac{1}{2}}, \\
u(\epsilon, (0,1)) &= u^{\frac{1}{2}}, \\
u(\epsilon, (1,1)) &= u.
\end{aligned}$$

Recall that the defining relations for the cocycles $\eta, \zeta$ are

$$\begin{aligned}
u(h, n_1n_2) &= \eta_h(n_1, n_2)u(h, n_1)\beta'_{n_1}(u(h, n_2)) \\
u(h_1h_2, n) &= \overline{\zeta_n(h_1, h_1)}\alpha_{h_1}(u(h_2, n^{h_1}))u(h_1, n)
\end{aligned}$$

for $n, n_i \in N = \mathbf{Z}_2 \times \mathbf{Z}_2$ and $h, h_i \in H = \mathbf{Z}_2 = \{e, \epsilon\}$. Thanks to $\alpha(u^x) = u^x$ and $u^{-\frac{1}{2}}u^{\frac{1}{2}} = u^2 = 1$, we obviously get $\zeta = 1$. Notice

$$\begin{aligned}
\theta(u^x) &= Ad(u^{-\frac{1}{2}})\alpha\beta\alpha\beta(u^x) \\
&= Ad(u^{-\frac{1}{2}})\alpha\beta\alpha(e^{ix\pi}u^{-x}) \\
&= Ad(u^{-\frac{1}{2}})(e^{ix\pi}e^{-ix\pi}u^x) \\
&= u^x, \\
\theta\beta(u^x) &= e^{ix\pi}\theta(u^{-x}) \\
&= e^{ix\pi}u^{-x}.
\end{aligned}$$

Having this rule in our mind, we compute

$$
\begin{aligned}
u(\epsilon,(1,0))\beta'_{(1,0)}(u(\epsilon,(1,0))) &= u^{-\frac{1}{2}}\beta(u^{-\frac{1}{2}}) = -iu^{-\frac{1}{2}}u^{\frac{1}{2}} = -i \\
( &= \overline{\eta_\epsilon((1,0),(1,0))}u(\epsilon,(0,0))\ ), \\
u(\epsilon,(1,0))\beta'_{(1,0)}(u(\epsilon,(0,1))) &= u^{-\frac{1}{2}}\beta(u^{\frac{1}{2}}) = iu^{-\frac{1}{2}}u^{-\frac{1}{2}} = iu^{-1} = iu, \\
u(\epsilon,(1,0))\beta'_{(1,0)}(u(\epsilon,(1,1))) &= u^{-\frac{1}{2}}\beta(u) = -u^{-\frac{1}{2}}u = -u^{\frac{1}{2}}, \\
u(\epsilon,(0,1))\beta'_{(0,1)}(u(\epsilon,(1,0))) &- u^{\frac{1}{2}}\theta\beta(u^{-\frac{1}{2}}) - -iu^{\frac{1}{2}}u^{\frac{1}{2}} - -iu, \\
u(\epsilon,(0,1))\beta'_{(0,1)}(u(\epsilon,(0,1))) &= u^{\frac{1}{2}}\theta\beta(u^{\frac{1}{2}}) = iu^{\frac{1}{2}}u^{-\frac{1}{2}} = i, \\
u(\epsilon,(0,1))\beta'_{(0,1)}(u(\epsilon,(1,1))) &= u^{\frac{1}{2}}\theta\beta(u) = -u^{\frac{1}{2}}u = -u^{-\frac{1}{2}}, \\
u(\epsilon,(1,1))\beta'_{(1,1)}(u(\epsilon,(1,0))) &= u\theta(u^{-\frac{1}{2}}) = uu^{-\frac{1}{2}} = u^{\frac{1}{2}}, \\
u(\epsilon,(1,1))\beta'_{(1,1)}(u(\epsilon,(0,1))) &= u\theta(u^{\frac{1}{2}}) = uu^{\frac{1}{2}} = u^{-\frac{1}{2}}, \\
u(\epsilon,(1,1))\beta'_{(1,1)}(u(\epsilon,(1,1))) &= u\theta(u) = uu = 1.
\end{aligned}
$$

By taking the complex conjugates of the coefficients, we get the following table of values of the cocycle $\eta_\epsilon((a,b),(c,d))$:

| | | $(c,d)$ | | | |
|---|---|---|---|---|---|
| | | $(0,0)$ | $(1,0)$ | $(0,1)$ | $(1,1)$ |
| | $(0,0)$ | 1 | 1 | 1 | 1 |
| $(a,b)$ | $(1,0)$ | 1 | $i$ | $-i$ | $-1$ |
| | $(0,1)$ | 1 | $i$ | $-i$ | $-1$ |
| | $(1,1)$ | 1 | 1 | 1 | 1 |

Let $\xi(\cdot,\cdot)$ be given by

$$\xi(\epsilon,(0,0)) = 1, \xi(\epsilon,(1,0)) = e^{-\frac{i\pi}{4}}, \xi(\epsilon,(0,1)) = e^{\frac{i\pi}{4}}, \xi(\epsilon,(1,1)) = 1,$$

and $\xi(e,\cdot) = 1$. Then, $(\partial^N\xi)_\epsilon((a,b)(c,d))$ is as follows:

| | | $(c,d)$ | | | |
|---|---|---|---|---|---|
| | | $(0,0)$ | $(1,0)$ | $(0,1)$ | $(1,1)$ |
| | $(0,0)$ | 1 | 1 | 1 | 1 |
| $(a,b)$ | $(1,0)$ | 1 | $-i$ | 1 | $-i$ |
| | $(0,1)$ | 1 | 1 | $i$ | $i$ |
| | $(1,1)$ | 1 | $-i$ | $i$ | 1 |

The product of the two tables is exactly the one for the Kac-Paljutkin algebra (i.e., the one in §2 of Chapter 7). Therefore, if one perturbs $\eta_\epsilon$ to $\eta_\epsilon(\partial^N\xi)$, then the new $\eta_\epsilon((a,b),(c,d))$ is $i^{2ad-ab-cd+(a+c)(b+d)}$. Notice that after this perturbation we still have $\zeta = 1$ because of $(\partial^H\xi) = 1$ (note $\overline{e^{\frac{i\pi}{4}}} = e^{-\frac{i\pi}{4}}$). Thus, Theorem 8.3 has been proved.

CHAPTER 9

# Structure theorems

Let $\mathcal{P} \supseteq \mathcal{Q}$ be a depth 2 inclusion with finite index, and $\mathcal{A}, \hat{\mathcal{A}}$ be the associated Kac and dual Kac algebras. Recall that the Galois and Weyl groups of the inclusion $\mathcal{P} \supseteq \mathcal{Q}$ are

$$\begin{aligned} \text{Galois } (\mathcal{P} \supseteq \mathcal{Q}) &= \{\alpha \in Aut(\mathcal{P});\ \alpha|_{\mathcal{Q}} = id_{\mathcal{Q}}\}, \\ \text{Weyl } (\mathcal{P} \supseteq \mathcal{Q}) &= \mathbf{N}(\mathcal{Q})/\mathbf{U}(\mathcal{Q}) \end{aligned}$$

(see the end of Chapter 6). Let $\rho : \mathcal{Q} \hookrightarrow \mathcal{P}$ be the inclusion map so that there is a bijective correspondence between the set of simple components of $\mathcal{A}$ (resp. $\hat{\mathcal{A}}$) and that of irreducibles contained in $\overline{\rho}\rho$ (resp. $\rho\overline{\rho}$). As was shown in Proposition 6.8, the intrinsic groups $G(\mathcal{A})$ and $G(\hat{\mathcal{A}})$ can be identified with Galois $(\mathcal{P} \supseteq \mathcal{Q})$ and Weyl $(\mathcal{P} \supseteq \mathcal{Q}) \simeq$ Galois $(\mathcal{P}_1 \supseteq \mathcal{P})$ respectively, where $\mathcal{P}_1$ is the basic extension of $\mathcal{P} \supseteq \mathcal{Q}$.

In this chapter some structure results for Kac algebras are obtained (Theorem 9.8). Namely, when the underlying algebra structure of a Kac algebra is of specific form, we will show that the Kac algebra in question arises from a certain semi-direct product $N \rtimes H$. This means that these Kac algebras can be classified by $H^2((H, K), \mathbf{T})/\sim$ (see Definition 9.2 below).

Our strategy here is as follows: Suppose we are given a finite-dimensional Kac algebras $\mathcal{A}$ with certain prescribed properties, and let $\mathcal{P} \supseteq \mathcal{Q}$ be the corresponding depth 2 inclusion. We identify the Weyl group of the inclusion with $G(\hat{\mathcal{A}})$, which can be described in terms of the set of 1-dimensional representations of $\mathcal{A}$. Let $G_0 \subseteq G(\hat{\mathcal{A}})$ be an abelian characteristic (see Definition 9.1 below) subgroup so that the crossed product $\mathcal{R} = \mathcal{Q} \rtimes G_0$ can be regarded as an intermediate subfactor $\mathcal{Q} \subseteq \mathcal{R} \subseteq \mathcal{P}$. Let $N$ be the dual group of $G_0$ (considered as a subgroup of $Aut(\mathcal{R})$ through the dual action) so that we have $\mathcal{R}^N = \mathcal{Q}$ by the duality. In many cases such a group $G_0$ can be taken with the following additional requirement: (i) a finite group $H \subseteq Aut(\mathcal{R})$ can be found in such a way that $\mathcal{P} = \mathcal{R} \rtimes H$, (ii) $N$ and $H$ form a semi-direct product group, either $N \rtimes H$ or $H \rtimes N$, in $Out(\mathcal{R})$.

Definition 9.1. *Let $\mathcal{A}$ be a Kac algebra. We say that a subgroup $H \subseteq G(\mathcal{A})$ is a* **characteristic subgroup** *of $\mathcal{A}$ if $H$ is globally invariant under every Kac algebra automorphism of $\mathcal{A}$. In particular, $G(\mathcal{A})$ itself is a characteristic subgroup.*

Definition 9.2. *Let $G = N \rtimes H$ be a semi-direct product group, and assume $[(\eta, \zeta)], [(\eta', \zeta')] \in H^2((N, H), \mathbf{T})$, where $[(\eta, \zeta)]$ denotes the class of $(\eta, \zeta)$. We say that $[(\eta, \zeta)]$ and $[(\eta', \zeta')]$ are equivalent (denoted by $\sim$) if there exists an automorphism $\theta \in Aut(G)$ that globally preserves $N, H$ and satisfies*

$$[(\eta_{\theta(\cdot)}(\theta(\cdot), \theta(\cdot)), \zeta_{\theta(\cdot)}(\theta(\cdot), \theta(\cdot))] = [(\eta', \zeta')].$$

## 1. Preliminaries

At first we record some basic results which will be repeatedly used in the rest of the chapter (and the article).

LEMMA 9.3. ([**34**]) *Let $\mathcal{P} \supseteq \mathcal{Q}$ be an irreducible inclusion of factors with finite index, and $\gamma : \mathcal{P} \longrightarrow \mathcal{Q}$ be the canonical endomorphism with the irreducible decomposition*

$$[\gamma] = \oplus_{i \in I} m_i [\pi_i]$$

*(with multiplicities $m_i$). We assume that a subset $J \subseteq I$ satisfies the following* (1) *and* (2):

$$\begin{cases} (1) \text{ For every } j \in J, \text{ there exists } \bar{j} \in J \text{ satisfying } \overline{[\pi_j]} = [\pi_{\bar{j}}], \\ (2) \text{ If an irreducible } [\nu] \text{ is contained in } [\pi_j][\pi_k] \text{ for some } j,k \in J, \\ \qquad \text{then either } [\nu] = [\pi_l] \text{ for some } l \in J \text{ or } [\nu] \text{ does not appear in } [\gamma]. \end{cases}$$

*Then, there exists a unique intermediate subfactor $\mathcal{L}$ between $\mathcal{P}$ and $\mathcal{Q}$ such that the canonical endomorphism*

$$\gamma_{\mathcal{P} \supseteq \mathcal{L}} : \mathcal{P} \longrightarrow \mathcal{L}$$

*has the following irreducible decomposition:*

$$[\gamma_{\mathcal{P} \supseteq \mathcal{L}}] = \oplus_{j \in J} m_j [\pi_j].$$

LEMMA 9.4. *Let $\mathcal{P} \supseteq \mathcal{Q}$ be an irreducible inclusion of factors with integer index and finite depth.*
*(i) (Bisch [**2**]) For any intermediate subfactor $\mathcal{L}$ between $\mathcal{P}$ and $\mathcal{Q}$, $[\mathcal{P} : \mathcal{L}]$ is also an integer dividing $[\mathcal{P} : \mathcal{Q}]$.*
*(ii) If $p = [\mathcal{P} : \mathcal{Q}]$ is prime, then either the canonical endomorphism $\gamma$ $(= \gamma_{\mathcal{P} \supseteq \mathcal{Q}})$ contains no non-trivial automorphism or $\mathcal{P} = \mathcal{Q} \rtimes \mathbf{Z}_p$.*

PROOF. There is no intermediate subfactor of $\mathcal{P} \supseteq \mathcal{Q}$ thanks to (i) so that either Galois($\mathcal{P} \supseteq \mathcal{Q}$) is trivial or $\mathcal{Q}$ is the fixed-point algebra under Galois($\mathcal{P} \supseteq \mathcal{Q}$). Thus we get (ii) since the latter case means $\mathcal{P} = \mathcal{Q} \rtimes \mathbf{Z}_p$ with the only group $\mathbf{Z}_p$ of order $p$. □

Using this lemma, we can show the following: Let $\mathcal{B}$ be a left co-ideal $*$-subalgebra of a finite-dimensional Kac algebra $\mathcal{A}$ (see [**34**]). Then, the dimension of $\mathcal{B}$ divides that of $\mathcal{A}$. See [**50**] for more general and purely Hopf algebraic arguments. In particular, both of the dimension of a Kac subalgebra and the order of the intrinsic group divide the dimension of $\mathcal{A}$.

LEMMA 9.5. *Let $\mathcal{P}$ be a factor, $G$ be a finite group, and $\{[\alpha_g]\}_{g \in G}$ $(\subseteq Out(\mathcal{P}))$ be an injective $G$-kernel. Assume that a family $\{[\rho_i]\}_{i \in I} \in Sect(\mathcal{P})$ of irreducible sectors satisfies $[\alpha_g \rho_i] = [\rho_i]$ (for each $g \in G, i \in I$). If $H^2(G, \mathbf{T})$ is trivial, then we can take representatives $\{\alpha_g\}$ and $\{\rho_i\}$ in such a way that $\{\alpha_g\}_{g \in G}$ is an action of $G$ and $\alpha_g \cdot \rho_i = \rho_i$ $(g \in G, i \in I)$.*

PROOF. After fixing $i_0 \in I$, we choose representatives $\{\alpha_g\}$ satisfying $\alpha_g \cdot \rho_{i_0} = \rho_{i_0}$ $(g \in G)$. Then, $\{\alpha_g\}_{g \in G}$ is an action of $G$. In fact, $\alpha_g \alpha_h = Adu\alpha_{gh}$ and

$$Adu\rho_{i_0} = Adu\alpha_{gh} \cdot \rho_{i_0} = \alpha_g \alpha_h \cdot \rho_{i_0} = \rho_{i_0}.$$

Then, the irreducibility of $\rho_{i_0}$ says that $u$ is a scalar and hence $\alpha_g \alpha_h = \alpha_{gh}$.

For general $i \in I$ (and $g \in G$) there exists a unitary $u_g^i \in \mathcal{P}$ satisfying $Ad(u_g^i) \cdot \alpha_g \cdot \rho_i = \rho_i$ by the assumption. Since $\{Ad(u_g^i) \cdot \alpha_g\}_{g \in G}$ is also an action of $G$ by the same reason, there exists $\omega^i \in Z^2(G, \mathbf{T})$ satisfying $u^i_g \alpha_h(u^i_h) = \omega^i(g,h)u^i_{gh}$, $g, h \in G$. Since $H^2(G, \mathbf{T})$ is trivial, we may and do assume that $\omega^i(g,h) = 1$ (after modifying $\{u^i_g\}$ if necessary) and hence $\{u^i_g\}_{g\in G}$ is an $\alpha$-cocycle. However, since every cocycle of an outer action of a finite group on a factor is trivial ([**4**]), one finds unitaries $v^i \in \mathcal{P}$ satisfying $u^i_g = v^{i*}\alpha_g(v^i)$ for each $g \in G$. Thus, we get $\alpha_g \cdot Ad(v^i) \cdot \rho_i = Ad(v^i) \cdot \rho_i$, i.e., the representatives $Ad(v^i)\rho_i$ do the job. □

We also need following purely group-theoretical results:

PROPOSITION 9.6. *(i) (Frobenius-Thompson) Let $G$ be a finite group with a subgroup $H$. If $gHg^{-1} \cap H = \{e\}$ for all $g \in G \setminus H$, then there exists a normal subgroup $N$, which is nilpotent, satisfying $G = N \rtimes H$. Such $G$ is called a Frobenius group.*
*(ii) (Burnside) Let $G$ be a finite group and assume that there is a $p$-Sylow subgroup $H$ such that it is contained in the center of the normalizer of $H$ in $G$. Then there exists a normal subgroup $N$ satisfying $G = N \rtimes H$.*
*(iii) Let $H$ be a finite group acting on another finite group $N$. If the action of $H$ has exactly two orbits and $\#N = \#H + 1$, then $N$ is a direct sum of $n$ copies of the cyclic group of prime order $p$ and $H \simeq \mathbf{Z}_{pn-1}$.*

PROOF. (i) See 8.5.5 and 10.5.6 of [**59**]. (ii) See 10.1.8 of [**59**]. (iii) See Exercise 10.5.6 of [**59**]. □

LEMMA 9.7. *(i) Let $G = N \rtimes H$ be a semi-direct product of finite groups $N$ and $H$, and we assume that $\#N$ and $\#H$ are relatively prime. If $K$ is a subgroup of $G$ satisfying $G = K \cdot H$ and $K \cap H = \{e\}$, then $K = N$.*
*(ii) Let $H, K$ be subgroups of a finite group $G$ satisfying $G = K \cdot H$ and $K \cap H = \{e\}$. If $H$ is a $p$-Sylow subgroup of $G$ contained in the center of its normalizer, then $K$ is normal (and hence $G = K \rtimes H$).*

PROOF. (i) Since $\#K = \frac{\#G}{\#H} = \#N$, it suffices to see $K \subseteq N$. If we had $k = nh$ with $k \in K, n \in N$, and $h \neq e \in H$, then we would get $e = (nh)^m \in Nh^m$. Here, $m$ (dividing $\#K = \#N$) is the order of $k$. But, this is a contradiction since $\#N, \#H$ are relatively prime and $h^m \neq e$. (ii) It is a direct consequence of (i) and Burnside's theorem above. □

## 2. Main theorem

We are now ready to state the main result in this chapter, whose proof will be presented in the next section.

THEOREM 9.8. *Let $\mathcal{A}$ be a finite-dimensional Kac algebra and $\hat{\mathcal{A}}$ its dual Kac algebra.*
*(i) If*

$$\mathcal{A} = \mathbf{C}^m \oplus M_m(\mathbf{C}),$$

*then $m = p^k - 1$ with $p$ prime and the intrinsic group $G(\hat{\mathcal{A}})$ of $\hat{\mathcal{A}}$ is the cyclic group $\mathbf{Z}_m$. Furthermore, $\mathcal{A}$ is classified by an element in $H^2((\underbrace{\mathbf{Z}_p \times \cdots \times \mathbf{Z}_p}_{k}, \mathbf{Z}_m), \mathbf{T})/\!\sim$ satisfying*

$$\ell^\infty(\mathbf{Z}_p \times \cdots \times \mathbf{Z}_p) \rtimes_\zeta \mathbf{Z}_m \cong \mathbf{C}^m \oplus M_m(\mathbf{C}).$$

*(ii) If*

$$\mathcal{A} = \mathbf{C}^{m^2} \oplus M_m(\mathbf{C}),$$

*then $G(\hat{\mathcal{A}})$ is an abelian group (with the dual group $N$) and $\mathcal{A}$ is classified by an element in $H^2((N, \mathbf{Z}_2), \mathbf{T})/\sim$ satisfying*

$$\mathbf{C}(N) \oplus \mathbf{C}_{\eta_1}(N) \cong \mathbf{C}^{m^2} \oplus M_m(\mathbf{C}).$$

*(iii) If*

$$\mathcal{A} = \mathbf{C}^p \oplus \underbrace{M_p(\mathbf{C}) \oplus \cdots \oplus M_p(\mathbf{C})}_{n}$$

*with $p$ prime, then $\mathcal{A}$ is classified by $N \rtimes \mathbf{Z}_p$, where $N$ is a nilpotent group with $\#N = 1 + np$, and an element in $H^2((N, \mathbf{Z}_p), \mathbf{T})/\sim$ satisfying*

$$\ell^\infty(N) \rtimes_\zeta \mathbf{Z}_p \cong \mathbf{C}^p \oplus \underbrace{M_p(\mathbf{C}) \oplus \cdots \oplus M_p(\mathbf{C})}_{n}.$$

*(iv) Assume*

$$\mathcal{A} = \mathbf{C}^{p^2} \oplus \overbrace{M_p(\mathbf{C}) \oplus \cdots \oplus M_p(\mathbf{C})}^{q-1},$$
$$\hat{\mathcal{A}} = \mathbf{C}^{p^2} \oplus \underbrace{M_p(\mathbf{C}) \oplus \cdots \oplus M_p(\mathbf{C})}_{q-1}$$

*with distinct primes $p, q$. Then, $\mathcal{A}$ is classified by $N \rtimes \mathbf{Z}_p$ with $N = \mathbf{Z}_q \rtimes \mathbf{Z}_p$ non-commutative, and an element in $H^2((N, \mathbf{Z}_p), \mathbf{T})/\sim$ satisfying*

$$\ell^\infty(N) \rtimes_\zeta \mathbf{Z}_p \cong \oplus_{h \in \mathbf{Z}_p} \mathbf{C}_{\eta_h}(N) \cong \mathbf{C}^{p^2} \oplus \underbrace{M_p(\mathbf{C}) \oplus \cdots \oplus M_p(\mathbf{C})}_{q-1}.$$

*(v) If*

$$\mathcal{A} = \mathbf{C}^{pq} \oplus \underbrace{M_q(\mathbf{C}) \oplus \cdots \oplus M_q(\mathbf{C})}_{np}$$

*with distinct primes $p, q$, then the $q$-Sylow subgroup $G_0$ of $G(\hat{\mathcal{A}})$ is normal and characteristic. Let $\mathcal{P} \supseteq \mathcal{Q}$ be the corresponding depth 2 inclusion, and set $\mathcal{R} = \mathcal{Q} \rtimes G_0$, which we regard as an intermediate subfactor of $\mathcal{P} \supseteq \mathcal{Q}$. We set $H = \hat{G}_0$, the dual group, and $\mathcal{R}^{(\alpha, H)} = \mathcal{Q}$ with the dual action $\alpha$. Then, there exists a group $N$ with $\#N = p(1 + nq)$ and its outer action $\beta$ on $\mathcal{R}$ satisfying $\mathcal{P} = \mathcal{R} \rtimes_\beta N$.*

1. *Under the additional assumption that $G(\hat{\mathcal{A}})$ is abelian, $\mathcal{A}$ is classified by $N \rtimes H$ and an element in $H^2((N, H), \mathbf{T})/\sim$ satisfying*

$$\ell^\infty(N) \rtimes_\zeta H \cong \mathbf{C}^{pq} \oplus \underbrace{M_q(\mathbf{C}) \oplus \cdots \oplus M_q(\mathbf{C})}_{np}.$$

2. *Moreover, $G(\hat{\mathcal{A}})$ is automatically abelian when $q \not\equiv 1 \pmod p$ or $r = 1 + nq$ is prime with $pq \not\equiv 1 \pmod r$.*

COROLLARY 9.9. *A finite-dimensional Kac algebra whose underlying algebra structure is given by*

$$\mathbf{C}^2 \oplus \underbrace{M_2(\mathbf{C}) \oplus \cdots \oplus M_2(\mathbf{C})}_{n}$$

*is co-commutative.*

PROOF. Thanks to (iii) in the above theorem, a Kac algebra $\mathcal{A}$ described in the corollary is classified by a Frobenius group $N \rtimes \mathbf{Z}_2$ and an element in $H^2((N,\mathbf{Z}_2),\mathbf{T})/\!\sim$ satisfying

$$\ell^\infty(N) \rtimes_\zeta \mathbf{Z}_2 \cong \mathbf{C}^2 \oplus \underbrace{M_2(\mathbf{C}) \oplus \cdots \oplus M_2(\mathbf{C})}_{n}.$$

Let $\tau$ be the generator of $\mathbf{Z}_2$. It is considered as a period 2 automorphism of $N$, and we at first note that $\tau$ is fixed-point free. Indeed, if there were fixed points (other than the unit $e \in N$), one would obtain more than two copies of $\mathbf{C}$ in $\mathcal{A}$. This fact implies that $N$ is an abelian group of odd order and $n^\tau = n^{-1}$ for each $n \in N$ (see, for example, 10.5 of [**59**]). Note that the order of $H^2(N,\mathbf{T})$ is odd and the induced automorphism of $\tau$ on $H^2(N,\mathbf{T})$ is trivial (see [**69**]).

Let $(\eta,\zeta) \in Z^2((N,\mathbf{Z}_2),\mathbf{T})$ be the element corresponding to $\mathcal{A}$. Then, the fundamental relation between $\eta$ and $\zeta$ says

$$1 = [\eta_\tau][\eta_\tau(\tau(\cdot),\tau(\cdot))] = [\eta_\tau]^2.$$

Since the group $H^2(N,\mathbf{T})$ (of odd order) cannot admit a subgroup of even order, we conclude $\eta_\tau \in B^2(N,\mathbf{T})$. Therefore, the dual Kac algebra of $\mathcal{A}$ is

$$\hat{\mathcal{A}} = \mathbf{C}(N) \oplus \mathbf{C}_{\eta_\tau}(N) \cong \mathbf{C}(N) \oplus \mathbf{C}(N),$$

which is commutative. □

It is not so difficult to find examples of non-co-commutative Kac algebras satisfying the conditions in the above theorem. We present an example for (i) for instance. Let $N = \mathbf{Z}_2 \times \mathbf{Z}_2 \times \mathbf{Z}_2$. We define $\tau \in Aut(N)$ by

$$\begin{pmatrix} 0 & 0 & 1 \\ 1 & 0 & 1 \\ 0 & 1 & 0 \end{pmatrix} \in GL(2,3).$$

Then, a direct computation shows that $\tau$ is a fixed-point free automorphism of order 7. Moreover, we can show $\sum_{i=0}^{6}(\tau \otimes \tau)^i = 0$ on $N \otimes N$, where the tensor product is over $\mathbf{Z}$. We define $\eta \in Z^2(N,\mathbf{T}) \setminus B^2(N,\mathbf{T})$ by

$$\eta((a,b,c),(a',b',c')) = (-1)^{ab'}.$$

We have

$$\prod_{i=0}^{6} \eta^{\tau^i} = 1 \text{ with } \eta^{\tau^i}(n_1,n_2) = \eta(\tau^i(n_1),\tau^i(n_2)),$$

and set

$$\eta_{\tau^i} = \prod_{j=0}^{i-1} \eta^{\tau^j} \quad (\in Z^2(N,\mathbf{T})) \quad \text{for } i = 0,1,\cdots,6.$$

The above identity guarantees $\eta_{\tau^i} \times \eta_{\tau^j}(\tau^i(\cdot),\tau^i(\cdot)) = \eta_{\tau^{i+j}}$, where $i+j$ is understood in mod 7. Thus, we get a $\mathbf{Z}_7$-equivariant family $\{\eta_{\tau_i}\}$ of cocycles, that is, $(\eta,\zeta) \in Z^2((N,\mathbf{Z}_7),\mathbf{T})$ with $\zeta_n(g,h) = 1$. We easily observe

$$\mathcal{A} = \ell^\infty(N) \rtimes \mathbf{Z}_7 \cong \mathbf{C}^7 \oplus M_7(\mathbf{C}),$$
$$\hat{\mathcal{A}} = \oplus_{i=0}^{6} \mathbf{C}_{\eta_{\tau^i}}(N) \cong \mathbf{C}^8 \oplus \underbrace{M_2(\mathbf{C}) \oplus \cdots \oplus M_2(\mathbf{C})}_{12},$$

and hence $\mathcal{A}$ is non-co-commutative.

## 3. Proof

Here the proof of Theorem 9.8 is presented. In what follows, $\mathcal{P} \supseteq \mathcal{Q}$ is the depth 2 inclusion corresponding to $\mathcal{A}$, and

$$\mathcal{P}_1 \supseteq \mathcal{P} \supseteq \mathcal{Q}$$

is its basic extension. Also,

$$\rho : \mathcal{Q} \hookrightarrow \mathcal{P} \quad \text{and} \quad \sigma : \mathcal{P} \hookrightarrow \mathcal{P}_1$$

denote the inclusion maps.

(i) Let $\pi$ be the sector corresponding to the simple component $M_m(\mathbf{C})$. Then, the multiplicity of $\pi$ is $m$, $d(\pi) = m$, and the irreducible decomposition of $[\sigma\bar{\sigma}]$ is given by

$$[\sigma\bar{\sigma}] = \oplus_{g \in G(\hat{\mathcal{A}})}[\tau_g] \oplus m[\pi]$$

with a $G(\hat{\mathcal{A}})$-kernel $\{[\tau_g]\}_{g \in G(\hat{\mathcal{A}})}$ (on $\mathcal{P}_1$). Since the system $\{[\tau_g]\}_{g \in G(\hat{\mathcal{A}})} \cup \{[\pi]\}$ is closed under conjugation and irreducible decomposition of products, we have the following fusion rule:

$$[\tau_g][\pi] = [\pi][\tau_g] = [\bar{\pi}] = [\pi],$$
$$[\pi]^2 = \oplus_{g \in G(\hat{\mathcal{A}})}[\tau_g] \oplus (m-1)[\pi].$$

We can obviously take representatives $\{\tau_g\}_{g \in G(\hat{\mathcal{A}})}$ satisfying $\tau_g \cdot \pi = \pi$. Then, $\tau_g$'s form a group (see the first part of the proof of Lemma 9.5). We denote by $\mathcal{R} = \mathcal{P}_1^{\tau}$ the fixed-point algebra of $\mathcal{P}_1$ under $\tau$, and by $\nu : \mathcal{R} \hookrightarrow \mathcal{P}_1$ the inclusion map. At this stage it is not known (but will be shown) if $\mathcal{R}$ is an intermediate subfactor of $\mathcal{P} \subseteq \mathcal{P}_1$ because of a possible 2 cohomology obstruction of $G(\hat{\mathcal{A}})$. However, as in the proof of Lemma 3.1, (iii) of [**30**], we can show that $G(\hat{\mathcal{A}})$ is abelian and there exists $\theta \in Aut(\mathcal{R})$ satisfying $[\pi] = [\nu\theta\bar{\nu}]$. Let $H$ be the dual group of $G(\hat{\mathcal{A}})$ and $\alpha$ be the action of $H$ on $\mathcal{R}$ such that $\mathcal{P}_1 = \mathcal{R} \rtimes_\alpha H$ and $\tau$ is the dual action of $\alpha$.

Let $G$ ($\subseteq Out(\mathcal{R})$) be the group generated by $[\alpha_H]$ and $[\alpha_H][\theta][\alpha_H]$, and we claim that

$$G = [\alpha_H] \cup [\alpha_H][\theta][\alpha_H] \text{ with } \#G = m + m^2.$$

Indeed, since $[\nu\theta\bar{\nu}][\nu\theta\bar{\nu}] = [\nu\theta\alpha_H\theta\bar{\nu}]$, each $\nu\theta\alpha_h\theta\bar{\nu}$ ($h \in H$) is contained in $\pi^2$. Thanks to the above fusion rule, we have either $[\nu\theta\alpha_h\theta\bar{\nu}] = [\alpha_H]$ or $[\nu\theta\alpha_h\theta\bar{\nu}] = [\pi]$ by comparing statistical dimensions. In the first case, the Frobenius reciprocity implies $[\nu\theta\alpha_h\theta] = [\nu]$ and hence $[\theta\alpha_h\theta]$ appears in $[\bar{\nu}\nu] = [\alpha_H]$. Thus, there exists $h' \in H$ satisfying $[\theta\alpha_h\theta] = [\alpha_{h'}]$. In the second case, the Frobenius reciprocity implies

$$1 = \dim(\pi, \pi) = \dim(\nu\theta\alpha_h\theta\bar{\nu}, \nu\theta\bar{\nu}) = \dim(\theta\alpha_h\theta, \bar{\nu}\nu\theta\bar{\nu}\nu) = \dim(\theta\alpha_h\theta, \alpha_H\theta\alpha_H),$$

showing $[\theta\alpha_h\theta] = [\alpha_{h_1}\theta\alpha_{h_2}]$ for some $h_1, h_2 \in H$. Consequently, in either case we have $([\alpha_H][\theta][\alpha_H])^2 \subseteq [\alpha_H] \cup [\alpha_H][\theta][\alpha_H]$. It remains to prove $\#G = m + m^2$. To this end, it suffices to observe $[\alpha_H] \cap [\alpha_H][\theta][\alpha_H] = \emptyset$ and $[\alpha_{h_1}\theta\alpha_{h_2}] = [\theta]$ only when $h_1 = h_2 = e$, both of which are consequences of the irreducibility of $\pi$. For the first fact, if we had $[\alpha_H] \cap [\alpha_H][\theta][\alpha_H] \neq \emptyset$ (hence $[\theta] = [\alpha_h]$ for some $h$ and $[\pi] = [\nu\alpha_h\bar{\nu}]$), then we would get the contradiction

$$1 = \dim(\pi, \pi) = \dim(\nu\alpha_h\bar{\nu}, \nu\alpha_h\bar{\nu}) = \dim(\bar{\nu}\nu\alpha_h, \alpha_h\bar{\nu}\nu) = \dim(\alpha_H, \alpha_H).$$

For the second fact, we just notice

$$1 = \dim(\pi, \pi) = \dim(\nu\theta\overline{\nu}, \nu\theta\overline{\nu}) = \dim(\overline{\nu}\nu, \theta\overline{\nu}\nu\theta^{-1}) = \dim(\alpha_H, \theta\alpha_H\theta^{-1}),$$

i.e., $[\alpha_H] \cap [\theta][\alpha_H][\theta^{-1}] = \{[id]\}$.

Next we show that $G$ is a Frobenius group (see Proposition 9.6,(i)). Let $g \in G \setminus H$ (with the identification $[\alpha_H] \cong H$). The above argument shows that there exist $h_1, h_2 \in H$ satisfying $g = [\alpha_{h_1}\theta\alpha_{h_2}]$ so that we have

$$[\alpha_H] \cap g[\alpha_H]g^{-1} = [\alpha_{h_1}]\left([\alpha_H] \cap [\theta][\alpha_H][\theta^{-1}]\right)[\alpha_{h_1}^{-1}],$$

implying $[\alpha_H] \cap g[\alpha_H]g^{-1} = \{[id]\}$ thanks to the above second fact. This means that $G$ is a Frobenius group and there exists a normal subgroup $N \subseteq G$ satisfying $G = N \rtimes H$. Note that $G = [\alpha_H] \cup [\alpha_H][\theta][\alpha_H]$ is the $(H, H)$-double coset decomposition of $G$, which means the action of $H$ on $N$ has just two orbits. Since $\#H = m$ and $\#N = \frac{m+m^2}{m} = m+1$, Lemma 9.6, (iii) implies that $N$ is isomorphic to $\underbrace{\mathbf{Z}_p \times \cdots \times \mathbf{Z}_p}_{k}$ with $p$ prime and $H \cong \mathbf{Z}_m$. Recall that $G(\hat{\mathcal{A}})$ is the dual group of $H$, and hence we have $G(\hat{\mathcal{A}}) \cong \mathbf{Z}_m$. Since $H^2(\mathbf{Z}_m, \mathbf{T})$ is trivial (and $[\tau_g][\sigma] = [\sigma]$), we can now assume

$$\mathcal{P}_1 \ (= \mathcal{R} \rtimes_\alpha H) \supseteq \mathcal{R} \supseteq \mathcal{P}$$

thanks to Lemma 9.5.

To deal with the above second inclusion, we let $\sigma_1 : \mathcal{P} \hookrightarrow \mathcal{R}$ be the inclusion map. Then, $\sigma = \nu\sigma_1$ so that $\sigma\overline{\sigma} = \nu\sigma_1\overline{\sigma}_1\overline{\nu}$ is the sum of $\nu\mu\overline{\nu}$'s over the irreducible components $[\mu]$ appearing in $\sigma_1\overline{\sigma}_1$. When $[\mu] = [id]$, we have $\nu\mu\overline{\nu} = \oplus_{g \in G(\hat{\mathcal{A}})}\tau_g$ where all the one-dimensional components in $\sigma\overline{\sigma}$ are already exhausted. Thus, for $[\mu] \neq [id]$ instead $\nu\mu\overline{\nu}$ can contain none of them, which forces $\dim(\pi, \nu\mu\overline{\nu}) = \dim(\nu\theta\overline{\nu}, \nu\mu\overline{\nu}) \neq 0$. Since $\pi = \nu\theta\overline{\nu}$, the Frobenius reciprocity implies

$$0 \neq \dim(\nu\theta\overline{\nu}, \nu\mu\overline{\nu}) = \dim(\overline{\nu}\nu\theta\overline{\nu}\nu, \mu) = \dim(\oplus_{h_1, h_2 \in H}\alpha_{h_1}\theta\alpha_{h_2}, \mu),$$

showing $[\mu] = [\alpha_{h_1}\theta\alpha_{h_2}] \in G$ for some $h_1, h_2 \in H$ and $d(\mu) = 1$. We have seen $\sigma_1\overline{\sigma}_1$ contains just one-dimensional irreducible components so that we have $\mathcal{P} = \mathcal{R}^{(\beta, K)}$, the fixed-point subalgebra, with a finite group $K$ and its outer action $\beta$ on $\mathcal{R}$. Finally, we remark that Lemma 9.7,(i) shows $[\beta_K] = N$ and we get the result.

(ii) Let

$$[\overline{\rho}\rho] = \oplus_{g \in G(\hat{\mathcal{A}})}[\tau_g] \oplus m[\pi]$$

be the irreducible decomposition of $[\overline{\rho}\rho]$ ($\in Sect(\mathcal{Q})$). Then, as before we easily see the following fusion rule:

$$[\tau_g][\pi] = [\pi][\tau_g] = [\overline{\pi}] = [\pi],$$
$$[\pi]^2 = [\pi][\overline{\pi}] = [\overline{\pi}][\pi] = \oplus_{g \in G(\hat{\mathcal{A}})}[\tau_g].$$

We consider the inclusion $\mathcal{Q} \supseteq \pi(\mathcal{Q})$. The second set of rules means that $\pi(\mathcal{Q})$ and the "dual" subfactor $\overline{\pi}(\mathcal{Q})$ are the fixed point algebras of finite group ($= G(\hat{\mathcal{A}})$) actions. Therefore, $G(\hat{\mathcal{A}})$ must be abelian.

We set $\mathcal{R} = \mathcal{Q} \rtimes G(\hat{\mathcal{A}}) \subseteq \mathcal{P}$. We denote by $N$ the dual group of $G(\hat{\mathcal{A}})$ and $\alpha$ the dual action of $G(\hat{\mathcal{A}})$. Thanks to $[\mathcal{P} : \mathcal{R}] = 2$, we have $\mathcal{P} = \mathcal{R} \rtimes_\beta \mathbf{Z}_2$ with a $\mathbf{Z}_2$ action $\beta$ on $\mathcal{R}$, and we are in the following situation:

$$\mathcal{Q} = \mathcal{R}^{(\alpha, N)} \subseteq \mathcal{R} \subseteq \mathcal{P} = \mathcal{R} \rtimes_\beta \mathbf{Z}_2.$$

The inclusion $\mathcal{P} \supseteq \mathcal{Q}$ being of depth 2, $[\alpha_H]$ and $[\beta_{\mathbf{Z}_2}]$ must form a product group $G$ in $Out(\mathcal{R})$. However, since $[G : [\alpha_H]] = 2$, the subgroup $[\alpha_H]$ is normal in $G$ and we are done.

(iii) Let

$$[\sigma\overline{\sigma}] = \oplus_{g\in G(\hat{\mathcal{A}})}[\tau_g] \oplus (\oplus_{i\in I} p[\pi_i])$$

be the irreducible decomposition of $[\sigma\overline{\sigma}]$. Here, $\{[\tau_g]\}_{g\in G(\hat{\mathcal{A}})}$ is a $G(\hat{\mathcal{A}})$-kernel (on $\mathcal{P}_1$) and $d(\pi_i) = p$. We take representatives $\{\tau_g\}_{g\in G(\hat{\mathcal{A}})}$ satisfying $\tau_g \cdot \sigma = \sigma$ (and hence $\tau_g$'s determine an action on $\mathcal{P}_1$ as before). Thus, we can set $\mathcal{R} = \mathcal{P}_1^{\tau} \supseteq \mathcal{P}$ and denote by $\nu : \mathcal{R} \hookrightarrow \mathcal{P}_1$ the inclusion map. Since $G(\hat{\mathcal{A}}) \cong \mathbf{Z}_p$ admits no 2-cohomology obstruction, as in the proof of Lemma 3.1 of [**30**] we can find $\theta_i \in Aut(\mathcal{R})$ satisfying $[\pi_i] = [\nu\theta_i\overline{\nu}]$. Let $\alpha$ be the action of $\mathbf{Z}_p$ on $\mathcal{R}$ whose dual action is $\tau$ so that we have

$$\mathcal{P} \subseteq \mathcal{R} \subseteq \mathcal{P}_1 = \mathcal{R} \rtimes_\alpha \mathbf{Z}_p.$$

Let $G$ ($\subseteq Out(\mathcal{R})$) be the group generated by $[\alpha_{\mathbf{Z}_p}]$ and $\cup_{i\in I}[\alpha_{\mathbf{Z}_p}\theta_i\alpha_{\mathbf{Z}_p}]$. Then, as in the proof of (i), we can show that $G$ is a Frobenius group (of order $p + mp^2$) and there exists a normal subgroup $N \subseteq G$ satisfying $G = N \rtimes H$. Note that $\#N = \frac{p+mp^2}{p} = 1 + mp$ is relatively prime to $p$. By repeating the arguments in the last paragraph of the proof of (i), we see $\mathcal{P} = \mathcal{R}^{(\beta,K)}$ (relative to a $K$ action $\beta$ on $\mathcal{R}$) and $[\beta_K] = N$.

(iv) Let

$$[\rho\overline{\rho}] = \oplus_{g\in G(\mathcal{A})}[\tau_g] \oplus (\oplus_{i\in I} p[\pi_i])$$

be the irreducible decomposition of $[\rho\overline{\rho}]$ ($\in Sect(\mathcal{P})$), where $\{[\tau_g]\}_{g\in G(\mathcal{A})}$ is a $G(\mathcal{A})$-kernel and $d(\pi_i) = p$. We set

$$\mathcal{Q} \subseteq \mathcal{R}_0 = \mathcal{Q} \rtimes G(\hat{\mathcal{A}}) \subseteq \mathcal{P},$$

and denote by $\rho_1 : \mathcal{R}_0 \hookrightarrow \mathcal{P}$ and $\rho_2 : \mathcal{Q} \hookrightarrow \mathcal{R}_0$ the inclusion maps. Notice that $\rho = \rho_1\rho_2$ and $\rho_1\overline{\rho}_1$ is contained in $\rho\overline{\rho}$. In particular, one-dimensional sectors in $\rho_1\overline{\rho}_1$ form a subgroup in the group $G(\mathcal{A})$. Also notice $[\mathcal{P} : \mathcal{R}_0] = \frac{p^2q}{p^2} = q$ is prime. Since $G(\mathcal{A})$ (of order $p^2$) cannot admit a subgroup of order $q$, Lemma 9.4,(ii) shows that $\rho_1\overline{\rho}_1$ contains no non-trivial automorphism. Thus, there exists a subset $J \subseteq I$ and natural numbers (multiplicities) $\{m_j\}_{j\in J}$ satisfying

$$[\rho_1\overline{\rho}_1] = [id] \oplus (\oplus_{j\in J} m_j[\pi_j]) \,.$$

Note that this shows $q = 1 + pn$ with $n = \sum_{j\in J} m_j$. Let $K$ be the dual group of $G(\hat{\mathcal{A}})$ and $\alpha$ the dual action of $G(\hat{\mathcal{A}})$ on $\mathcal{R}_0$. Since $\mathcal{Q} = \mathcal{R}_0^{(\alpha,K)}$, we get

$$[\rho\overline{\rho}] = [\rho_1\rho_2\overline{\rho}_2\overline{\rho}_1] = \oplus_{k\in K}[\rho_1\alpha_k\overline{\rho}_1].$$

Comparing the two expressions of $[\rho\overline{\rho}]$ so far, we observe that each $\tau_g$ ($g \in G(\mathcal{A})$) appears in some $\rho_1\alpha_k\overline{\rho}_1$ ($k \in K$). Note that the Frobenius reciprocity shows

$$\tau_g \text{ appears in } \rho_1\alpha_k\overline{\rho}_1 \iff [\tau_g][\rho_1] = [\rho_1][\alpha_k].$$

From the assumption on the underlying algebra structure of $\hat{\mathcal{A}}$, we also know that $\overline{\rho}_1\rho_1$ contains no non-trivial automorphism by the same reasoning as above. Therefore, the above characterization shows that $k = k(g)$ is uniquely determined by $g$ and that the map $g \in G(\mathcal{A}) \longrightarrow k = k(g) \in K$ is a group homomorphism. This homomorphism is actually an isomorphism between $G(\mathcal{A})$ and $K$ due to the non-existence of (non-trivial) automorphisms in $\rho_1\bar{\rho}_1$. We identify these two groups ($G(\mathcal{A}) = K$) and write $[\tau_g\rho_1] = [\rho_1\alpha_g]$ (so that $[\rho_1\alpha_g\bar{\rho}_1] = [\tau_g\rho_1\bar{\rho}_1]$). The discussions so far show

$$[\rho\bar{\rho}] = \oplus_{g\in G(\mathcal{A})}[\tau_g][\rho_1\bar{\rho}_1] = \oplus_{g\in G(\mathcal{A})}[\tau_g] \oplus \left(\oplus_{j\in J, g\in G(\mathcal{A})} m_j[\tau_g\pi_j]\right).$$

Now, we introduce subgroups $\{G_i\}_{i\in I}$ of $G(\mathcal{A})$ by

$$G_i = \{g \in G(\mathcal{A});\ [\tau_g][\pi_i] = [\pi_i]\},$$

which is a subgroup of $G(\mathcal{A})$ of order either $1, p$ or $p^2$ because of $\#G(\mathcal{A}) = p^2$. Also note that $g \in G_i$ if and only if $[\tau_g]$ appears in $[\pi_i\overline{\pi_i}]$. Note that $[\pi_i\overline{\pi_i}]$ is decomposed into the $[\tau_g]$'s ($g \in G_i$) and some other $[\pi_j]$'s of dimension $p$. Since $[\pi_i\overline{\pi_i}]$ contains $[id]$, by comparing dimensions we conclude that $\#G_i$ must be either $p$ or $p^2$. The latter is actually impossible. In fact, it would imply that the multiplicity of $\pi_i$ in $[\rho\bar{\rho}]$ is at least $p^2$ by the expression of $[\rho\bar{\rho}]$ in the second paragraph, which is impossible. Thus, we have shown $\#G_i = p$ for each $i \in I$ (and $G_i \cong \mathbf{Z}_p$).

We next show that $G_i$ does not depend on $i \in I$. For simplicity we write $\overline{[\pi_i]} = [\pi_{\bar{i}}]$ ($i, \bar{i} \in I$). We take representatives $\{\tau_g\}_{g\in G(\mathcal{A})}$ satisfying $\tau_g \cdot \rho = \rho$, and let $\nu_i$ be the inclusion map $\mathcal{P}^{(G_i,\tau)} \hookrightarrow \mathcal{P}$. Since $H^2(G_i, \mathbf{T})$ is trivial, we can show the existence of an isomorphism $\theta_i : \mathcal{P}^{G_{\bar{i}}} \longrightarrow \mathcal{P}^{G_i}$ satisfying $[\pi_i] = [\nu_i\theta_i\overline{\nu_{\bar{i}}}]$ (as in Lemma 3.1 of [**30**]). We have $[\pi_i\overline{\pi_i}] = [\nu_i\theta_i\overline{\nu_{\bar{i}}}\nu_{\bar{i}}\overline{\theta_i}\overline{\nu_i}]$ and notice that $[\overline{\nu_{\bar{i}}}\nu_{\bar{i}}]$ decomposes into automorphisms Therefore, by considering an irreducible component of $[\pi_i\overline{\pi_i}]$, we can find $i_0 \in I$ such that $[\pi_{i_0}] = [\nu_{i_0}\theta_{i_0}\overline{\nu}_{i_0}]$. This expression clearly shows $G_{i_0} = G_{\bar{i}_0}$. We set

$$I_0 = \{i \in I;\ G_i = G_{\bar{i}} = G_{i_0}\},$$

and note that $i \in I_0$ if and only if $[\pi_i]$ is of the form $[\nu_{i_0}\theta\overline{\nu}_{i_0}]$ with some automorphism $\theta$. The product of two such $\pi_i$'s is decomposed into the sum of $[\nu_{i_0}\theta\overline{\nu}_{i_0}]$'s again since $\overline{\nu}_{i_0}\nu_{i_0}$ is decomposed into automorphisms. Also, if $[\tau_g\pi_i] = [\pi_j]$ ($g \in G_{i_0}$) (or $[\pi_i\tau_g] = [\pi_j]$), then $i \in I_0$ imply $j \in I_0$ since $G(\mathcal{A})$ is abelian. Therefore, the irreducible sectors $\{\tau_g\}_{g\in G(\mathcal{A})} \cup \{\pi_i\}_{i\in I_0}$ form a system closed under conjugation and irreducible decomposition of products in the sense of Lemma 9.3, and there exists an intermediate subfactor $\mathcal{L}$ of $\mathcal{P} \supseteq \mathcal{Q}$ such that

$$\oplus_{g\in G(\mathcal{A})}[\tau_g] \oplus (\oplus_{i\in I_0} p[\pi_i]) = \gamma_{\mathcal{P}\supseteq\mathcal{Q}}, \text{ the canonical endomorphism of } \mathcal{P} \supseteq \mathcal{Q}.$$

Since $[\mathcal{P} : \mathcal{L}] > p^2$ (because of $\#G(\mathcal{A}) = p^2$) and $[\mathcal{P} : \mathcal{Q}] = p^2q$, we actually have to have $\mathcal{L} = \mathcal{Q}$ thanks to Lemma 9.4,(i). Therefore, we conclude $I_0 = I$, that is, $G_i = G_{i_0}$ for each $i \in I$ as desired.

Note that $G_{i_0}$ is a characteristic subgroup of $G(\mathcal{A})$ because it is canonically determined by the fusion rule of the irreducible representations of $\hat{\mathcal{A}}$. We write $\nu = \nu_{i_0}$ and set

$$\mathcal{Q} \subseteq \mathcal{R} = \mathcal{P}^{(G_{i_0},\tau)} \subseteq \mathcal{P}.$$

Since $G_{i_0} \cong \mathbf{Z}_p$, we have $\mathcal{P} = \mathcal{R} \rtimes_\beta \mathbf{Z}_p$ with a $\mathbf{Z}_p$ outer action $\beta$ on $\mathcal{R}$. As in the proof of (i), we can find a group $N$ of order $pq$ and its outer action $\mu$ on $\mathcal{R}$ such that $\mathcal{Q} = \mathcal{R}^{(\mu,N)}$ and $[\beta_{\mathbf{Z}_p}], [\mu_N]$ form a product group $G$ (in $Out(\mathcal{R})$ of order $p^2q$. We claim that $N$ is non-commutative. This follows from the fact that $[\mu_N]$ sits in $G(\hat{\mathcal{A}})$, the group of one-dimensional sectors in $\bar{\rho}\rho$. In fact, if $N$ (of order $pq$) were abelian, then $[\mu_N]$ admits an element of order $q$ and hence $G(\hat{\mathcal{A}})$ (of order $p^2$) contains a subgroup of order $q$, a contradiction. Therefore, $N$ is non-commutative and hence $N = \mathbf{Z}_q \rtimes \mathbf{Z}_p$ with a non-trivial action of $\mathbf{Z}_p$ on $\mathbf{Z}_q$. It remains to show that $[\mu_N]$ is a normal subgroup in $G$. If $p = 2$, then $[\mu_N]$ is a subgroup of index 2 so that it is normal. Thus, we can assume $p \neq 2$ in the rest. Since $q = 1 + np > p$, the Sylow theorem guarantees that either a $q$-Sylow subgroup is normal or $p^2 \equiv 1 \pmod q$ (and hence $q$ divides $p + 1$). The latter case does not occur because of $p \neq 2$, and so $G = \mathbf{Z}_q \rtimes H$ for any $p$-Sylow subgroup $H$. Since $H$ is abelian as $\#H = p^2$, it is easy to see that any subgroup of order $pq$, which always contains the $q$-Sylow subgroup $\mathbf{Z}_q$ of $G$, is normal in $G$.

(v) Let

$$[\sigma\bar{\sigma}] = \oplus_{g \in G(\hat{\mathcal{A}})}[\tau_g] \oplus (\oplus_{i \in I} q[\pi_i])$$

be the irreducible decomposition of $[\sigma\bar{\sigma}]$ $(\in Sect(\mathcal{P}_1))$. Since $\#G(\hat{\mathcal{A}}) = pq$ and $d(\pi_i) = q$, as before we observe that $[\pi_i\bar{\pi}_i]$ contains exactly $q$ automorphisms and set

$$G_i = \{g \in G(\hat{\mathcal{A}});\ [\tau_g][\pi_i] = [\pi_i]\},$$

which is a $q$-Sylow subgroup of $G(\hat{\mathcal{A}})$.

We show (in the next few paragraphs) that $G_i$ is normal and it does not depend on $i$. Suppose that a $q$-Sylow group of $G(\hat{\mathcal{A}})$ is not normal. Then, $G(\hat{\mathcal{A}})$ has the following presentation:

$$G(\hat{\mathcal{A}}) = \langle a, b;\ a^p = b^q = 1, bab^{-1} = a^k\rangle \text{ with } k \not\equiv 0, 1 \pmod p.$$

Note that $\{\langle a^l b a^{-l}\rangle;\ l = 0, 1, \dots p-1\}$ exhaust all the $q$-Sylow subgroups of $G(\hat{\mathcal{A}})$. We set

$$J = \{i \in I;\ G_i = G_{\bar{i}} = \langle b\rangle\}$$

(where the meaning of $\bar{i}$ is as in the proof of (iv)). As was remarked above, for each $i \in I$ one finds $0 \le l_1, l_2 \le p-1$ such that

$$G_i = \langle a^{l_1} b a^{-l_1}\rangle \quad \text{and} \quad G_{\bar{i}} = \langle a^{l_2} b a^{-l_2}\rangle,$$

which means $\tau_b\tau_{a^{-l_1}}\pi_i = \tau_{a^{-l_1}}\pi_i$, and $\tau_b\tau_{a^{-l_2}}\pi_{\bar{i}} = \tau_{a^{-l_2}}\pi_{\bar{i}}$. Therefore, $j \in I$ satisfying $[\pi_j] = [\tau_{a^{-l_1}}\pi_i\tau_{a^{l_2}}]$ (and hence, $[\pi_{\bar{j}}] = [\tau_{a^{-l_2}}\pi_{\bar{i}}\tau_{a^{l_1}}]$) belongs to $J$, and $J$ is not empty. Remark that this argument also shows that every $[\pi_i]$ is expressed as $[\tau_{a^s}\pi_j\tau_{a^t}]$ for some $j \in J$ and $0 \le s, t \le p-1$. We claim that this expression is unique. To prove the claim, it suffices to show that $[\tau_{a^s}\pi_{j_1}] = [\pi_{j_2}\tau_{a^t}]$ $(j_1, j_2 \in J$, $0 \le s, t \le p-1)$ implies $s = t = 0$. We take $i \in I$ satisfying $[\pi_i] = [\tau_{a^s}\pi_{j_1}] = [\pi_{j_2}\tau_{a^t}]$. The equation $[\tau_b][\pi_{j_2}] = [\pi_{j_2}]$ $(j_2 \in J)$ multiplied by $[\tau_{a^t}]$ from the right gives us $[\tau_b][\pi_i] = [\pi_i]$ while $[\tau_b][\pi_{j_1}] = [\pi_{j_1}]$ $(j_1 \in J)$ and $[\pi_i] = [\tau_{a^s}][\pi_{j_1}]$ imply $[\tau_{a^s b a^{-s}}][\pi_i] = [\pi_i]$. Therefore, we have $\langle b, a^s b a^{-s}\rangle \subseteq G_i$, which forces $s = 0$ due to $\#G_i = q$. The symmetric argument together with the conjugate $[\pi_{\bar{i}}] = \overline{[\pi_i]}$ also gives us $t = 0$.

Let $\nu : \mathcal{P}_1^{\tau_b} \hookrightarrow \mathcal{P}_1$ be the inclusion map. Then, since $H^2(\mathbf{Z}_q, \mathbf{T})$ is trivial, as before there exists $\theta_j \in Aut(\mathcal{P}_1^{\tau_b})$ for each $j \in J$ satisfying $[\pi_j] = [\nu\theta_j\bar{\nu}]$. Thus, every $[\pi_i]$ is expressed as $[\tau_{a^s}\nu\theta_j\bar{\nu}\tau_{a^t}]$, and this expression is unique. We fix $j \in J$ and consider the irreducible decomposition of $[\nu\theta_j\bar{\nu}][\tau_a\nu\theta_j\bar{\nu}]$. We claim that

$$[\nu\theta_{j'}\bar{\nu}] \ (j' \in J) \text{ is not contained in } [\nu\theta_j\bar{\nu}][\tau_a\nu\theta_j\bar{\nu}].$$

Indeed, by the Frobenius reciprocity we get

$$\dim(\nu\theta_j\bar{\nu}\tau_a\nu\theta_j\bar{\nu}, \nu\theta_{j'}\bar{\nu}) = \dim(\bar{\nu}\tau_a\nu, \theta_j^{-1}\bar{\nu}\nu\theta_{j'}\bar{\nu}\nu\theta_j^{-1}).$$

Note that $\theta_j^{-1}\bar{\nu}\nu\theta_{j'}\bar{\nu}\nu\theta_j^{-1}$ is decomposed into 1-dimensional sectors. On the other hand, $\bar{\nu}\tau_a\nu$ (of dimension $q$) is irreducible because of

$$\dim(\bar{\nu}\tau_a\nu, \bar{\nu}\tau_a\nu) = \dim(\tau_a\nu\bar{\nu}\tau_{a^{-1}}, \nu\bar{\nu}) = \dim(\oplus_s\tau_{ab^sa^{-1}}, \oplus_t\tau_{b^t}) = 1,$$

which proves the claim. We also claim that

$$[\nu\theta_j\bar{\nu}\tau_a\nu\theta_j\bar{\nu}] \text{ cannot contain } [\tau_{a^s}\nu\theta_{j'}\bar{\nu}\tau_{a^t}] \ (j' \in J,\ 0 < s,t \leq p-1).$$

Note that $[\nu\theta_j\bar{\nu}\tau_a\nu\theta_j\bar{\nu}]$ is stable under the multiplication of $[\tau_b]$ from the both sides due to $[\tau_b][\nu] = [\nu]$. Therefore, if the above containment occurred for some $s,t$, then $[\nu\theta_j\bar{\nu}\tau_a\nu\theta_j\bar{\nu}]$ would contain all of $[\tau_{b^ua^s}\nu\theta_{j'}\bar{\nu}\tau_{a^tb^v}] = [\tau_{a^sk^u}\nu\theta_{j'}\bar{\nu}\tau_{a^{tk^{-u}}}]$. This is impossible because the dimension of $[\nu\theta_j\bar{\nu}\tau_a\nu\theta_j\bar{\nu}]$ is $q^2$.

The discussions in the preceding paragraph show that the irreducible decomposition of $[\nu\theta_j\bar{\nu}\tau_a\nu\theta_j\bar{\nu}]$ is either of the following forms:

$$\oplus_l[\tau_{a^{sk^l}}\nu\theta_{j'}\bar{\nu}] \quad \text{or} \quad \oplus_l\,[\nu\theta_{j'}\bar{\nu}\tau_{a^{sk^l}}] \quad \text{for some } j' \in J \text{ and } \ 0 < s \leq p-1.$$

We may assume the former by taking the conjugate sector if necessary. Let $[\bar{\nu}\nu] = \oplus_{l=0}^{q-1}[\alpha^l]$ be the irreducible decomposition, and we have

$$\begin{aligned} 1 &= \dim(\nu\theta_j\bar{\nu}\tau_a\nu\theta_j\bar{\nu}, \tau_{a^s}\nu\theta_{j'}\bar{\nu}) = \dim(\tau_{a^{-s}}\nu\theta_j\bar{\nu}\tau_a, \nu\theta_{j'}\bar{\nu}\nu\theta_j^{-1}\bar{\nu}) \\ &= \dim(\tau_{a^{-s}}\nu\theta_j\bar{\nu}\tau_a, \oplus_l\nu\theta_{j'}\alpha^l\theta_j^{-1}\bar{\nu}). \end{aligned}$$

This shows $[\tau_{a^{-s}}\nu\theta_j\bar{\nu}\tau_a] = [\nu\theta_{j'}\alpha^l\theta_j^{-1}\bar{\nu}]$ for some $l$. But, since $[\tau_b][\nu] = [\nu]$, this would imply $[\tau_b][\tau_{a^{-s}}\nu\theta_j\bar{\nu}\tau_a] = [\tau_{a^{-s}}\nu\theta_j\bar{\nu}\tau_a]$, which is impossible ($j \in J$). Thus the $q$-Sylow subgroup of $G(\hat{\mathcal{A}})$ is normal and $G_i$ does not depend on $i$.

We denote by $G_0$ the $q$-Sylow subgroup of $G(\hat{\mathcal{A}})$. We set

$$\mathcal{P} \subseteq \mathcal{R} = \mathcal{P}_1^{(G_0,\tau)} \subseteq \mathcal{P}_1,$$

and denote by $\nu : \mathcal{R} \hookrightarrow \mathcal{P}_1$ the inclusion map. Let $\alpha$ be an action of $\mathbf{Z}_q$ on $\mathcal{R}$ satisfying $\mathcal{P}_1 = \mathcal{R} \rtimes_\alpha \mathbf{Z}_q$. Since $G_i = G_{\bar{i}} = G_0$, there exists $\theta_i \in Aut(\mathcal{R})$ for each $i \in I$ satisfying $[\pi_i] = [\nu\theta_i\bar{\nu}]$. Thus, in the same way as before, we can show $\mathcal{P} = \mathcal{R}^{(\beta,K)}$ with a group $K$ (of order $\frac{pq+npq^2}{q} = p(1+nq)$) and its outer action $\beta$ on $\mathcal{R}$, and $[\alpha_{\mathbf{Z}_q}], [\beta_K]$ form a product group $G$ in $Out(\mathcal{R})$. This finishes the proof of the first part.

Since $G_0$ is normal in $G(\hat{\mathcal{A}})$, we have $G(\hat{\mathcal{A}}) = G_0 \rtimes G_1$ with with a subgroup $G_1$ of order $p$. We set $H = [\alpha_{\mathbf{Z}_q}]$, which is a $q$-Sylow subgroup of $G$ because of $\#K = p(1+nq)$. Since $G$ is a product group, there exists a subgroup $K_0 \subset K$ satisfying $N_G(H) = H \rtimes [\beta_{K_0}]$. For $k \in K_0 \setminus \{e\}$ we have

$$\dim(\nu\beta_k\bar{\nu}, \nu\beta_k\bar{\nu}) = \dim(\beta_k\bar{\nu}\nu\beta_{k^{-1}}, \bar{\nu}\nu) = \dim(\oplus_s\beta_k\alpha_s\beta_{k^{-1}}, \oplus_t\alpha_t) = q.$$

Thus, $[\nu\beta_k\bar{\nu}]$ (of dimension $q$) is not irreducible and decomposed into 1-dimensional sectors. The Frobenius reciprocity implies that there exists a unique element $g \in G_1$ satisfying $[\tau_g][\nu] = [\nu][\beta_k]$. Since this correspondence from $k$ to $g$ is unique, $G_1$ is isomorphic to $K_0$, and we get $\#K_0 = p$. We identify $G_1$ with $K_0$, and write $[\tau_g][\nu] = [\nu][\beta_g]$. Note that $\tau_g$ ($g \in G_1$) globally preserves $\mathcal{R}$ because $g$ normalizes $G_0$. Thus, there exists a normalizer $u_g \in \mathbf{N}(\mathcal{R})$ of $\mathcal{R}$ in $\mathcal{P}_1$ satisfying $\beta_g = Ad(u_g)\cdot\tau_g|_{\mathcal{R}}$. We fix a generator $\alpha \in \alpha_{\mathbf{Z}_q}$, and denote by $\lambda$ the implementing unitary of $\alpha$ in $\mathcal{P}_1$. We also fix generators $g_0 \in G_0$, $g_1 \in G_1$ satisfying $\tau_{g_0}(\lambda) = e^{\frac{2\pi i}{q}}\lambda$, and set

$$\hat{\alpha} = \tau_{g_0}, \ \mu = \tau_{g_1}, \text{ and } \beta = \beta_{g_1}.$$

Since $\mu$ normalizes $\langle\hat{\alpha}\rangle$, there exists $0 < k \leq q-1$ satisfying $\hat{\alpha}\mu = \mu\hat{\alpha}^k$. Thus, there exists a unitary $w \in \mathcal{R}$ satisfying $\mu(\lambda) = w\lambda^k$. Take $u \in \mathbf{N}(\mathcal{R})$ satisfying $\beta = Ad(u)\cdot\mu|_{\mathcal{R}}$. Since $\mathbf{N}(\mathcal{R})/\mathbf{U}(\mathcal{R}) = \langle[\lambda]\rangle$, there exist $0 \leq s \leq q-1$ and a unitary $v \in \mathcal{R}$ satisfying $u = v\lambda^s$. We show $[\beta][\alpha] = [\alpha^k][\beta]$. Indeed, for $x \in \mathcal{R}$ we get

$$\begin{aligned}\beta\cdot\alpha(x) &= Ad(v\lambda^s)\cdot\mu(\lambda x\lambda^*) = Ad(v\lambda^s w\lambda^k)\cdot\mu(x) = Ad(v\alpha^s(w)\lambda^{s+k})\cdot\mu(x)\\ &= Ad(v\alpha^s(w)\lambda^k v^*)\cdot\beta(x) = Ad(v\alpha^s(w)\alpha^k(v^*))\cdot\alpha^k\cdot\beta(x).\end{aligned}$$

Now, we show the two criteria in (2) for $G(\hat{\mathcal{A}})$ to be abelian. At first, if $q \not\equiv 1 \pmod{p}$, then $\mathbf{Z}_p$ acts on $\mathbf{Z}_q$ trivially and so $G(\hat{\mathcal{A}})$ is abelian. Next, we assume that $r = 1 + nq$ is prime and $pq \not\equiv 1 \pmod{r}$. Then, $G$ is a product group of $H$ and $[\beta_K]$ with $\#H = q$, $\#K = pr$. We may further assume $q \equiv 1 \pmod{p}$ (by the above), i.e., there exists a positive integer $m$ satisfying $q = 1 + mp$. Therefore, we have $pq \not\equiv 1 \pmod{r}$, $p \not\equiv 1 \pmod{r}$, and $q \not\equiv 1 \pmod{r}$. The Sylow theorem implies that a $r$-Sylow group of $G$, denoted by $S_r$, is unique and normal. Thus, $S_r$ is a normal subgroup of $[\beta_K]$ as well, and we get $[\beta_K] = S_r \rtimes [\beta_{K_0}]$. Since $S_r \cap N_G(H) = \{e\}$, the action of $H$ on $S_r$ is non-trivial and consequently faithful. Since $S_r$ is the cyclic group of order $r$, $Aut(S_r) \cong \mathbf{Z}_{r-1}$. This means that the action of $H$ and that of $[\beta_{K_0}]$ on $S_r$ commute: i.e. for any pair $h \in H$, $k \in [\beta_{K_0}]$, $khk^{-1}h^{-1} \in H$ acts on $S_r$ trivially. Thus, we get $khk^{-1}h^{-1} = e$, i.e., $H$ and $[\beta_{K_0}]$ commute. This shows that $k = 1$ and $G(\hat{\mathcal{A}})$ is abelian.

Finally, we show that $[\beta_K]$ is normal in $G$ under the assumption that $G(\hat{\mathcal{A}})$ is abelian, which will prove (1). To do so, it suffices to prove that $H$ is contained in the center of the normalizer $N_G(H)$ of $H$ in $G$ thanks to Lemma 9.7, (ii). However, the above computation shows that $H$ and $[\beta_{K_0}]$ commute when $G(\hat{\mathcal{A}})$ is abelian. Thus, $N_G(H)$ is abelian as desired.

CHAPTER 10

# Classification of certain Kac algebras

The dimension of a given Kac algebra imposes certain restriction on the underlying algebra structure (see §1), giving rise to quite useful information to classify Kac algebras of a fixed dimension. It was shown by Masuoka ([**51**]) that a Kac algebra (more precisely Hopf algebra) of dimension $2p$ with $p$ prime is trivial (i.e., either commutative or co-commutative). In §2 a similar result for Kac algebras of dimension $3p$ is shown. In §3 we determine structure of Kac algebras $\mathcal{A}$ of dimension $p^2q$ under the assumption that the rank of each component of $\mathcal{A}$ and $\hat{\mathcal{A}}$ divides $\dim \mathcal{A}$. In the next §4 we see that this additional assumption is automatic as long as the dimension is less than 60. The semi-direct product $D_{14} \rtimes \mathbf{Z}_3$ (no cocycle involved here) gives us a non-trivial Kac algebra of dimension $42 = 2 \cdot 3 \cdot 7$. Within the above range of dimensions, we also prove that all the other Kac algebras of dimension $pq$ or $pqr$ are trivial.

## 1. Useful criteria

First, we collect some useful criteria for classification of finite-dimensional Kac algebras, which will be freely used in later sections.

PROPOSITION 10.1. *Let $\mathcal{A}$ be a finite-dimensional Kac algebra with the following algebra structure:*

$$\mathbf{C}^{m_0} \oplus \underbrace{M_{n_1}(\mathbf{C}) \oplus \cdots \oplus M_{n_1}(\mathbf{C})}_{m_1} \oplus \underbrace{M_{n_2}(\mathbf{C}) \oplus \cdots \oplus M_{n_2}(\mathbf{C})}_{m_2} \oplus \cdots$$
$$\oplus \underbrace{M_{n_k}(\mathbf{C}) \oplus \cdots \oplus M_{n_k}(\mathbf{C})}_{m_k}$$

*with $2 \leq n_1 < n_2 < \cdots < n_k$. Then, we have*
*(i) $m_0$ is a divisor of $\dim \mathcal{A}$.*
*(ii) For every $i = 1, 2, \ldots, k$, $\frac{m_0}{(m_0, n_i^2)}$ is a divisor of $m_i$.*
*(iii) When $m_0$ is odd and $n_1 = 2$, $\dim \mathcal{A}$ is a multiple of 12. If in addition $m_0 \not\equiv 0$ (mod 3), then $\dim \mathcal{A}$ is a multiple of 60.*

PROOF. Let $\hat{\mathcal{A}}$ be the dual Kac algebra of $\mathcal{A}$ as usual. We obviously get (i) thanks to $m_0 = \#G(\hat{\mathcal{A}})$.
(ii) Let $\mathcal{P} \supseteq \mathcal{Q}$ be the inclusion of factors corresponding to the dual Kac algebra and $\rho : \mathcal{Q} \hookrightarrow \mathcal{P}$ be the inclusion map. Then, the irreducible decomposition of $[\rho\bar{\rho}]$ is

$$\oplus_{g \in G(\hat{\mathcal{A}})}[\tau_g] \oplus \oplus_{i=1}^{k}(\oplus_{a \in I_i} n_i[\pi_a^i]))$$

with $\#I_i = m_i$. We set

$$G_a^i = \{g \in G(\hat{\mathcal{A}});\ [\tau_g][\pi_a^i] = [\pi_a^i]\}.$$

Since $G_a^i$ is a subgroup of $G(\hat{\mathcal{A}})$, $\#G_a^i$ divides $\#G(\hat{\mathcal{A}}) = m_0$. On the other hand, since $G_a^i$ is the Galois group for the inclusion $\mathcal{Q} \supseteq \pi_a^i(\mathcal{Q})$, $\#G_a^i$ divides $[\mathcal{Q} : \pi_a^i(\mathcal{Q})] = n_i^2$ as well, and so we conclude that $\#G_a^i$ divides $(m_0, n_i^2)$.

Choose and fix $i \in \{1, 2, \cdots, k\}$. Note that $G(\hat{\mathcal{A}})$ acts on $\{[\pi_a^i]\}_{a \in I_i}$ by multiplication from the left and $\{[\pi_a^i]\}_{a \in I_i}$ is decomposed into several orbits. Let us consider the orbit containing $[\pi_a^i]$, and note that $G_a^i$ is the isotropy group at $[\pi_a^i]$ under the current action. The cardinality of the orbit of $[\pi_a^i]$ (i.e., the homogeneous space $G(\hat{\mathcal{A}})/G_a^i$) is $\frac{m_0}{\#G_a^i}$, which is a multiple of $\frac{m_0}{(m_0, n_i^2)}$ thanks to what was seen in the previous paragraph. Since this is the case for each orbit, we see that $m_i$, the sum of these cardinalities, is also a multiple of $\frac{m_0}{(m_0, n_i^2)}$.

(iii) Let $\pi$ be a (unique) irreducible component of $\rho\bar{\rho}$ with $d(\pi) = 2$, and $\{\rho_j\}_{j \in J}$ be the set of irreducibles generated by $\pi\bar{\pi}$ (i.e., the ones corresponding to vertices at even levels of the principal graph of $\mathcal{P} \supseteq \pi(\mathcal{P})$). Note that the automorphisms appearing in $\pi\bar{\pi}$ form a subgroup in the group $G(\hat{\mathcal{A}})$ of odd ($= m_0$) order. Therefore, this subgroup cannot contain an element of even order. This means that the principal graph of $\mathcal{P} \supseteq \pi(\mathcal{P})$ must be either $E_6^{(1)}$ or $E_8^{(1)}$ thanks to the classification (see [**16**]) of subfactors of index 4 (or graphs with norm 2).

Figure 4 (Coxeter-Dynkin graph $E_6^{(1)}$)

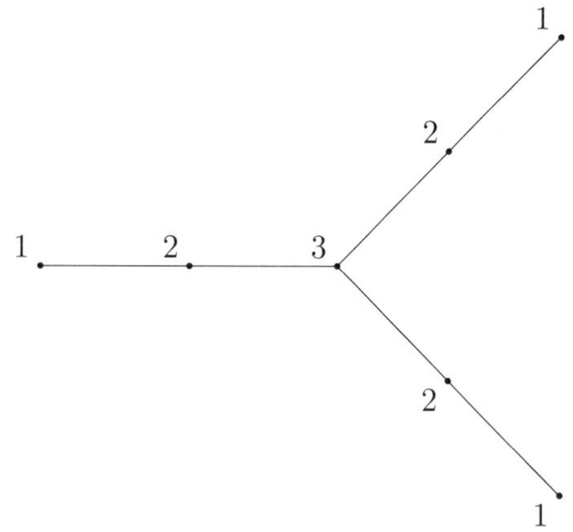


Figure 5 (Coxeter-Dynkin graph $E_8^{(1)}$)

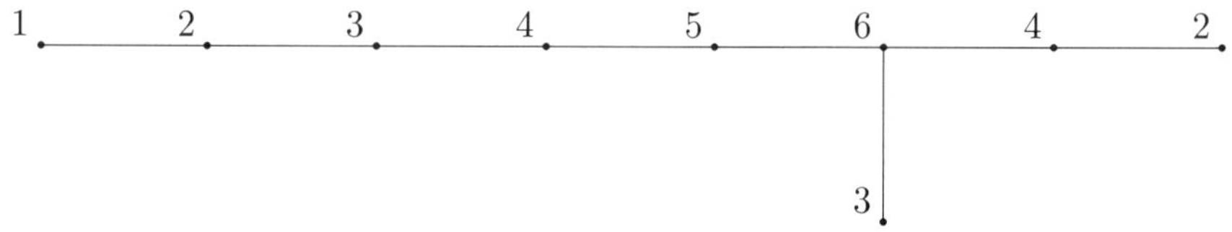

On the other hand, since $\{\rho_j\}_{j\in J}$ is closed under conjugation and irreducible decomposition of products by definition, there exists an intermediate subfactor $\mathcal{P} \supseteq \mathcal{L} \supseteq \mathcal{Q}$ satisfying $[\mathcal{P} : \mathcal{L}] = \sum_{j\in J} d(\rho_j)^2$. Note

$$\sum_{j\in J} d(\rho_j)^2 = \begin{cases} 12 & \text{when the principal graph of } \mathcal{P} \supseteq \pi(\mathcal{P}) \text{ is } E_6^{(1)}, \\ 60 & \text{when the principal graph of } \mathcal{P} \supseteq \pi(\mathcal{P}) \text{ is } E_8^{(1)}. \end{cases}$$

Since $[\mathcal{P} : \mathcal{L}]$ divides $[\mathcal{P} : \mathcal{Q}] = \dim \mathcal{A}$, 12 divides $\dim \mathcal{A}$ in either case. When the graph is $E_6^{(1)}$, the group of the automorphisms appearing in $\pi\overline{\pi}$ is $\mathbf{Z}_3$. Therefore, if we further assume $m_0 \not\equiv 0 \pmod 3$, then possibility of the graph $E_6^{(1)}$ is eliminated and the principal graph of $\mathcal{P} \supseteq \pi(\mathcal{P})$ must be $E_8^{(1)}$. Thus, in this case 60 divides $\dim \mathcal{A}$. □

PROPOSITION 10.2. *(i) If the algebra*

$$\mathbf{C}^r \oplus \underbrace{M_s(\mathbf{C}) \oplus \cdots M_s(\mathbf{C})}_{m}$$

*admits a Kac algebra structure, then s is a divisor of r.*
*(ii) Let r, s, t be natural numbers. If any pair of these is relatively prime, then there exists no Kac algebra with the following algebra structure:*

$$\mathbf{C}^r \oplus \underbrace{M_s(\mathbf{C}) \oplus \cdots M_s(\mathbf{C})}_{m} \oplus \underbrace{M_t(\mathbf{C}) \oplus \cdots M_t(\mathbf{C})}_{n}.$$

PROOF. (i) Assume that $\mathcal{A}$ is such a Kac algebra described here. Let $\mathcal{P} \supseteq \mathcal{Q}$ be the inclusion of factors corresponding the dual Kac algebra $\hat{\mathcal{A}}$, and $\rho : \mathcal{Q} \hookrightarrow \mathcal{P}$ the inclusion map. Then, the irreducible decomposition of $[\rho\overline{\rho}]$ is

$$[\rho\overline{\rho}] = \oplus_{g\in G(\hat{\mathcal{A}})}[\theta_g] \oplus (\oplus_{i\in I} s[\pi_i])$$

with $\#I = m$, $d(\theta_g) = 1$, and $d(\pi_i) = s$. For each $i \in I$ we set

$$G_i = \{g \in G(\hat{\mathcal{A}});\ [\tau_g][\pi_i] = [\pi_i]\}.$$

Then, $G_i$ is a subgroup of $G(\hat{\mathcal{A}})$ so that $\#G_i$ divides $r = \#G(\hat{\mathcal{A}})$.

By the Frobenius reciprocity the irreducible decomposition of $[\pi_i][\overline{\pi}_i]$ is of the form

$$[\pi_i][\overline{\pi}_i] = \oplus_{g\in G_i}[\theta_g] \oplus (\oplus_{j\in I} m_j[\pi_j]).$$

Then, by counting dimensions, we observe $s^2 = \#G_i + \sum_{j\in I} m_j s$. This shows that $s$ divides $\#G_i$ and hence divides $r$.
(ii) We may and do assume $s > t$. To show the result by contradiction let us suppose that a Kac algebra $\mathcal{A}$ described above exists. Let $\mathcal{P} \supseteq \mathcal{Q}$ be the inclusion of factors corresponding to the dual Kac algebra $\hat{\mathcal{A}}$, and $\rho : \mathcal{Q} \hookrightarrow \mathcal{P}$ be the inclusion map. Then, the irreducible decomposition of $[\rho\overline{\rho}]$ is of the form

$$[\rho\overline{\rho}] = \oplus_{g\in G(\hat{\mathcal{A}})}[\theta_g] \oplus (\oplus_{i\in I} s[\pi_i]) \oplus (\oplus_{j\in J} t[\sigma_j])$$

with $\#I = m$, $\#J = n$, $d(\theta_g) = 1$, $d(\pi_i) = s$, and $d(\sigma_j) = t$. Note that

$$\begin{cases} \pi_{i_0}\pi_{i_1} \text{ contains at most one automorphism,} \\ \sigma_{j_0}\sigma_{j_1} \text{ contains at most one automorphism,} \\ \pi_{i_0}\sigma_{j_0} \text{ contains no automorphism.} \end{cases}$$

In fact, the order of the group $\{g \in G(\hat{\mathcal{A}});\ [\theta_g][\pi_i] = [\pi_i]\}$ and that of the group $\{g \in G(\hat{\mathcal{A}});\ [\theta_g][\sigma_j] = [\sigma_j]\}$ divide $(r, s^2)$ and $(r, t^2)$ respectively as was seen in the proof of Proposition 10.1,(i). Since $r, s, t$ are mutually relatively prime, we see that the two groups are trivial, from which the first two facts easily follow. For example, if the product $\pi_{i_0}\pi_{i_1}$ contained two different automorphisms $\theta, \theta'$, then $\pi_{i_0}\pi_{i_1}\theta^{-1}$ would contain $id$ and $\theta'\theta^{-1}$. Thus, $\pi_{i_1}\theta^{-1}$ is the conjugate of $\pi_{i_0}$ and $\pi_{i_0}\overline{\pi}_{i_0}$ would contain a non-trivial automorphism, a contradiction. The third fact is obvious since $[\theta_g] \prec [\pi_{i_0}][\sigma_{j_0}]$ would imply the contradiction $[\theta_g][\overline{\sigma}_{j_0}] = [\pi_{i_0}]$.

We show

$$\begin{cases} (1)\ \pi_i\overline{\pi_i} \text{ always contains some } \sigma_j, \\ (2)\ \sigma_j\overline{\sigma_j} \text{ always contains some } \pi_i, \\ (3) \text{ the irreducible components of } \pi_i\sigma_j \text{ consist of either } \pi_{i'}\text{'s or } \sigma_{j'}\text{'s}, \\ (4) \text{ the irreducible components of } \sigma_{j_1}\sigma_{j_2} \text{ consist of } \sigma_{j'}\text{'s unless } [\sigma_{j_1}] = [\theta_g\overline{\sigma}_{j_2}]. \end{cases}$$

Since $\pi_i\overline{\pi_i}$ and $\sigma_j\overline{\sigma_j}$ contain the identity exactly once, we get (1) and (2) by counting dimensions. To show (3), we at first recall that $\pi_i\sigma_j$ contains no automorphism. Therefore, a priori it is decomposed into $\pi_{i'}$'s and $\sigma_{j'}$'s. However, if two kinds of sectors were mixed, then we would get $st = ks + \ell t$ for some integers $k, \ell > 0$, which is a contradiction. Finally assume $[\sigma_{j_1}] \neq [\theta_g\overline{\sigma}_{j_2}]$ for each $g \in G(\hat{\mathcal{A}})$. This means that $\sigma_{j_1}\sigma_{j_2}$ contains no automorphism. If it were decomposed into two kinds of sectors, then we would get $t^2 = ks + \ell t$ with $k > 0$. Since $t, s$ are relatively prime, $k$ must be a multiple of $t$ and hence $t^2 \geq ks \geq ts$, contradicting $s > t$.

We claim that

$$[\pi_i][\sigma_j] \text{ is decomposed into } [\pi_{i'}]\text{'s for each } i \in I \text{ and } j \in J.$$

If this claim is established, we are done. In fact, after fixing $j \in J$ we choose $\pi_i$ appearing in $\sigma_j\overline{\sigma}_j$. The Frobenius reciprocity says that $\pi_i\sigma_j$ contains $\sigma_j$, which contradicts the above claim. Thus, it remains to prove the above claim. To this end, let us assume that the statement in the above claim is not valid for some $[\pi_{i_0}][\sigma_{j_0}]$. Thanks to (3), it means that $[\pi_{i_0}][\sigma_{j_0}]$ is decomposed into $[\sigma_j]$'s. Notice that $[\sigma_j] \prec [\pi_{i_0}][\sigma_{j_0}]$ implies $[\pi_{i_0}] \prec [\sigma_j][\overline{\sigma_{j_0}}]$ so that the above (4) guarantees $[\sigma_j] = [\theta_g][\sigma_{j_0}]$ for some $g \in G(\hat{\mathcal{A}})$. Since every irreducible component of $[\pi_{i_0}\sigma_{j_0}]$ is of this form, we get

$$[\pi_{i_0}\sigma_{j_0}] = \oplus_{g\in G(\hat{\mathcal{A}})} m_g[\theta_g\sigma_{j_0}].$$

Counting dimensions, we get $st = \sum_{g\in G(\hat{\mathcal{A}})} m_g t$, i.e., $s = \sum_{g\in G(\hat{\mathcal{A}})} m_g$. On the other hand, we have

$$m_g = \dim(\theta_g\sigma_{j_0}, \pi_{i_0}\sigma_{j_0}) = \dim(\sigma_{j_0}\overline{\sigma}_{j_0}, \theta_{g^{-1}}\pi_{i_0}).$$

Also notice that sectors $[\theta_{g^{-1}}\pi_{i_0}]$'s are all distinct (as $g$ varies) because the group $\{g \in G(\hat{\mathcal{A}});\ [\theta_g][\pi_{i_0}] = [\pi_{i_0}]\}$ is trivial. Hence, $\sigma_{j_0}\overline{\sigma}_{j_0}$ contains $\oplus_{g\in G(\hat{\mathcal{A}})} m_g[\theta_{g^{-1}}\pi_{i_0}]$ (and the identity), and hence we get $t^2 > \sum m_g s = s^2$ by comparing dimensions, which is a contradiction. □

## 2. $3p$ theorem

In [**51**] Masuoka showed that a finite-dimensional Hopf algebra of dimension $2p$ with $p$ prime is either commutative or co-commutative under certain conditions. This result was generalized by Kobayashi and Masuoka ([**39**]) to the case of $3p$ where $p$ is a prime with $p \equiv 2 \pmod 3$. In this section, we generalize Masuoka's result to Kac algebras with dimension $3p$ by following his arguments in [**51**] together with Proposition 10.1. See [**66**] for more detailed related results.

LEMMA 10.3. *Let $\mathcal{A}$ be a Kac algebra of dimension $pq$ with primes $p > q > 2$. We further assume that the order $\#G(\mathcal{A})$ of the intrinsic group is either 1 or $q$.*
*(i) The Kac algebra $\mathcal{A}$ admits no simple component of rank 2.*
*(ii) If the dual Kac algebra $\hat{\mathcal{A}}$ has no simple component of rank $r$ with $2 \leq r < q$, then $\mathcal{A}$ is either commutative or co-commutative.*

PROOF. (i) is an easy consequence of Proposition 10.1. To prove (ii), we make use of Masuoka's argument in [**51**]. Suppose $\mathcal{A}$ is neither commutative nor co-commutative. Let $\{u(\pi)_{ij}\}_{\pi\in I}$ be the set of irreducible unitary co-representations of $\mathcal{A}$, and we set $\chi_\pi = \sum_i u(\pi)_{ii}$ (known as a character). We denote by $\mathcal{B}$ the subalgebra of $\mathcal{A}$ generated by the characters, and by $e_{\mathcal{A}}$ the integral of $\mathcal{A}$. Let $e_{\mathcal{A}}, e_1, e_2, \ldots, e_l$ be the minimal central projections of $\mathcal{B}$. In a similar way as in Proposition 3,(i) of [**51**], we know that $\dim \mathcal{A}e_i$ is either $q$ or $p$ for each $i = 1, 2, \ldots l$, and we may assume

$$\dim \mathcal{A}e_i = \begin{cases} q & \text{for } i = 1, 2, \ldots, m \\ p & \text{for } i = m+1, m+2, \ldots m+n \end{cases}$$

with $l = m + n$. Note that $n$ is uniquely determined by $pq = 1 + mq + np$ and $1 \leq n < q$. We set $N = \dim \mathcal{B} - 1$ $(\geq m + n)$ so that the number of simple components of $\hat{\mathcal{A}}$ is $N + 1$.

If $\#G(\hat{\mathcal{A}})$ were trivial, we would have

$$\begin{aligned} pq &= \dim \hat{\mathcal{A}} \geq 1 + q^2 N \geq 1 + q^2(m+n) \\ &= 1 + q^2\left(n + \frac{pq - np - 1}{q}\right) = 1 + q(nq - np + pq - 1). \end{aligned}$$

By dividing the both sides by $q$, we would get

$$p \geq 1 + nq - np + pq - 1 = nq - np + pq > -np + pq,$$

This would mean $n + 1 > q$, contradicting $n < q$.

Thus, we conclude $\#G(\hat{\mathcal{A}}) = q$ and

$$pq = \dim \hat{\mathcal{A}} \geq q + q^2(N - q + 1) \geq q + q^2(m + n - q + 1),$$

and so $(p - q)(n + 1 - q) \geq 0$. Since $n < q$, we have $n = q - 1$ and the equalities hold in the above, which implies that the rank of every non-commutative simple component of $\hat{\mathcal{A}}$ is $q$. Now, we can apply Theorem 9.8 to get

$$\mathcal{A} \cong \oplus_{i=0}^{p-1} \mathbf{C}_{\eta_i}(\mathbf{Z}_q) \cong \oplus_{i=0}^{p-1} \mathbf{C}(\mathbf{Z}_q).$$

This is a contradiction and the lemma is proved. □

In the next two results one knows that the order of $G(\mathcal{A})$ is not $p$ (thanks to Proposition 10.1 and dimension counting). Applying Lemma 10.3 to the case $q = 3$, we get the following:

THEOREM 10.4. *Every Kac algebra of dimension $3p$ with prime $p$ is either commutative or co-commutative.*

**Remark.** The only additional facts needed here (compared with the case of $2p$) are (ii) and (iii) of Proposition 10.1, which likely hold for Hopf algebras with reasonable representation theory. Therefore, we believe that the theorem holds for semisimple Hopf algebras over a field with characteristic 0 as well.

Using the same argument as in the proof of Lemma 10.3 and Proposition 10.2, we get the following:

PROPOSITION 10.5. *Let $\mathcal{A}$ be a Kac algebra of dimension $5p$ with prime $p > 5$. If $p \equiv 2 \pmod 5$ or $p \equiv 4 \pmod 5$, then $\mathcal{A}$ is isomorphic to the group algebra of $\mathbf{Z}_{5p}$.*

PROOF. Note that $\mathbf{Z}_{5p}$ is a unique group of order $5p$ because of $p \not\equiv 1 \pmod 5$ so that $\mathcal{A}$ is commutative if and only if so is the dual Kac algebra. To show the result by contradiction, let us assume that there exists a simple component of $\hat{\mathcal{A}}$ of rank either 3 or 4 (otherwise we are done thanks to (ii) of Lemma 10.3). We use the same notations as in the proof of Lemma 10.3,(ii) with $q = 5$. If $\#G(\mathcal{A}) = 1$, thanks to Proposition 10.2,(ii), we would have

$$\begin{aligned} 5p = \dim \hat{\mathcal{A}} &\geq 1 + 9(N-2) + 16 + 25 \\ &\geq 9(m+n-2) + 42 = 9\left(\frac{5p - np - 1}{5} + n\right) + 24. \end{aligned}$$

However, we have either $n = 1$ or $n = 2$ by the assumption and the above inequality is impossible.

Thus, we can suppose $\#G(\hat{\mathcal{A}}) = 5$. If $\hat{\mathcal{A}}$ were of the form $\mathbf{C}^5 \oplus M_r(\mathbf{C}) \oplus M_r(\mathbf{C}) \oplus \cdots \oplus M_r(\mathbf{C})$, then $r$ must be 5 thanks to Proposition 10.2,(i). Thus, the dimension would be $5 + 5^2k = 5(1+5k)$ which is impossible. If we had only two different ranks of simple components of $\hat{\mathcal{A}}$ (other than 1), then we would get the following thanks to Proposition 10.2,(ii):

$$5p \geq 5 + 9(N-5) + 25 \geq 5 + 9(m+n-5) + 25 = 9\left(n + \frac{5p - np - 1}{5}\right) - 15,$$

which is again impossible. Finally, if we had three different ranks of simple components of $\hat{\mathcal{A}}$ (other than 1), we would get

$$5p \geq 5 + 9(N-6) + 16 + 25 > 5 + 9(N-5) + 25$$

and this is also impossible. Therefore, contradiction was obtained in all the cases. □

## 3. $p^2q$ case

In this section, we classify Kac algebras $\mathcal{A}$ of dimension $p^2q$ with distinct primes $p, q$ **under the assumption** that

the rank of every simple component of $\mathcal{A}$ and $\hat{\mathcal{A}}$ divides $\dim \mathcal{A}$.

In the Hopf algebra theory such an algebra is referred to as a Hopf algebra of Frobenius type. In the next section this additional requirement will be shown automatic for Kac algebras of small dimensions.

LEMMA 10.6. *Let $\mathcal{A}$ be a non-commutative Kac algebra of dimension $p^2q$ with distinct primes $p, q$. If the rank of every simple component of $\mathcal{A}$ divides $\dim \mathcal{A}$, then the algebra structure of $\mathcal{A}$ is one of the following:*

*(i)* $p \equiv \pm 1 \pmod q$ *and* $\mathcal{A} \cong \mathbf{C}^q \oplus \underbrace{M_q(\mathbf{C}) \oplus \cdots M_q(\mathbf{C})}_{\frac{p^2-1}{q}}$,

*(ii)* $p \equiv 1 \pmod q$ *and* $\mathcal{A} \cong \mathbf{C}^{pq} \oplus \underbrace{M_q(\mathbf{C}) \oplus \cdots M_q(\mathbf{C})}_{\frac{p(p-1)}{q}}$,

(*iii*) $q \equiv 1 \pmod{p^2}$ *and* $\mathcal{A} \cong \mathbf{C}^{p^2} \oplus \underbrace{M_{p^2}(\mathbf{C}) \oplus \cdots M_{p^2}(\mathbf{C})}_{\frac{q-1}{p^2}}$,

(*iv*) $\mathcal{A} \cong \mathbf{C}^{p^2} \oplus \underbrace{M_p(\mathbf{C}) \oplus \cdots M_p(\mathbf{C})}_{q-1}$.

PROOF. Thanks to Proposition 10.1(i),(ii), the only possible underlying algebras of $\mathcal{A}$ are as follows:

$$(10.1)\ \mathbf{C} \oplus \underbrace{M_p(\mathbf{C}) \oplus \cdots M_p(\mathbf{C})}_{m} \oplus \underbrace{M_{p^2}(\mathbf{C}) \oplus \cdots M_{p^2}(\mathbf{C})}_{n} \oplus \underbrace{M_q(\mathbf{C}) \oplus \cdots M_q(\mathbf{C})}_{l},$$

$$(10.2)\ \mathbf{C}^p \oplus \underbrace{M_p(\mathbf{C}) \oplus \cdots M_p(\mathbf{C})}_{m} \oplus \underbrace{M_{p^2}(\mathbf{C}) \oplus \cdots M_{p^2}(\mathbf{C})}_{n} \oplus \underbrace{M_q(\mathbf{C}) \oplus \cdots M_q(\mathbf{C})}_{lp},$$

$$(10.3)\ \mathbf{C}^{p^2} \oplus \underbrace{M_p(\mathbf{C}) \oplus \cdots M_p(\mathbf{C})}_{m} \oplus \underbrace{M_{p^2}(\mathbf{C}) \oplus \cdots M_{p^2}(\mathbf{C})}_{n} \oplus \underbrace{M_q(\mathbf{C}) \oplus \cdots M_q(\mathbf{C})}_{lp^2},$$

$$(10.4)\ \mathbf{C}^q \oplus \underbrace{M_p(\mathbf{C}) \oplus \cdots M_p(\mathbf{C})}_{mq} \oplus \underbrace{M_{p^2}(\mathbf{C}) \oplus \cdots M_{p^2}(\mathbf{C})}_{nq} \oplus \underbrace{M_q(\mathbf{C}) \oplus \cdots M_q(\mathbf{C})}_{l},$$

$$(10.5)\ \mathbf{C}^{pq} \oplus \underbrace{M_p(\mathbf{C}) \oplus \cdots M_p(\mathbf{C})}_{mq} \oplus \underbrace{M_{p^2}(\mathbf{C}) \oplus \cdots M_{p^2}(\mathbf{C})}_{nq} \oplus \underbrace{M_q(\mathbf{C}) \oplus \cdots M_q(\mathbf{C})}_{lp}.$$

Case (10.1): We obviously have $p^2q = 1 + mp^2 + np^4 + lq^2$. In particular, $l \neq 0$ and $p^2q > q^2$, that is, $p^2 > q$. Thus, $p^2q > np^4 \geq np^2q$ and hence $n = 0$. However, Proposition 10.2,(ii) says that such a Kac algebra does not exist.

Case (10.2): We have $p^2q = p + mp^2 + np^4 + lpq^2$, that is, $pq = 1 + mp + np^3 + lq^2$. As above we get $l \neq 0$ so that we observe $pq > q^2$, that is, $p > q$. This then implies $n = 0$, and $m \neq 0$. We also observe $p > q > m$. Let $\mathcal{P} \supseteq \mathcal{Q}$ be the inclusion of factors corresponding to the dual Kac algebra of $\mathcal{A}$ and $\rho : \mathcal{Q} \hookrightarrow \mathcal{P}$ the inclusion map. Then, the irreducible decomposition of $\rho\bar{\rho}$ is

$$[\rho\bar{\rho}] = \oplus_{g \in G(\hat{\mathcal{A}})}[\tau_g] \oplus (\oplus_{i \in I} p[\pi_i]) \oplus (\oplus_{j \in J} q[\sigma_j])$$

with $\#I = m$, $\#J = lp$, $d(\pi_i) = p$, and $d(\sigma_j) = q$. The inequality $\#G(\hat{\mathcal{A}}) = p > m$ guarantees

$$[\tau_g][\pi_i] = [\pi_i][\tau_g] = [\pi_i] \quad \text{for each } i \in I \text{ and } g \in G(\hat{\mathcal{A}}).$$

In fact, the order of $\{g \in G(\hat{\mathcal{A}});\ [\tau_g\pi_i] = [\pi_i]\}$ divides $\#G(\hat{\mathcal{A}}) = p$. If it were 1, then the cardinality of the orbit of $[\pi_i]$ under the (left) $G(\hat{\mathcal{A}})$ action would be $p$, which is a contradiction. Thus, the order must be $p$, which means $[\tau_g][\pi_i] = [\pi_i]$ for each $g \in G(\hat{\mathcal{A}})$. Considering the left action of $G(\hat{\mathcal{A}})$, we also get $[\pi_i][\tau_g] = [\pi_i]$ for each $g \in G(\hat{\mathcal{A}})$.

We take representatives $\{\tau_g\}_{g \in G(\hat{\mathcal{A}})}$ satisfying $\tau_g \cdot \rho = \rho$ and denote by $\nu : \mathcal{P}^\tau \hookrightarrow \mathcal{P}$ the inclusion map. Then, since $H^2(\mathbf{Z}_p, \mathbf{T})$ is trivial, as usual we can show $[\pi_i] = [\nu\theta_i\bar{\nu}]$ $(i \in I)$ with some $\theta_i \in Aut(\mathcal{P}^\tau)$. We claim that $\{\tau_g\}_{G(\hat{\mathcal{A}})} \cup \{\pi_i\}_{i \in I}$ is a system closed under conjugation and irreducible decomposition of products, i.e., for each $i_1, i_2 \in I$ a sector $\sigma_j$ does not appear in the product $\pi_{i_1}\pi_{i_2}$. Let $[\bar{\nu}\nu] = \oplus_{k=0}^{p-1}[\alpha^k]$ be the irreducible decomposition $(d(\alpha) = 1)$. If $[\pi_{i_1}][\pi_{i_2}] = \oplus_{k=0}^{p-1}[\nu\theta_{i_1}\alpha^k\theta_{i_2}\bar{\nu}]$ contained $\sigma_j$ for some, i.e., $\sigma_j \prec \nu\theta_{i_1}\alpha^k\theta_{j_2}\bar{\nu}$ for some $k$, then $\nu\theta_{i_1}\alpha^k\theta_{i_2}\bar{\nu}$ would contain all of the $p$ sectors $\tau_g\sigma_j$ ($g \in G(\hat{\mathcal{A}})$) (because of $[\tau_g][\nu] = [\nu]$). On the other hand, the dimension of $\nu\theta_{i_1}\alpha^k\theta_{i_2}\bar{\nu}$ is $p$ and it is easy to see that the above $p$ sectors of dimension $q$ are all distinct. This contradiction proves the claim. The claim guarantees the existence of an intermediate subfactor $\mathcal{P} \supseteq \mathcal{L} \supseteq \mathcal{Q}$ with $[\mathcal{P} : \mathcal{L}] = p + mp^2 = p(1 + mp)$. Since this index divides $p^2q$, we must have $1 + mp = q$. However, this would mean $q > p$, a contradiction.

Case (10.3): We have $p^2q = p^2 + mp^2 + np^4 + lp^2q^2$. Note $l = 0$ and hence $q = 1 + m + np^2$. The case $m = 0$ corresponds to (iii). We assume $m \neq 0$ and will show $n = 0$ in the rest, which corresponds to (iv).

Let $\mathcal{P} \supseteq \mathcal{Q}$ be the inclusion of factors corresponding to the dual Kac algebra of $\mathcal{A}$, and $\rho : \mathcal{Q} \hookrightarrow \mathcal{P}$ be the inclusion map. Then, the irreducible decomposition of $\rho\bar{\rho}$ is

$$[\rho\bar{\rho}] = \oplus_{g\in G(\hat{\mathcal{A}})}[\tau_g] \oplus (\oplus_{i\in I} p[\pi_i]) \oplus (\oplus_{j\in J} p^2[\sigma_j])$$

with $\#I = m$, $\#J = n$, $d(\pi_i) = p$, and $d(\sigma_j) = p^2$ (and $n = 0$ is to be shown). We take representatives $\{\tau_g\}$ satisfying $\tau_g \cdot \rho = \rho$. Note that $G(\hat{\mathcal{A}})$ is isomorphic to either $\mathbf{Z}_{p^2}$ or $\mathbf{Z}_p \times \mathbf{Z}_p$. Let

$$G_i = \{g \in G(\hat{\mathcal{A}});\ [\tau_g][\pi_i] = [\pi_i]\}.$$

The identity appears in $\pi_i\bar{\pi}_i$ so that $\sigma_j$ (of dimension $p^2$) cannot appear, i.e., $\pi_i\bar{\pi}_i$ is decomposed into $\tau_g$'s and $\pi_i$'s. By counting dimensions, we see that a non-trivial automorphism has to appear in $\pi_i\bar{\pi}_i$. This means that $G_i$ is not trivial and consequently the order $\#G$ is either $p$ or $p^2$. We also remark

$$G_i = G(\hat{\mathcal{A}}) \quad \text{if and only if} \quad G_{\bar{i}} = G(\hat{\mathcal{A}}).$$

Indeed, if $G_i = G(\hat{\mathcal{A}})$, then $\pi_i(\mathcal{P})$ (a subfactor of index $p^2$) is the fixed-point algebra of $\mathcal{P}$ under an abelian group ($= G(\hat{\mathcal{A}})$) action. This means that so is $\pi_{\bar{i}}(\mathcal{P})$ and hence $G_{\bar{i}} = G(\hat{\mathcal{A}})$ as well.

We set

$$I_0 = \{i \in I; \#G_i = p\}.$$

The above remark means that $\{\pi_i\}_{i\in I_0}$ is invariant under conjugation. Let $\nu_i : \mathcal{P}^{(G_i,\tau)} \hookrightarrow \mathcal{P}$ be the inclusion map. As before, we can show that for each $i \in I_0$ there exists an isomorphism $\theta_i : \mathcal{P}^{(G_{\bar{i}},\tau)} \longrightarrow \mathcal{P}^{(G_i,\tau)}$ satisfying $[\pi_i] = [\nu_i\theta_i\bar{\nu}_{\bar{i}}]$. Firstly, we consider the case $I_0 \neq \emptyset$ and choose $i_0 \in I_0$. By taking an irreducible component of $\pi_{i_0}\bar{\pi}_{i_0}$ if necessary, we may and do assume $G_{i_0} = G_{\bar{i}_0}$. Let $I_1 = \{i \in I_0;\ G_i = G_{\bar{i}}\}$. Then, $\{\tau_g\}_{g\in G(\hat{\mathcal{A}})} \cup \{\pi_i\}_{i\in I_1}$ is closed under conjugation and irreducible decomposition of products. Thus there exists an intermediate subfactor $\mathcal{P} \supseteq \mathcal{L} \supseteq \mathcal{Q}$ with index $[\mathcal{P} : \mathcal{L}] = (p^2 + \#I_1 \times p^2)$. Since this index value divides $p^2q$, we must have $q = 1 + \#I_1$. On the other hand, since $q = 1 + m + nq^2$ and $\#I_1 \leq \#I = m$, we must have $n = 0$. Secondly, we consider the case $I_0 = \emptyset$ and claim that $\{\tau_g\}_{g\in G(\hat{\mathcal{A}})} \cup \{\pi_i\}_{i\in I}$ is a closed system in this case. Then, we would get an intermediate subfactor $\mathcal{L}$ with $[\mathcal{P} : \mathcal{L}] = p^2 + mp^2$, which forces $q = 1 + m$ and $n = 0$ as above. To show the claim, it suffices to see that $\pi_{i_1}\pi_{i_2}$ is reducible for each $i_1, i_2 \in I$ (due to $d(\sigma_j) = p^2 = d(\pi_{i_1})d(\pi_{i_2})$). But, the assumption $I_0 = \emptyset$ means $\bar{\pi}_{i_1}\pi_{i_1} = \pi_{i_2}\bar{\pi}_{i_2} = \oplus_{g\in G(\hat{\mathcal{A}})}\tau_g$ so that the Frobenius reciprocity implies

$$\dim(\pi_{i_1}\pi_{i_2}, \pi_{i_1}\pi_{i_2}) = \dim(\bar{\pi}_{i_1}\pi_{i_1}, \pi_{i_2}\bar{\pi}_{i_2}) = p^2$$

as desired.

Case (10.4): We have $p^2q = q + mp^2q + np^4q + lq^2$. This forces $m = n = 0$ and $p^2 = 1 + lq$ (note that $\frac{p^2-1}{q}$ is an integer so that $p \equiv \pm 1 \pmod q$), which corresponds to (i).

Case (10.5): We have $p^2q = pq + mp^2q + np^4q + lpq^2$. Thus, we get $m = n = 0$, $p = 1 + lq$ and this corresponds to (ii). □

The following gives a complete classification of Kac algebras with the assumption described at the beginning of this section:

THEOREM 10.7. *Let $\mathcal{A}$ be a Kac algebra of dimension $p^2q$ with distinct primes $p, q$ with the above additional requirement.*
*(i) If $\mathcal{A}$ is not co-commutative and*

$$\mathcal{A} \cong \mathbf{C}^q \oplus \underbrace{M_q(\mathbf{C}) \oplus \cdots M_q(\mathbf{C})}_{\frac{p^2-1}{q}},$$

*then, $p, q$ are odd primes with $p \equiv 1 \pmod q$ and*

$$\hat{\mathcal{A}} \cong \mathbf{C}^{p^2} \oplus \underbrace{M_p(\mathbf{C}) \oplus \cdots M_p(\mathbf{C})}_{q-1}.$$

*The intrinsic groups are*

$$G(\mathcal{A}) \cong \mathbf{Z}_p \times \mathbf{Z}_p \quad \textit{and} \quad G(\hat{\mathcal{A}}) \cong \mathbf{Z}_q.$$

*There are exactly $\frac{q-1}{2}$ Kac algebras of this type.*
*(ii) If $\mathcal{A}$ is not co-commutative and*

$$\mathcal{A} \cong \mathbf{C}^{pq} \oplus \underbrace{M_q(\mathbf{C}) \oplus \cdots M_q(\mathbf{C})}_{\frac{p(p-1)}{q}},$$

*then $p \equiv 1 \pmod q$ and*

$$\hat{\mathcal{A}} \cong \mathbf{C}^{p^2} \oplus \underbrace{M_p(\mathbf{C}) \oplus \cdots M_p(\mathbf{C})}_{q-1}.$$

*The intrinsic groups are*

$$G(\mathcal{A}) \cong \mathbf{Z}_p \times \mathbf{Z}_p \quad \textit{and} \quad G(\hat{\mathcal{A}}) \cong \mathbf{Z}_{pq}.$$

*There is only one Kac algebra of this type.*
*(iii) If*

$$\mathcal{A} \cong \mathbf{C}^{p^2} \oplus \underbrace{M_{p^2}(\mathbf{C}) \oplus \cdots M_{p^2}(\mathbf{C})}_{\frac{q-1}{p^2}},$$

*then $\hat{\mathcal{A}}$ is commutative.*
*(iv) If*

$$\mathcal{A} \cong \hat{\mathcal{A}} \cong \mathbf{C}^{p^2} \oplus \underbrace{M_p(\mathbf{C}) \oplus \cdots M_p(\mathbf{C})}_{q-1},$$

*then $q \equiv 1 \pmod p$ and $\mathcal{A}, \hat{\mathcal{A}}$ are self-dual. There are exactly $p$ Kac algebras of this type. The intrinsic groups are*

$$G(\mathcal{A}) \cong G(\hat{\mathcal{A}}) \cong \mathbf{Z}_p \times \mathbf{Z}_p$$

*in one case, and*

$$G(\mathcal{A}) \cong G(\hat{\mathcal{A}}) \cong \mathbf{Z}_{p^2}$$

*in the other $p-1$ cases.*

PROOF.
(i) If $p \equiv \pm 1 \pmod q$ and $\mathcal{A} \cong \mathbf{C}^q \oplus M_q(\mathbf{C}) \oplus \cdots M_q(\mathbf{C})$, Theorem 9.8 and its corollary imply that $q \neq 2$ and such a Kac algebra is classified by $N \rtimes \mathbf{Z}_q$ with $\#N = p^2$ and an element $[(\eta, \zeta)]$ in $H^2((N,(\mathbf{Z}_q)),\mathbf{T})/\sim$ giving $\mathcal{A} \cong \ell^\infty((\mathbf{Z}_p \times \mathbf{Z}_p)) \rtimes_\zeta \mathbf{Z}_q$. Note that $\{\eta_{\tau^k}\}$ is not trivial (otherwise $\hat{\mathcal{A}} \cong \oplus_{k=0}^{q-1}\mathbf{C}_{\eta_{\tau^k}}(N)$ would be abelian). Therefore, we must have $N = \mathbf{Z}_p \times \mathbf{Z}_p$. Here, we use the additive notation for $\mathbf{Z}_p \times \mathbf{Z}_p$ and denote by $\tau$ the generator of $\mathbf{Z}_q$, which we regard as $\tau \in Aut(\mathbf{Z}_p \times \mathbf{Z}_p) = GL(2,p)$. Note that $\ell^\infty((\mathbf{Z}_p \times \mathbf{Z}_p)) \rtimes_\zeta \mathbf{Z}_q \cong \mathcal{A}$ if and only if $\tau$ is a fixed-point free automorphism of order $q$. Therefore, we assume this condition.

We set $\eta = \eta_\tau$ and note that $\zeta = 1$ implies

$$[\eta_{\tau^k}] = [\eta\eta^\tau\eta^{\tau^2}\cdots\eta^{\tau^{k-1}}] \quad \text{and} \quad [1] = [\eta\eta^\tau\eta^{\tau^2}\cdots\eta^{\tau^{q-1}}] \quad \text{in } H^2(\mathbf{Z}_p \times \mathbf{Z}_p, \mathbf{T}),$$

Here, $\eta^{\tau^k}$ means $\eta(\tau^k(\cdot),\tau^k(\cdot))$. We firstly claim $p \neq 2$. Indeed, if $p = 2$, then $\#Aut(\mathbf{Z}_p \times \mathbf{Z}_p) = 6$ and the only possible case would be $q = 3$. Note that $H^2(\mathbf{Z}_2 \times \mathbf{Z}_2, \mathbf{T}) \cong \mathbf{Z}_2$ on which $\tau$ acts trivially. Thus, we would get

$$[1] = [\eta\eta^\tau\eta^{\tau^2}] = [\eta]^3 \in H^2(\mathbf{Z}_2 \times \mathbf{Z}_2, \mathbf{T}) \cong \mathbf{Z}_2,$$

i.e., $[\eta] = [1]$. This contradicts that $\hat{\mathcal{A}}$ is non-commutative, and hence the claim is established. Thus, in the rest we assume that $p$ is an odd prime. We may assume

$$\eta_\tau((a,b),(c,d)) = \omega^{ad-bc}$$

where $\omega$ is a primitive $p$-th root of unity. Let $\tau_0$ be the determinant of $\tau$ regarded as an element of $GL(2,p)$. We claim that $\tau_0 \not\equiv 1 \pmod p$. Indeed, if $\tau_0 \equiv 1 \pmod p$, $\tau$ would act on $\eta_\tau$ trivially and we would get $\eta_\tau^q \in B^2(\mathbf{Z}_p \times \mathbf{Z}_p, \mathbf{T})$, contradicting the non-cocommutativity again. Since $\tau_0^q \equiv 1 \pmod p$ and $\mathbf{Z}_p^\times \cong \mathbf{Z}_{p-1}$, we get $p \equiv 1 \pmod q$. Note that in this case we have

$$\begin{cases} \eta\eta^\tau\eta^{\tau^2}\cdots\eta^{\tau^{q-1}} = \eta^{\sum_{k=0}^{q-1}\tau_0^k} = 1, \\ [\eta_{\tau^k}] \neq [1] \text{ for } 1 \leq k \leq q-1. \end{cases}$$

Since $\mathbf{C}_{\eta_{\tau^k}}(\mathbf{Z}_p \times \mathbf{Z}_p) \cong M_p(\mathbf{C})$ for $1 \leq k \leq q-1$, we get

$$\hat{\mathcal{A}} \cong \mathbf{C}^{p^2} \oplus \underbrace{M_p(\mathbf{C}) \oplus \cdots M_p(\mathbf{C})}_{q-1}.$$

There are essentially two different choices for $\tau$. In one case, $\tau$ is given by a scalar matrix

$$\begin{pmatrix} a & 0 \\ 0 & a \end{pmatrix},$$

where $a$ satisfies $a^q \equiv 1 \pmod p$, $a \not\equiv 1 \pmod p$. In the other case, $\tau$ is not a scalar. Since $x^q - 1 = \prod_{k=0}^{q-1}(x - \tau_0^k)$ in $\mathbf{F}_p[x]$, the minimal polynomial of $\tau$, is given by $(x - \tau_0^k)(x - \tau_0^{1-k})$, where $0 \leq k \leq \frac{q-1}{2}$. Thus, we may assume that $\tau$ is diagonal. Furthermore, since $\tau$ is a fixed-point free automorphism, we see $k \neq 0, 1$. In both cases, the number of freedom coming from $\zeta$ is $\#H^2(\mathbf{Z}_q, \widehat{\mathbf{Z}_p \times \mathbf{Z}_p})$, where the action of $\mathbf{Z}_q$ is fixed-point free. Therefore, by considering the corresponding group extensions, we can see that $H^2(\mathbf{Z}_q, \widehat{\mathbf{Z}_p \times \mathbf{Z}_p})$ is trivial. Thus, we conclude that $\#H^2(((\mathbf{Z}_p \times \mathbf{Z}_p), \mathbf{Z}_q), \mathbf{T}) = p$ and $p-1$ of them correspond to our (non-cocommutative) Kac algebras.

Now we count the exact number of isomorphism classes. To do so, we need to determine which of the above can be identified by group isomorphisms preserving $\mathbf{Z}_p \times \mathbf{Z}_p$ and $\mathbf{Z}_q$. By taking a suitable generator of $\mathbf{Z}_q$, we may think of just one value for $\tau_0$. By using an automorphism of $\mathbf{Z}_p \times \mathbf{Z}_p$ expressed by a diagonal matrix, we may think of just one value for primitive $p$-th root of unity $\omega$. Then, the only automorphisms available now are those acting on $\mathbf{Z}_q$ trivially and acting on $\mathbf{Z}_p \times \mathbf{Z}_p$ by an element in $SL(2,p)$. Thus, in the case where $\tau$ is a scalar, there is only one possibility of $\tau$. In the other case, we have $\frac{q-3}{2}$ possibilities of $\tau$. Therefore, there are exactly $\frac{q-1}{2}$ Kac algebras of this type. Finally, it is easy to determine the intrinsic groups based on Theorem 6.1 and Theorem 6.2.

(ii) If $p \equiv 1 \pmod q$ and

$$\mathcal{A} \cong \mathbf{C}^{pq} \oplus \underbrace{M_q(\mathbf{C}) \oplus \cdots M_q(\mathbf{C})}_{\frac{p(p-1)}{q}},$$

Theorem 9.8 implies that such a Kac algebra is classified by $N \rtimes \mathbf{Z}_q$ with $\#N = p^2$ and an element $[(\eta, \zeta)]$ in $H^2((N, \mathbf{Z}_q), \mathbf{T})/\!\sim$ giving $\ell^\infty((\mathbf{Z}_p \times \mathbf{Z}_p)) \rtimes_\zeta \mathbf{Z}_q \cong \mathcal{A}$. By the same reason as in (i), we must have $N = \mathbf{Z}_p \times \mathbf{Z}_p$. We denote by $\tau$ a generator of $\mathbf{Z}_q$, which we regards as an automorphism $\tau \in Aut(\mathbf{Z}_p \times \mathbf{Z}_p) = GL(2,p)$. Note that $\ell^\infty((\mathbf{Z}_p \times \mathbf{Z}_p)) \rtimes_\zeta \mathbf{Z}_q \cong \mathcal{A}$ if and only if $\tau$ has exactly $p$ fixed points. Thus we may assume that $\tau$ is given by

$$\begin{pmatrix} 1 & b \\ 0 & a \end{pmatrix}.$$

Since the order of $\tau$ is $q$, we get $a^q \equiv 1 \pmod p$, $a \not\equiv 1 \pmod p$, and $b \equiv 0 \pmod p$. As above, we set $\eta = \eta_\tau$, which is not a coboundary. Since $p \neq 2$, we may assume

$$\eta_\tau((a,b),(c,d)) = \omega^{ad-bc},$$

where $\omega$ is a primitive $p$-th root of unity. Note

$$\begin{cases} \eta\eta^\tau\eta^{\tau^2}\cdots\eta^{\tau^{q-1}} = \eta^{\sum_{k=0}^{q-1} a^k} = 1 \\ [\eta_{\tau^k}] \neq [1] \text{ for } 1 \leq k \leq q-1 \end{cases}$$

again, and we get

$$\hat{\mathcal{A}} \cong \mathbf{C}^{p^2} \oplus \underbrace{M_p(\mathbf{C}) \oplus \cdots M_p(\mathbf{C})}_{q-1}.$$

Let $\chi_1$, $\chi_2$ be generators of the dual group $\widehat{\mathbf{Z}_p \times \mathbf{Z}_p}$ satisfying

$$\chi((1,0)) = \omega, \quad \chi((0,1)) = 1, \quad \chi((1,0)) = \omega, \quad \chi((0,1)) = \omega.$$

The number of freedom coming from $\zeta$ is $\#H^2(\mathbf{Z}_q, \widehat{\mathbf{Z}_p \times \mathbf{Z}_p})$, where the action of $\mathbf{Z}_q$ is given by the same matrix as $\tau$ in terms of $\chi_1$, $\chi_2$. Therefore, by considering the corresponding group extensions, we can see that $H^2(\mathbf{Z}_q, \widehat{\mathbf{Z}_p \times \mathbf{Z}_p})$ is trivial. Therefore, we conclude that $\#H^2(((\mathbf{Z}_p \times \mathbf{Z}_p), (\mathbf{Z}_q)), \mathbf{T}) = p$ and $p-1$ of them correspond to our Kac algebras. In the same way as in (i), we may think of just one value for $a$ and $\omega$, and so there is only one Kac algebra of this type. It is easy to determine the intrinsic groups based on Theorem 6.1 and Theorem 6.2.

(iii) We assume $q \equiv 1 \pmod{p^2}$ and

$$\mathcal{A} \cong \mathbf{C}^{p^2} \oplus \underbrace{M_{p^2}(\mathbf{C}) \oplus \cdots M_{p^2}(\mathbf{C})}_{\frac{q-1}{p^2}}.$$

Let $\mathcal{P} \supseteq \mathcal{Q}$ be the inclusion of subfactor corresponding to the dual Kac algebra of $\mathcal{A}$, and $\rho : \mathcal{Q} \hookrightarrow \mathcal{P}$ be the inclusion map. Then, the irreducible decomposition of $\rho\bar{\rho}$ is

$$[\rho\bar{\rho}] = \oplus_{g \in G(\hat{\mathcal{A}})}[\tau_g] \oplus (\oplus_{j \in J} p^2[\sigma_j])$$

with $\#J = \frac{q-1}{p^2}$ and $d(\sigma_j) = p^2$. We take representatives $\{\tau_g\}$ satisfying $\tau_g \cdot \rho = \rho$. By counting argument dimension, we see that $\sigma_j\overline{\sigma_j}$ contains the $p^2$ automorphisms $\tau_g$ $(g \in G(\hat{\mathcal{A}}))$ so that we conclude

$$[\tau_g][\sigma_j] = [\sigma_j][\tau_g] = [\sigma_j] \quad \text{for each pair } j \in J \text{ and } g \in G(\hat{\mathcal{A}}).$$

Therefore, we have $Ad(u_g^j) \cdot \tau_g \cdot \sigma_j = \sigma_j$ with unitaries $\{u_g^j\}_{j \in J, g \in G(\hat{\mathcal{A}})} \subseteq \mathcal{P}$. We set $\tau_g^j = Ad(u_g^j) \cdot \tau_g$ $(g \in G(\hat{\mathcal{A}}))$, which gives rise to an action. Let $\omega_j$ be the 2-cocycle on $G(\hat{\mathcal{A}})$ carried by $\{u_g^j\}_{g \in G(\hat{\mathcal{A}})}$ and $\nu_j : \mathcal{P}^{\tau^j} \hookrightarrow \mathcal{P}$ be the inclusion map. Then, as before we have $[\sigma_j] = [\nu_j \theta_j \overline{\nu}_{\bar{j}}]$ with isomorphisms $\theta_j : \mathcal{P}^{\tau^{\bar{j}}} \longrightarrow \mathcal{P}^{\tau^j}$.

We claim that the class $[\omega_j] \in H^2(G(\hat{\mathcal{A}}), \mathbf{T})$ does not depend upon $j$. We fix $j_0 \in J$. By taking an irreducible component of $\sigma_{j_0}\overline{\sigma}_{j_0}$ if necessary, we may and do assume $[\omega_{j_0}] = [\omega_{\bar{j}_0}]$. We set

$$J_0 = \{j \in J;\ [\omega_j] = [\omega_{\bar{j}}] = [\omega_{j_0}]\}.$$

Then, $\{\tau_g\}_{g \in G(\hat{\mathcal{A}})} \cup \{\sigma_j\}_{j \in J_0}$ is closed under conjugation and irreducible decomposition of products. Thus there exists an intermediate subfactor $\mathcal{P} \supseteq \mathcal{L} \supseteq \mathcal{Q}$ with index

$$[\mathcal{P} : \mathcal{L}] = p^2 + \#J_0 \times p^4 = p^2(1 + \#J_0 \times p^2).$$

Since this index divides $p^2q$, that is, $1+\#J_0 \times p^2$ divides $q$, we conclude $1+\#J_0 \times p^2 = q$ $(= 1 + \#J \times p^2)$. Therefore, we get $J_0 = J$ and the claim is proved.

Let $H$ be the dual group of the (abelian) group $G(\hat{\mathcal{A}})$ (of order $p^2$), and we set $\mathcal{R} = \mathcal{P}^{\tau^{j_0}}$. Note $\mathcal{P} = \mathcal{R} \rtimes_\alpha H$ with an action $\alpha$ whose dual is $\tau^{j_0}$. Let $G$ be the subgroup in $Aut(\mathcal{R})$ generated by $[\alpha_H]$ and $\cup_{j \in J}[\alpha_H \theta_j \alpha_H]$. Then, the arguments in the proof of Theorem 9.8,(i) show that $G$ is a Frobenius group with the following properties: there exists a normal subgroup $N$ of order $q$ such that $G = N \rtimes H$ and the length of each non-trivial orbit of $H$ action on $N$ is $p^2$. Since $Aut(N) \cong \mathbf{Z}_{q-1}$, we get $H = \mathbf{Z}_{p^2}$ and $H^2(G(\hat{\mathcal{A}}), \mathbf{T})$ is actually trivial. Thus, we can identify $\tau$ with $\tau^{j_0}$ and $\mathcal{R}$ is an intermediate subfactor of $\mathcal{P} \supseteq \mathcal{Q}$. As in the proof of Theorem 9.8,(i), we can show that there exists a $\mathbf{Z}_q$ action $\beta$ such that $\mathcal{Q}$ is the fixed-point algebra of $\mathcal{R}$ under $\beta$ and $N = [\beta_{\mathbf{Z}_q}]$. Therefore, we have

$$\hat{\mathcal{A}} \cong \oplus_{i=0}^{p^2-1} \mathbf{C}_{\eta_i}(\mathbf{Z}_q) \cong \oplus_{i=0}^{p^2-1} \mathbf{C}(\mathbf{Z}_q),$$

which is commutative.

(iv) Since

$$\mathcal{A} \cong \hat{\mathcal{A}} \cong \mathbf{C}^{p^2} \oplus \underbrace{M_p(\mathbf{C}) \oplus \cdots M_p(\mathbf{C})}_{q-1},$$

Theorem 9.8 says that such a Kac algebra is classified by $N \rtimes \mathbf{Z}_p$, where $N \cong \mathbf{Z}_q \rtimes \mathbf{Z}_p$ is non-commutative, and an element $[(\eta, \zeta)]$ in $H^2(((N), (\mathbf{Z}_p)), \mathbf{T})/\sim$ giving $\ell^\infty(N) \rtimes_\zeta \mathbf{Z}_p \cong \mathcal{A}$.

Note at first that we must have $q \equiv 1 \pmod p$ since $N$ is non-commutative. Let $\tau$ be a generator of $\mathbf{Z}_p$, which we regards as an automorphism of $N$. Also note that $\ell^\infty(N) \rtimes \mathbf{Z}_p \cong \mathcal{A}$ if and only if the order of the fixed-point subgroup $K$ of $N$ under $\tau$ is $p$. Thus, we have $N = S_q \rtimes K$ with a $q$-Sylow subgroup $S_q$ of $N$. Therefore, we can take generators $a \in S_q$, $b \in K$ satisfying

$$bab^{-1} = a^r \quad \text{with } r^p \equiv 1 \pmod q \text{ and } r \not\equiv 1 \pmod q.$$

By changing generators if necessary, we may also assume $\tau(a) = a^r$, $\tau(b) = b$. Since $H^2(N, \mathbf{T})$ is trivial, we may assume $\eta_i = 1$, and so we certainly have

$$\hat{\mathcal{A}} \cong \oplus_{i=0}^{p-1} \mathbf{C}(N) \cong \mathbf{C}^{p^2} \oplus \underbrace{M_p(\mathbf{C}) \oplus \cdots M_p(\mathbf{C})}_{q-1}.$$

We define $\chi \in Hom(N, \mathbf{T}) \cong \mathbf{Z}_p$ by $\chi(a) = 1$, $\chi(b) = \omega$ where $\omega$ is a primitive $p$-th root of unity. Then, $\chi$ is a generator of $Hom(N, \mathbf{T})$. Note that $\tau$ acts on $\chi$ trivially. Since $\eta_i = 1$, we have $H^2((N, \mathbf{Z}_p), \mathbf{T}) \cong H^2(\mathbf{Z}_p, Hom(N, \mathbf{T}))$ with the trivial $\mathbf{Z}_p$ action on $Hom(N, \mathbf{T})$. By considering the corresponding extensions, we thus conclude $H^2((N, \mathbf{Z}_p), \mathbf{T}) \cong \mathbf{Z}_p$. We define $(1, \zeta^s) \in Z^2((N, \mathbf{Z}_p), \mathbf{T})$, $s = 0, 1, \ldots, p-1$, which represent the classes in $H^2((N, \mathbf{Z}_p), \mathbf{T})$, by

$$\zeta^s_{a^i b^j}(k, l) = \begin{cases} 1, & \text{when } k + l < p, \\ \omega^{js}, & \text{when } k + l \geq p, \end{cases}$$

where $0 \leq k, l \leq p-1$.

It is easy to see

$$G(\mathcal{A}) \cong \begin{cases} \mathbf{Z}_p \times \mathbf{Z}_p & \text{when } s = 0, \\ \mathbf{Z}_{p^2} & \text{when } s = 1, 2, \cdots, p-1. \end{cases}$$

Note that every automorphism $\varphi$ of $N \rtimes \mathbf{Z}_p$ preserving $N$ and $\mathbf{Z}_p$ globally is given by $\varphi(a) = a^m$, $\varphi(b) = b$, and $\varphi(\tau) = \tau$, and it acts on $\zeta^s$ trivially. Therefore, there are $p$ non-isomorphic Kac algebras of this type.

It remains to show that all of them are self-dual. To see this, we make use of near actions of $(N, \mathbf{Z}_p)$ on a factor $\mathcal{R}$. Let $(\alpha, \beta)$ be a near action of $(N, \mathbf{Z}_p)$ on a factor $\mathcal{R}$ with the invariant $(1, \zeta^s)$; more precisely there exist unitaries $\{u(\tau^k, n)\} \subseteq \mathcal{R}$ such that

$$\beta_{\tau^k} \cdot \alpha_{n^{\tau^k}} \cdot \beta_{\tau^{-k}} = Ad(u(\tau^k, n)) \cdot \alpha_n \quad \text{carrying} \quad (1, \zeta^s).$$

Since the near action $(\alpha\, \beta)$ restricted to $(S_q, \mathbf{Z}_p)$ has trivial invariant, there exists a $\alpha|_{S_q}$-cocycle $\{v_{a^i}\}$ satisfying $\beta_\tau \cdot Ad(v_{a^i}) \cdot \alpha_{a^i} \cdot \beta_{\tau^{-1}} = Ad(v_{a^{ir}}) \cdot \alpha_{a^{ir}}$. Taking $v \in \mathcal{R}$ satisfying $v_{a^i} = v\alpha_{a^i}(v^*)$ and considering $Ad(v) \cdot \alpha_{a^i b^j} \cdot Ad(v^*)$ instead of $\alpha_{a^i b^j}$, we may assume $u(\tau^k, a^i) = 1$ from the beginning. Since $a^i b^j = b^j a^{r_0^k}$, where $r_0$ is an integer satisfying $rr_0 \equiv 1 \pmod q$, and $\{u(\tau^k, n)\}_{n \in N}$ is an $\alpha$-cocycle for fixed $\tau^k$, we have $u(\tau^k, a^i b^j) = \alpha_{a^i}(u(\tau^k, b^j)) = u(\tau^k, b^j)$. We set $u = u(\tau, b) \in \mathcal{R}^{(S_q, \alpha)}$. Then, our near action is characterized by the following properties:

$$\beta_\tau \cdot \alpha_a \cdot \beta_\tau^{-1} = \alpha_a^r, \qquad \beta_\tau \cdot \alpha_b \cdot \beta_\tau^{-1} = Ad(u) \cdot \alpha_b,$$
$$u\alpha_b(u)\alpha_b^2(u) \cdots \alpha_b^{p-1}(u) = 1, \quad u^* \beta_\tau(u^*) \beta_\tau^2(u^*) \cdots \beta_\tau^{p-1}(u^*) = \omega^s.$$

We set

$$\mathcal{P} = \mathcal{R} \rtimes_\alpha N \supseteq \mathcal{R} \supseteq \mathcal{Q} = \mathcal{R}^\beta.$$

This corresponds to $\mathcal{A}$, and we denote by $\{\lambda_n\}_{n\in N}$ the implementing unitary in $\mathcal{R}$ for $\alpha$. To show that $\mathcal{A}$ is self-dual (as a Kac algebra), we consider

$$\mathcal{P} \supseteq \mathcal{L} = \mathcal{R} \rtimes_\alpha S_q \ (= \langle \mathcal{R}, \lambda_a \rangle'') \supseteq \mathcal{Q}$$

Then, we have $\mathcal{P} = \mathcal{L} \rtimes K$ with the restricted action of $Ad(\lambda_b)$ to $\mathcal{L}$. Let $\theta$ be a non-trivial automorphism of $\mathcal{L}$ coming from the dual action of $\alpha|_K$, and $\mu$ be the extension of $\beta_\tau$ to $\mathcal{L}$ defined by

$$\mu(x) = \begin{cases} \beta_\tau(x) & \text{if } x \in \mathcal{R}, \\ \lambda_a^r & \text{if } x = \lambda_a. \end{cases}$$

By direct computation we see that

$$\tilde{\alpha}_a = \theta, \quad \tilde{\alpha}_b = \mu^{-1}, \quad \tilde{\beta}_\tau = Ad(\lambda_b^*)|_{\mathcal{L}}$$

determine a near action $(N, \mathbf{Z}_p)$ on $\mathcal{L}$, which we denote by $(\tilde{\alpha}, \tilde{\beta})$. We compute

$$\tilde{\beta}_\tau \cdot \tilde{\alpha}_a \cdot \tilde{\beta}_\tau^{-1} = \tilde{\alpha}_a{}^r, \qquad \tilde{\beta}_\tau \cdot \tilde{\alpha}_b \cdot \tilde{\beta}_\tau{}^{-1} = Ad(v) \cdot \tilde{\alpha}_b,$$

with $v = \beta_\tau^{-1} \cdot \alpha_b^{-1}(u^*) = \alpha_b^{-1} \cdot \beta_\tau^{-1}(u^*)$. Since we have $\mathcal{P} = \mathcal{L} \rtimes_{\tilde{\beta}} \mathbf{Z}_p$ and $\mathcal{Q} = \mathcal{L}^{\tilde{\alpha}}$, the dual inclusion of $\mathcal{P} \supseteq \mathcal{Q}$ can be identified with

$$\mathcal{L} \rtimes_{\tilde{\alpha}} N \supseteq \mathcal{L}^{\tilde{\beta}},$$

which corresponds to the dual Kac algebra $\hat{\mathcal{A}}$. Therefore, $\hat{\mathcal{A}}$ is determined by the invariant of $(\tilde{\alpha}, \tilde{\beta})$, and direct computation shows

$$v\tilde{\alpha}_b(v)\tilde{\alpha}_b{}^2(v) \cdots \tilde{\alpha}_b{}^{p-1}(v) = \omega^s, \quad v^* \tilde{\beta}_\tau(v^*) \tilde{\beta}_\tau{}^2(v^*) \cdots \tilde{\beta}_\tau{}^{p-1}(v^*) = 1.$$

Thus, after changing a phase of $v$ by $c \in \mathbf{T}$ satisfying $c^p = \omega^{-s}$, the above two qualities are switched to 1 and $\omega^s$. This means that the invariants of $(\alpha, \beta)$ and $(\tilde{\alpha}, \tilde{\beta})$ are the same, and we are done. □

## 4. Classification of low dimensional Kac algebras

Let $p, q, r$ be distinct primes. In this section, we classify Kac algebras of dimension $pq$, $p^2q$, $pqr$ less than 60. Note that general classification results already exist for the dimensions $p$, $2p$, $p^2$, $p^3$ (see [**51, 53, 54, 77**]). Here we do not deal with Kac algebras of dimension 16, 24 and so on. Instead, those of dimension 16 and 24 will be considered in later chapters.

THEOREM 10.8. *A Kac algebra of dimension $pq < 60$ with distinct primes $p, q$ is either commutative or co-commutative.*

PROOF. Thanks to Theorem 10.4 and Proposition 10.5, the only case we need to examine is the case of dimension 55. Let $\mathcal{A}$ be a Kac algebra with $\dim \mathcal{A} = 55$. Lemma 10.3,(i) shows that either $\#G(\hat{\mathcal{A}}) = 5$ or $\#G(\hat{\mathcal{A}}) = 1$ (if $\mathcal{A}$ is non-commutative). In the first case, the only algebra allowed by Proposition 10.1 is $\mathbf{C}^5 \oplus M_5(\mathbf{C}) \oplus M_5(\mathbf{C})$. Theorem 9.8,(iii) shows that the dual Kac algebra $\hat{\mathcal{A}}$ is given by

$$\mathbf{C}(\mathbf{Z}_{11}) \oplus \mathbf{C}(\mathbf{Z}_{11}) \oplus \mathbf{C}(\mathbf{Z}_{11}) \oplus \mathbf{C}(\mathbf{Z}_{11}) \oplus \mathbf{C}(\mathbf{Z}_{11})$$

so that $\mathcal{A}$ is co-commutative. In the second case, the only algebra allowed by Proposition 10.1 is $\mathbf{C} \oplus M_3(\mathbf{C}) \oplus M_3(\mathbf{C}) \oplus M_6(\mathbf{C})$. But it is easy to show that no consistent fusion rule for sectors corresponding to this Kac algebra is possible. (Consider the irreducible decomposition of $\pi\overline{\pi}$ with $d(\pi) = 3$). $\square$

Elimination of the case $\dim \mathcal{A} = 55$ was also done in §10.3 of [**66**].

To deal with the $p^2q$ case, we need the next general facts on Kac algebras containing simple components of rank 2. For the proof classification of subfactors of index 4 plays an important role (as in the proof of Proposition 10.1,(iii)).

LEMMA 10.9. *Let $\mathcal{A}$ be a finite-dimensional Kac algebra with algebra structure*

$$\mathbf{C}^{m_0} \oplus \underbrace{M_{n_1}(\mathbf{C}) \oplus \cdots \oplus M_{n_1}(\mathbf{C})}_{m_1} \oplus \underbrace{M_{n_2}(\mathbf{C}) \oplus \cdots \oplus M_{n_2}(\mathbf{C})}_{m_2} \oplus \cdots$$
$$\oplus \underbrace{M_{n_k}(\mathbf{C}) \oplus \cdots \oplus M_{n_k}(\mathbf{C})}_{m_k}$$

*with $n_1 < n_2 < \cdots < n_k$, and we assume $n_1 = 2$.*
*(i) If $m_0 = 2$ and $\dim \mathcal{A}$ is not a multiple of 12, then there exists an integer $1 \leq m \leq m_1$ such that $2(1+2m)$ divides $\dim \mathcal{A}$.*
*(ii) If $m_0 = 2$ and $m_1 = 1$, then $\dim \mathcal{A}$ is a multiple of 6.*
*(iii) If $m_0 = 4$ and $m_1 = 1$, then $\dim \mathcal{A}$ is a multiple of 8.*
*(iv) If $n_2 > 4$, then $(m_0 + 4m_1)$ divides $\dim \mathcal{A}$.*
*(v) If $k = 2$, $n_2 = 3$ and $\dim \mathcal{A}$ is not a multiple of 12, then $(m_0 + 4m_1)$ divides $\dim \mathcal{A}$.*
*(vi) If $m_0 = 2$ and $n_2 \neq 3$, then $(2 + 4m_1)$ divides $\dim \mathcal{A}$.*

PROOF. Let $\mathcal{P} \supseteq \mathcal{Q}$ be the inclusion of factors corresponding to the dual Kac algebra and $\rho : \mathcal{Q} \hookrightarrow \mathcal{P}$ be the inclusion map. Then, the irreducible decomposition of $[\rho\overline{\rho}]$ is

$$\oplus_{g \in G(\hat{\mathcal{A}})}[\tau_g] \oplus \oplus_{i=1}^{k}(\oplus_{a \in I_i} n_i [\pi_a^i]))$$

with $\# I_i = m_i$.
(i) We fix $a \in I_1$ and consider the inclusion $\mathcal{P} \supseteq \pi_a^1(\mathcal{P})$ of index 4. Since 12 is not a divisor of $\dim \mathcal{A}$, the principal graph of $\mathcal{P} \supseteq \pi_a^1(\mathcal{P})$ is none of $E_6^{(1)}$, $E_7^{(1)}$ and $E_8^{(1)}$ by the argument in the proof of Proposition 10.1,(iii). See the proof of Proposition 10.1 for the graphs $E_6^{(1)}$, $E_8^{(1)}$ (i.e., Figure 4 and Figure 5) while the graph $E_7^{(1)}$ is as follows:

Figure 6 (Coxeter-Dynkin graph $E_7^{(1)}$)

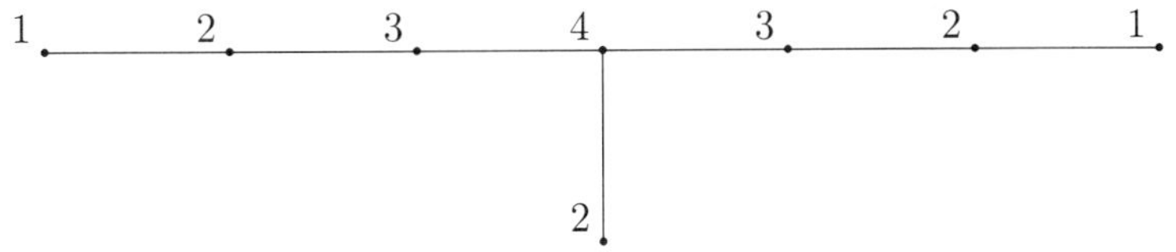


Also it is impossible that $\pi\overline{\pi}$ (of dimension 4) decompose into the sum of the identity and an irreducible sector of dimension 3 (so that $\pi\overline{\pi}$ must contain the non-trivial automorphism $\tau_1$). This means that the principal graph of $\mathcal{P} \supseteq \pi_a^1(\mathcal{P})$ must be $D_k^{(1)}$

with some $k \geq 5$ even. Then, the set of sectors generated by $\pi_a^1\overline{\pi}_a^1$ is of the following form: $\{id, \tau_1\} \cup \{\pi_a^1\}_{a\in J}$ where $J$ is a subset of $I_1$. Since this system is closed under conjugation and irreducible decomposition, $2 + 4m$ (with $m = \#J_1 \leq \#J = m_1$) divides $\dim \mathcal{A}$.
(ii) Let $\pi^1$ be the unique sector with $d(\pi^1) = 2$ contained in $\rho\bar{\rho}$ so that we have $[\tau_1][\pi^1] = [\pi^1][\tau_1] = [\pi^1] = [\overline{\pi}^1]$. We also note $[\pi^1][\overline{\pi}^1] = [id] \oplus [\tau_1] \oplus [\pi^1]$. Thus, $\{id, \tau_1, \pi^1\}$ is closed under conjugation and irreducible decomposition of products and we get the result.
(iii) In this case $[\pi^1][\overline{\pi}^1]$ (where $\pi^1$ is as in (ii)) decomposes into the sum of the four automorphisms $\{\tau_g\}_{g\in G(\hat{\mathcal{A}})}$. Thus, by considering $\{\tau_g\}_{g\in G(\hat{\mathcal{A}})} \cup \{\pi^1\}$, we get the result.
(iv) Since sectors of dimension 3 or 4 are not available, $[\pi_a^1][\pi_b^1]$ (of dimension 4) must decompose into automorphisms and sectors of dimension 2. Therefore, $\{\tau_g\}_{g\in G(\hat{\mathcal{A}})} \cup \{\pi_a^1\}_{a\in I_1}$ is closed under conjugation and irreducible decomposition of products so that $m_0 + 4m_1$ divides $\dim \mathcal{A}$.
(v) The usual argument shows the result if we can check that $\{\tau_g\}_{g\in G(\hat{\mathcal{A}})} \cup \{\pi_a^1\}_{a\in I_1}$ is closed under conjugation and irreducible decomposition. To see this by contradiction, let us assume $[\pi_a^1\pi_b^1] = [\tau_g] \oplus [\pi_c^2]$ for some $c \in I_1$, i.e., $[\pi_a^1\pi_b^1\tau_{g^{-1}}] = [id] \oplus [\pi_{c'}^2]$. Then, $[\pi_b^1\tau_{g^{-1}}]$ is the conjugate sector of $[\pi_a^1]$ and we have $[\pi_a^1\overline{\pi}_a^1] = [id] \oplus [\pi_{c'}^2]$. In particular, the principal graph of $\mathcal{P} \supseteq \pi_a^1(\mathcal{P})$ would be one of $E_6^{(1)}$, $E_7^{(1)}$, $E_8^{(1)}$. However, this would mean that 12 divides $\dim \mathcal{A}$ (as in the proof of Proposition 10.1,(iii)), a contradiction. (The graphs $E_6^{(1)}, E_8^{(1)}$ were handled in the proof of Proposition 10.1,(iii). If we have $E_7^{(1)}$, then $\dim \mathcal{A}$ is a multiple of 24.)
(vi) Since sectors of dimension 3 do not exist, each $\pi_a^1\overline{\pi}_a^1$ is decomposed into $id$, $\tau_1$ and a sector of dimension 2. Thus, the Frobenius reciprocity implies

$$\dim(\pi_a^1\pi_b^1, \pi_a^1\pi_b^1) = \dim(\overline{\pi}_a^1\pi_a^1, \pi_b^1\overline{\pi}_b^1) \geq \dim(id \oplus \tau_1, id \oplus \tau_1) = 2$$

for each $\pi_a^1, \pi_b^1$, i.e., $\pi_a^1\pi_b^1$ is not irreducible. Therefore, by the same reason as above $\pi_a^1\pi_b^1$ is also decomposed into $id$, $\tau^1$ and a sector of dimension 2. Therefore, $\{id, \tau_1\} \cup \{\pi_a^1\}_{a\in I_1}$ is closed under conjugation and irreducible decomposition of products, and we get the result. □

THEOREM 10.10. *All non-trivial Kac algebras of dimension $p^2q < 60$ with distinct primes $p$, $q$ satisfy the additional requirement in Theorem* 10.7, *Therefore, Theorem* 10.7 *actually gives us the complete classification* (*as long as $p^2q < 60$*).

Note that Kac algebras $\mathcal{A}$ covered in this theorem are those of the following dimensions:

$$\text{(i) } \dim \mathcal{A} = 12, \text{ (ii) } \dim \mathcal{A} = 18, \text{ (iii) } \dim \mathcal{A} = 20,$$
$$\text{(iv) } \dim \mathcal{A} = 28, \text{ (v) } \dim \mathcal{A} = 44, \text{ (vi) } \dim \mathcal{A} = 45,$$
$$\text{(vii) } \dim \mathcal{A} = 50, \text{ and (viii) } \dim \mathcal{A} = 52.$$

Thanks to Theorem 10.7, it suffices to show that the rank of every simple component of $\mathcal{A}$ divides $p^2q$. In the proof below, "possible algebras" mean those allowed by Proposition 10.1 and Proposition 10.2. Assume that $\mathcal{P} \supseteq \mathcal{Q}$ denotes the inclusion of factors corresponding to the dual Kac algebra with the inclusion map $\rho : \mathcal{P} \hookrightarrow \mathcal{Q}$.

(i) $\dim \mathcal{A} = 12$
The only possible algebras are

$$\begin{aligned}&\mathbf{C}^4 \oplus M_2(\mathbf{C}) \oplus M_2(\mathbf{C}),\\&\mathbf{C}^3 \oplus M_3(\mathbf{C})\end{aligned}$$

and indeed 1,2 and 3 divide 12.

In fact, the second case is eliminated by Theorem 10.7,(i) and the first case is handled in Theorem 10.7,(iv).

(ii) $\dim \mathcal{A} = 18$
The only possible algebras are

$$\begin{aligned}&\mathbf{C}^9 \oplus M_3(\mathbf{C}),\\&\mathbf{C}^6 \oplus M_2(\mathbf{C}) \oplus M_2(\mathbf{C}) \oplus M_2(\mathbf{C}),\\&\mathbf{C}^2 \oplus M_2(\mathbf{C}) \oplus M_2(\mathbf{C}) \oplus M_2(\mathbf{C}) \oplus M_2(\mathbf{C}),\end{aligned}$$

and 1,2 and 3 divide 12. (This case is treated in [**52**].)

(iii) $\dim \mathcal{A} = 20$
The only possible algebras are

$$\begin{aligned}&\mathbf{C}^4 \oplus M_4(\mathbf{C}),\\&\mathbf{C}^4 \oplus M_2(\mathbf{C}) \oplus M_2(\mathbf{C}) \oplus M_2(\mathbf{C}) \oplus M_2(\mathbf{C}),\end{aligned}$$

and 1,2 and 4 divide 20.

In fact, only the second case occurs thanks to Theorem 10.7,(iii) and this case is handled in Theorem 10.7,(iv).

(iv) $\dim \mathcal{A} = 28$
The only possible algebras are

$$\begin{aligned}&\mathbf{C}^4 \oplus M_2(\mathbf{C}) \oplus M_2(\mathbf{C}) \oplus M_4(\mathbf{C}),\\&\mathbf{C}^4 \oplus \underbrace{M_2(\mathbf{C}) \oplus \cdots \oplus M_2(\mathbf{C})}_{6},\\&\mathbf{C}^2 \oplus M_2(\mathbf{C}) \oplus M_2(\mathbf{C}) \oplus M_3(\mathbf{C}) \oplus M_3(\mathbf{C}).\end{aligned}$$

Lemma 10.9,(i) shows that the third case does not occur, and 1,2 and 4 divide 28.

In fact, only the second case occurs thanks to Lemma 10.6 and this case is handled in Theorem 10.7, (iv).

(v) $\dim \mathcal{A} = 44$
The only possible algebras are

$$\begin{aligned}&\mathbf{C}^4 \oplus M_2(\mathbf{C}) \oplus M_6(\mathbf{C}),\\&\mathbf{C}^4 \oplus M_2(\mathbf{C}) \oplus M_2(\mathbf{C}) \oplus M_4(\mathbf{C}) \oplus M_4(\mathbf{C}),\\&\mathbf{C}^4 \oplus \underbrace{M_2(\mathbf{C}) \oplus \cdots \oplus M_2(\mathbf{C})}_{6} \oplus M_4(\mathbf{C}),\end{aligned}$$

$$\mathbf{C}^4 \oplus M_2(\mathbf{C}) \oplus \underbrace{M_3(\mathbf{C}) \oplus \cdots \oplus M_3(\mathbf{C})}_{4},$$
$$\mathbf{C}^4 \oplus \underbrace{M_2(\mathbf{C}) \oplus \cdots \oplus M_2(\mathbf{C})}_{10},$$
$$\mathbf{C}^2 \oplus M_2(\mathbf{C}) \oplus M_2(\mathbf{C}) \oplus M_3(\mathbf{C}) \oplus M_3(\mathbf{C}) \oplus M_4(\mathbf{C}),$$
$$\mathbf{C}^2 \oplus \underbrace{M_2(\mathbf{C}) \oplus \cdots \oplus M_2(\mathbf{C})}_{6} \oplus M_3(\mathbf{C}) \oplus M_3(\mathbf{C}).$$

The 1st, the 4th, the 6th and 7th cases do not occur thanks to (iii), (iii), (i) and (v) of Lemma 10.9 respectively and 1,2 and 4 divide 44.

In fact, the 2nd and the 3rd cases are eliminated due to Lemma 10.6 and the only the 5th case occurs. The 5th case is handled in Theorem 10.7, (iv).

(vi) $\dim \mathcal{A} = 45$
The only possible algebra is

$$\mathbf{C}^9 \oplus M_3(\mathbf{C}) \oplus M_3(\mathbf{C}) \oplus M_3(\mathbf{C}) \oplus M_3(\mathbf{C}),$$

and 1,3 divide 45.

However, since $5 \not\equiv 1 \pmod 3$, this case does not occur thanks to Theorem 10.7,(iv).

(vii) $\dim \mathcal{A} = 50$
The only possible algebras are

$$\mathbf{C}^{25} \oplus M_5(\mathbf{C}),$$
$$\mathbf{C}^{10} \oplus \underbrace{M_2(\mathbf{C}) \oplus \cdots \oplus M_2(\mathbf{C})}_{10},$$
$$\mathbf{C}^5 \oplus \underbrace{M_2(\mathbf{C}) \oplus \cdots \oplus M_2(\mathbf{C})}_{5} \oplus M_5(\mathbf{C}),$$
$$\mathbf{C}^2 \oplus M_2(\mathbf{C}) \oplus M_2(\mathbf{C}) \oplus M_2(\mathbf{C}) \oplus M_6(\mathbf{C}),$$
$$\mathbf{C}^2 \oplus \underbrace{M_2(\mathbf{C}) \oplus \cdots \oplus M_2(\mathbf{C})}_{4} \oplus M_4(\mathbf{C}) \oplus M_4(\mathbf{C}),$$
$$\mathbf{C}^2 \oplus \underbrace{M_2(\mathbf{C}) \oplus \cdots \oplus M_2(\mathbf{C})}_{8} \oplus M_4(\mathbf{C}),$$
$$\mathbf{C}^2 \oplus M_2(\mathbf{C}) \oplus M_2(\mathbf{C}) \oplus M_2(\mathbf{C}) \oplus \underbrace{M_3(\mathbf{C}) \oplus \cdots \oplus M_3(\mathbf{C})}_{4},$$
$$\mathbf{C}^2 \oplus \underbrace{M_2(\mathbf{C}) \oplus \cdots \oplus M_2(\mathbf{C})}_{12}.$$

The 4th, the 5th, the 6th and the 7th cases do not occur thanks to (iv), (vi), (vi), and (v) of Lemma 10.9 respectively, and 1,2 and 5 divide 50.

The 3rd case is eliminated by Lemma 10.6 while the 8th case is eliminated by Theorem 10.7,(i). The 1st and the 2nd cases are handled in Theorem 10.7,(ii).

(viii) $\dim \mathcal{A} = 52$

The only possible algebras are

$$\begin{aligned}
&\mathbf{C}^4 \oplus M_2(\mathbf{C}) \oplus M_2(\mathbf{C}) \oplus M_2(\mathbf{C}) \oplus M_6(\mathbf{C}),\\
&\mathbf{C}^4 \oplus M_4(\mathbf{C}) \oplus M_4(\mathbf{C}) \oplus M_4(\mathbf{C}),\\
&\mathbf{C}^4 \oplus \underbrace{M_2(\mathbf{C}) \oplus \cdots \oplus M_2(\mathbf{C})}_{4} \oplus M_4(\mathbf{C}) \oplus M_4(\mathbf{C}),\\
&\mathbf{C}^4 \oplus \underbrace{M_2(\mathbf{C}) \oplus \cdots \oplus M_2(\mathbf{C})}_{8} \oplus M_4(\mathbf{C}),\\
&\mathbf{C}^4 \oplus M_2(\mathbf{C}) \oplus M_2(\mathbf{C}) \oplus M_2(\mathbf{C}) \oplus \underbrace{M_3(\mathbf{C}) \oplus \cdots \oplus M_3(\mathbf{C})}_{4},\\
&\mathbf{C}^4 \oplus \underbrace{M_2(\mathbf{C}) \oplus \cdots \oplus M_2(\mathbf{C})}_{12},\\
&\mathbf{C}^2 \oplus M_3(\mathbf{C}) \oplus M_3(\mathbf{C}) \oplus M_4(\mathbf{C}) \oplus M_4(\mathbf{C}),\\
&\mathbf{C}^2 \oplus \underbrace{M_2(\mathbf{C}) \oplus \cdots \oplus M_2(\mathbf{C})}_{4} \oplus M_3(\mathbf{C}) \oplus M_3(\mathbf{C}) \oplus M_4(\mathbf{C}),\\
&\mathbf{C}^2 \oplus \underbrace{M_2(\mathbf{C}) \oplus \cdots \oplus M_2(\mathbf{C})}_{8} \oplus M_3(\mathbf{C}) \oplus M_3(\mathbf{C}).
\end{aligned}$$

The 1st, the 5th, the 8th and the 9th are eliminated thanks to (iv), (v), (i) and (v) of Lemma 10.9 respectively.

We have to eliminate the 7th case (then 1,2 and 4 indeed divide 52). Suppose $\mathcal{A}$ is as in the 7th case. Then, we have the following irreducible decomposition of $\rho\bar{\rho}$:

$$[\rho\bar{\rho}] = [id] \oplus [\tau] \oplus 3[\pi_1] \oplus 3[\pi_2] \oplus 4[\sigma_1] \oplus 4[\sigma_2],$$

where $\tau$ is an automorphism of order 2 and $d(\pi_1) = d(\pi_2) = 3,\ d(\sigma_1) = d(\sigma_2) = 4$. We claim

$$[\pi_2] = [\tau\pi_1] = [\pi_1\tau].$$

In fact, if we had $[\tau] \prec [\pi_1\bar{\pi}_1]$, then the inclusion $\mathcal{P} \supseteq \pi_1(\mathcal{P})$ of index 9 would contain an intermediate subfactor of index 2, a contradiction. Thus, we know $[\tau\pi_1] \neq [\pi_1]$ and $[\tau\pi_1] = [\pi_2]$. By considering $\mathcal{P} \supseteq \bar{\pi}_1(\mathcal{P})$, we also get $[\pi_1\tau] = [\pi_2]$. By counting dimensions, we can see that $\pi_1\bar{\pi}_1$ contains exactly two sectors of dimension 4. We first assume $[\pi_1\bar{\pi}_1] = [id] \oplus 2[\sigma_1]$. The Frobenius reciprocity implies that the multiplicity of $\pi_1$ in $\sigma_1\pi_1$ is 2. Since $d(\sigma_1\pi_1) = 12$ and $\sigma_1\pi_1$ cannot contain an automorphism, we are forced to have $[\sigma_1\pi_1] = 2[\pi_1] \oplus 2[\pi_2]$ (by counting dimensions). Then, we get

$$\begin{aligned}
4 = \dim(\sigma_1\pi_1, \sigma_1\pi_1) &= \dim(\bar{\sigma}_1\sigma_1, \pi_1\bar{\pi}_1) = \dim(\bar{\sigma}_1\sigma_1, id \oplus 2\sigma_1)\\
&= \dim(\bar{\sigma}_1\sigma_1, id) + 2\dim(\bar{\sigma}_1\sigma_1, \sigma_1) = 1 + 2\dim(\bar{\sigma}_1\sigma_1, \sigma_1),
\end{aligned}$$

which is a contradiction. Therefore, we must have $[\pi_1\bar{\pi}_1] = [id] \oplus [\sigma_1] \oplus [\sigma_2]$, and similarly we get

$$[\pi_1\bar{\pi}_1] = [\pi_2\bar{\pi}_2] = [id] \oplus [\sigma_1] \oplus [\sigma_2].$$

The Frobenius reciprocity implies that the multiplicity of $\pi_i$ in $\sigma_j\pi_i$ is 1 for $i = 1, 2$, $j = 1, 2$. Since $\sigma_j\pi_i$ cannot contain automorphisms, $\sigma_j\pi_i$ is decomposed into $\pi_1, \pi_2$ and sectors of dimension 4. This is obviously impossible so that we have shown that the 7th case does not occur.

The 3rd and 4th cases are eliminated by Lemma 10.6 while the 2nd case is eliminated by Theorem 10.7,(iii). The 6th case is the only possible case and it is handled in Theorem 10.7,(iv).

Thus, the proof of Theorem 10.10 is completed. In the rest of the section we deal with Kac algebras of dimension $pqr$ ($< 60$), i.e., $30 = 2\times 3\times 5$ and $42 = 2\times 3\times 7$.

THEOREM 10.11. *A Kac algebra of dimension* 30 *is always either commutative or co-commutative.*

PROOF. Let $\mathcal{A}$ be a non-commutative Kac algebra of dimension 30. Then, possible algebras are as follows:

$$\begin{aligned}
&\mathbf{C}^{10} \oplus \underbrace{M_2(\mathbf{C}) \oplus \cdots \oplus M_2(\mathbf{C})}_{5},\\
&\mathbf{C}^{6} \oplus \underbrace{M_2(\mathbf{C}) \oplus \cdots \oplus M_2(\mathbf{C})}_{6},\\
&\mathbf{C}^{5} \oplus M_5(\mathbf{C}),\\
&\mathbf{C}^{3} \oplus M_3(\mathbf{C}) \oplus M_3(\mathbf{C}) \oplus M_3(\mathbf{C}),\\
&\mathbf{C}^{2} \oplus M_2(\mathbf{C}) \oplus M_2(\mathbf{C}) \oplus M_2(\mathbf{C}) \oplus M_4(\mathbf{C}),\\
&\mathbf{C}^{2} \oplus \underbrace{M_2(\mathbf{C}) \oplus \cdots \oplus M_2(\mathbf{C})}_{7}.
\end{aligned}$$

The 5th case does not occur due to Lemma 10.9,(vi) while Theorem 9.8,(i) shows that the 3rd case is impossible. The 4th case is covered by Theorem 9,(iii). However, the group $N$ (of order 10) in this theorem is either the dihedral group $D_{10}$ or $\mathbf{Z}_2 \times \mathbf{Z}_5$. Note that $\mathbf{Z}_3$ admits just the trivial action on both of them so that $\ell^\infty(N) \rtimes_\varsigma \mathbf{Z}_3$ must be abelian, showing that the 4th case is impossible. Theorem 9.8 can be applicable to the remaining 3 cases, and we easily observe that the dual Kac algebras in these cases are all commutative. □

THEOREM 10.12. *The bicrossed product of the pair* $(D_{14}, \mathbf{Z}_3)$ *(where the dihedral group* $D_{14}$ *is normal in the product group* $D_{14} \cdot \mathbf{Z}_3$*) gives rise to a Kac algebra* $\mathcal{A}$ *with the following underlying algebra structure:*

$$\begin{cases}
\mathcal{A} \cong \mathbf{C}^6 \oplus \underbrace{M_3(\mathbf{C}) \oplus \cdots \oplus M_3(\mathbf{C})}_{4},\\
\hat{\mathcal{A}} \cong \mathbf{C}^6 \oplus \underbrace{M_2(\mathbf{C}) \oplus \cdots \oplus M_2(\mathbf{C})}_{9},
\end{cases}$$

*and the intrinsic groups* $G(\mathcal{A}), G(\hat{\mathcal{A}})$ *are both isomorphic to* $\mathbf{Z}_6$*. Moreover, the Kac algebra* $\mathcal{A}$ *and its dual* $\hat{\mathcal{A}}$ *are the only non-trivial Kac algebras of dimension* 42.

PROOF. Let $\mathcal{A}$ be a non-commutative Kac algebra of dimension 42. As before, $\mathcal{P} \supseteq \mathcal{Q}$ denotes the inclusion of factors corresponding to the dual Kac algebra and

$\rho : \mathcal{P} \hookrightarrow \mathcal{Q}$ denotes the inclusion map. Then, possible algebras are as follows:

$$\mathbf{C}^{14} \oplus \underbrace{M_2(\mathbf{C}) \oplus \cdots \oplus M_2(\mathbf{C})}_{7},$$
$$\mathbf{C}^6 \oplus M_6(\mathbf{C}),$$
$$\mathbf{C}^6 \oplus \underbrace{M_3(\mathbf{C}) \oplus \cdots \oplus M_3(\mathbf{C})}_{4},$$
$$\mathbf{C}^6 \oplus \underbrace{M_2(\mathbf{C}) \oplus \cdots \oplus M_2(\mathbf{C})}_{9},$$
$$\mathbf{C}^2 \oplus M_2(\mathbf{C}) \oplus M_6(\mathbf{C}),$$
$$\mathbf{C}^2 \oplus M_2(\mathbf{C}) \oplus M_2(\mathbf{C}) \oplus M_4(\mathbf{C}) \oplus M_4(\mathbf{C}),$$
$$\mathbf{C}^2 \oplus \underbrace{M_2(\mathbf{C}) \oplus \cdots \oplus M_2(\mathbf{C})}_{6} \oplus M_4(\mathbf{C}),$$
$$\mathbf{C}^2 \oplus M_2(\mathbf{C}) \oplus \underbrace{M_3(\mathbf{C}) \oplus \cdots \oplus M_3(\mathbf{C})}_{4},$$
$$\mathbf{C}^2 \oplus \underbrace{M_2(\mathbf{C}) \oplus \cdots \oplus M_2(\mathbf{C})}_{10}.$$

From Theorem 9.8 we see that the dual Kac algebra $\hat{\mathcal{A}}$ is commutative in the 2nd and the 9th cases. Neither the 6th nor the 7th case occurs thanks to Lemma 10.9,(vi). Hence we have to analyze the 1st, 3rd, 4th, 5th and 8th cases.

In the 1st case, Theorem 9.8 implies that such a Kac algebra is classified by $N \rtimes \mathbf{Z}_2$ with $\#N = 21$ and an element $[(\eta, \zeta)]$ in $H^2((N, \mathbf{Z}_2), \mathbf{T})/\!\sim$ giving $\mathcal{A} \cong \ell^\infty(N) \rtimes_\zeta \mathbf{Z}_2$. Let $\tau$ be the generator of $\mathbf{Z}_2$ which we regard as an automorphism of $N$. Then,

$$\ell^\infty(N) \rtimes_{\tau,\zeta} \mathbf{Z}_2 \cong \mathbf{C}^{14} \oplus \underbrace{M_2(\mathbf{C}) \oplus \cdots \oplus M_2(\mathbf{C})}_{7}$$

if and only if the cardinality of the fixed-point subgroup $N_0$ of $N$ under $\tau$ is 7. Since $\#N = 21$, $N_0$ is a normal subgroup of $N$. If $N$ is commutative, it is cyclic and $H^2(N, \mathbf{T})$ is trivial, which means that $\mathcal{A}$ is co-commutative. If $N$ is non-commutative, then we have the following presentation of $N$:

$$N = \langle a, b;\ a^7 = b^3 = 1,\ bab^{-1} = a^2 \rangle.$$

However, there is no period two automorphism whose fixed-point subgroup is $\langle a \rangle$, which means that this possibility does not occur.

In the 3rd case, Theorem 9.8 implies that such a Kac algebra is classified by $N \rtimes \mathbf{Z}_3$ with $\#N = 14$ and an element $[(\eta, \zeta)]$ in $H^2((N, \mathbf{Z}_3), \mathbf{T})/\!\sim$ giving $\mathcal{A} \cong \ell^\infty(N) \rtimes_\zeta \mathbf{Z}_2$. Let $\tau$ be a generator of $\mathbf{Z}_3$ which we regard as an automorphism of $N$. Then,

$$\ell^\infty(N) \rtimes_{\tau,\zeta} \mathbf{Z}_3 \cong \mathbf{C}^6 \oplus \underbrace{M_3(\mathbf{C}) \oplus \cdots \oplus M_3(\mathbf{C})}_{4}$$

if and only if the cardinality of the fixed-point subgroup $N_0$ of $N$ under $\tau$ is 2. If $N$ is commutative, $\mathcal{A}$ is co-commutative as before. If $N$ is non-commutative, it is isomorphic to the dihedral group $D_{14}$ so that we can take generators $a, b \in N$ satisfying

$$a^7 = b^2 = 1,\ bab^{-1} = a^{-1},\ a^\tau = a^{-1},\ b^\tau = b.$$

Since $H^2(N,\mathbf{T})$ is trivial, we may assume $\eta_i = 1$ and we have

$$\hat{\mathcal{A}} \cong \mathbf{C}(D_{12}) \oplus \mathbf{C}(D_{12}) \oplus \mathbf{C}(D_{12}) = \mathbf{C}^6 \oplus \underbrace{M_2(\mathbf{C}) \oplus \cdots \oplus M_2(\mathbf{C})}_{9}$$

(which is the algebra in the 4th case). Since $\eta$ is trivial, we have

$$H^2((N,\mathbf{Z}_3),\mathbf{T}) \cong H^2(\mathbf{Z}_3, Hom(N,\mathbf{T})) \cong H^2(\mathbf{Z}_3,\mathbf{Z}_2)$$

with the trivial $\mathbf{Z}_3$ action on $Hom(N,\mathbf{T})$. Thus, $H^2((N,\mathbf{Z}_3),\mathbf{T})$ is trivial, and there exists only one Kac algebra of this type.

In the 4th case, Theorem 9.8 implies that such a Kac algebra is classified by $N \rtimes \mathbf{Z}_2$ with $\#N = 21$ and an element $[(\eta,\zeta)]$ in $H^2((N,\mathbf{Z}_2),\mathbf{T})/\sim$ giving $\mathcal{A} \cong \ell^\infty(N) \rtimes_\zeta \mathbf{Z}_2$. Let $\tau$ be the generator of $\mathbf{Z}_2$ which we regard as an automorphism of $N$. Then,

$$\ell^\infty(N) \rtimes_{\tau,\zeta} \mathbf{Z}_2 \cong \mathbf{C}^6 \oplus \underbrace{M_2(\mathbf{C}) \oplus \cdots \oplus M_2(\mathbf{C})}_{9}$$

if and only if the cardinality of the fixed-point subgroup $N_0$ of $N$ under $\tau$ is 3. If $N$ is commutative, $\mathcal{A}$ is co-commutative as before. If $N$ is non-commutative, we can take generators $a, b \in N$ satisfying

$$a^7 = b^3 = 1,\ bab^{-1} = a^2,\ a^\tau = a^{-1},\ b^\tau = b.$$

Since $H^2(N,\mathbf{T})$ is trivial, we may assume $\eta_i = 1$ and we have

$$\hat{\mathcal{A}} \cong \mathbf{C}(N) \oplus \mathbf{C}(N) \oplus \mathbf{C}(N) = \mathbf{C}^6 \oplus \underbrace{M_3(\mathbf{C}) \oplus \cdots \oplus M_3(\mathbf{C})}_{4}.$$

Thus, this is the dual case of the 3rd case.

Only the 5th and 8th cases remain now. We will show that no consistent fusion rule is possible in the both cases so that they do not occur.

In the 5th case, we have the following irreducible decomposition of $\rho\bar{\rho}$:

$$[\rho\bar{\rho}] = [id] \oplus [\tau] \oplus 2[\pi] \oplus 6[\sigma],$$

where $\tau$ is an automorphism of order 2 and $d(\pi) = 2$, $d(\sigma) = 6$. We note that $\{id, \tau, \pi\}$ is closed under conjugation and irreducible decomposition of products by counting dimensions. Thus, there exists an intermediate subfactor $\mathcal{P} \supseteq \mathcal{R} \supseteq \mathcal{Q}$ such that $[\rho_1\bar{\rho}_1] = [id] \oplus [\tau] \oplus 2[\pi]$, where $\rho_1 : \mathcal{R} \hookrightarrow \mathcal{P}$ is the inclusion map. Since $\mathcal{P} \supseteq \mathcal{R}$ is a depth 2 subfactor of index 6, we know $\mathcal{P} = \mathcal{R} \rtimes_\alpha \mathfrak{S}_3$ with an action $\alpha$ of the symmetric group $\mathfrak{S}_3$ and so we have $[\bar{\rho}_1\rho_1] = \oplus_{g\in\mathfrak{S}_3}[\alpha_g]$. Let $\rho_2 : \mathcal{Q} \hookrightarrow \mathcal{R}$ be the inclusion map. The discussion so far means

$$[\bar{\rho}\rho] = [\bar{\rho}_2\bar{\rho}_1\rho_1\rho_2] = \oplus_{g\in\mathfrak{S}_3}[\bar{\rho}_2\alpha_g\rho_2].$$

Note that as a consequence of the discussions so far, the algebra structure of the dual Kac algebra must be also as in the 5th case or 8th case. In either case one has two automorphisms and hence so does $[\bar{\rho}\rho]$. Let $\{id, \tilde{\tau}\}$ be the automorphisms in $[\bar{\rho}\rho]$. Since the index of $\mathcal{R} \supseteq \mathcal{Q}$ is $= 7$, $\bar{\rho}_2\rho_2$ does not contain the automorphism $\tilde{\tau}$ (of period 2). This means that $\bar{\rho}_2\alpha_g\rho_2$ contains $\tilde{\tau}$ for a unique element $g \in \mathfrak{S}_3 \setminus \{e\}$. Therefore, we have a presentation of $\mathfrak{S}_3$ with the following property:

$$\begin{cases} \mathfrak{S}_3 = \langle a, b;\ a^3 = b^2 = e, bab = a^2\rangle, \\ [\alpha_b\rho_2] = [\rho_2\tilde{\tau}], \\ \text{neither } [\bar{\rho}_2\alpha_a\rho_2] \text{ nor } [\bar{\rho}_2\alpha_{a^2}\rho_2] \text{ contains an automorphism.} \end{cases}$$

Firstly we assume that the dual Kac algebra is as in the 5th case. Since $\overline{\rho}_2\alpha_a\rho_2$ does not contain an automorphism, it is decomposed into sectors with dimension either 2 or 6. But, this is obviously impossible because of $d(\overline{\rho}_2\alpha_a\rho_2) = 7$. Secondly we assume the dual Kac algebra is as in the 8th case. As above one finds an intermediate subfactor $\mathcal{P} \supseteq \mathcal{L} \supseteq \mathcal{Q}$ satisfying $\mathcal{Q} = \mathcal{L}^{\mathfrak{S}_3}$. By exchanging the roles of $\rho$ and $\overline{\rho}$, we get the same contradiction as in the preceding case.

We finally assume that $\mathcal{A}$ is as in the 8th case. If $\hat{\mathcal{A}}$ is as in the 5th case, then we have already had a contradiction (by applying the argument in the previous paragraph to $\hat{\mathcal{A}}$). Thus, we may and do assume that both of $\mathcal{A} \cong \hat{\mathcal{A}}$ are as in the 8th case. The irreducible decomposition of $\overline{\rho}\rho$ is as follows:

$$[\overline{\rho}\rho] = [id] \oplus [\tilde{\tau}] \oplus 2[\tilde{\pi}] \oplus 3[\tilde{\sigma}_1] \oplus 3[\tilde{\sigma}_2] \oplus 3[\tilde{\sigma}_3] \oplus 3[\tilde{\sigma}_4]$$

with $d(\tilde{\pi}) = 2, d(\tilde{\sigma}_i) = 3$ $(i = 1, 2, 3, 4)$. The argument on an intermediate subfactor of index 6 in the 5th case is also applicable to the current case and we use the same notations as before. Then, since $[\alpha_b\rho_2] = [\rho_2\tilde{\tau}]$, we have the following decomposition:

$$\begin{aligned}[\overline{\rho}\rho] &= \oplus_{g\in\mathfrak{S}_3}[\overline{\rho}_2\alpha_g\rho_2] \\ &= [\overline{\rho}_2\rho_2] \oplus [\tilde{\tau}][\overline{\rho}_2\rho_2] \oplus [\overline{\rho}_2\alpha_a\rho_2] \oplus [\tilde{\tau}][\overline{\rho}_2\alpha_a\rho_2] \oplus [\overline{\rho}_2\alpha_{a^2}\rho_2] \oplus [\tilde{\tau}][\overline{\rho}_2\alpha_{a^2}\rho_2].\end{aligned}$$

The first $[\overline{\rho}_2\rho_2]$ (of dimension 7) contains no non-trivial automorphism and the multiplicity of $[\tilde{\pi}]$ in $[\overline{\rho}\rho]$ is 2. Thus, by counting dimensions we observe that $[\tilde{\pi}]$ cannot appear in $[\overline{\rho}_2\rho_2]$. Note that it cannot appear in $[\tilde{\tau}\overline{\rho}_2\rho_2]$ either (because of $[\tilde{\pi}] = [\tilde{\tau}\tilde{\pi}]$). Thus, the remaining 4 sectors altogether must contain $[\tilde{\pi}]$ twice. However, it is impossible. Indeed, if $[\tilde{\pi}]$ appears in one of the 4 sectors, then it appears in all of them because $[\tilde{\pi}]$ is self-conjugate and $[\tilde{\tau}\tilde{\pi}] = [\tilde{\pi}]$. □

CHAPTER 11

# Classification of Kac algebras of dimension 16

In this chapter classification of Kac algebras of dimension 16 is presented. By counting dimensions, we see that the order of the intrinsic group $G(\mathcal{A})$ of a non-trivial Kac algebra $\mathcal{A}$ of dimension 16 is either 4 or 8. When $G(\mathcal{A})$ is an abelian group of order 8, it is relatively easy to see that $\mathcal{A}$ arises from $N \rtimes \mathbf{Z}_2$ with $N$ $(=\widehat{G(\mathcal{A})}$, the dual group) abelian (see Proposition 11.3). Moreover, the possibility of the cyclic group $N = \mathbf{Z}_8$ is eliminated (see Proposition 11.3,(i)) due to the fact that this group does not possess a non-trivial cocycle. More delicate cases are

$$\begin{cases} G(\mathcal{A}) \text{ is non-commutative group of order 8,} \\ G(\mathcal{A}) \text{ is a group of order 4,} \end{cases}$$

which will be treated in §3. We will show that $\mathcal{A}$ arises from $D_8 \rtimes \mathbf{Z}_2$ in both of these cases (see Proposition 11.15 and Proposition 11.16).

## 1. Case when $G(\mathcal{A})$ is an abelian group of order 8

To deal with the case when $G(\mathcal{A})$ is an abelian group of order 8 (see Proposition 11.3), we begin by determining all possible $\mathbf{Z}_2$ actions on the abelian groups $\mathbf{Z}_2 \times \mathbf{Z}_2 \times \mathbf{Z}_2$, $\mathbf{Z}_4 \times \mathbf{Z}_2$ as well as their cohomology groups for later purposes. It is routine to show

LEMMA 11.1.
*(i) Any period two automorphism of* $N_0 = \mathbf{Z}_2 \times \mathbf{Z}_2 \times \mathbf{Z}_2$ *is conjugate to* $\tau_0$ *defined by*

$$(x_1, x_2, x_3)^{\tau_0} = (x_1, x_3, x_2),$$

*and the centralizer* $C_{Aut(N_0)}(\tau_0)$ *of* $\tau_0$ *in* $Aut(N_0) = GL(3, 2)$ *is*

$$C_{Aut(N_0)}(\tau) = \left\{ \begin{pmatrix} 1 & a & a \\ b & 1 & 0 \\ b & 0 & 1 \end{pmatrix}, \begin{pmatrix} 1 & c & c \\ d & 0 & 1 \\ d & 1 & 0 \end{pmatrix} ; \ a, b, c, d = 0, 1 \right\}.$$

*(ii) The automorphism group* $Aut(N_1)$ *of* $N_1 = \mathbf{Z}_4 \times \mathbf{Z}_2$ *is isomorphic to the dihedral group* $D_8$ *with generators*

$$(x_1, x_2)^{\varphi} = (x_1 + 2x_2, x_1 + x_2) \quad \text{and} \quad (x_1, x_2)^{\psi} = (x_1, x_1 + x_2)$$

*satisfying*

$$\varphi^4 = \psi^2 = id, \ \psi\varphi\psi = \varphi^3.$$

*(iii) The conjugacy classes of period two automorphisms of* $N_1$ *consist of*

$$\{\varphi^2\}, \quad \{\psi, \varphi^2\psi\}, \quad \{\varphi\psi, \varphi^3\psi\},$$

*and the centralizers of their representatives are as follows:*

$$C_{Aut(N_1)}(\varphi^2) = Aut(N_1),$$
$$C_{Aut(N_1)}(\psi) = \{id, \varphi^2, \psi, \varphi^2\psi\},$$
$$C_{Aut(N_1)}(\varphi\psi) = \{id, \varphi^2, \varphi\psi, \varphi^3\psi\}.$$

*(iv) For*

$$\tau_1 = \varphi\psi, \ \tau_2 = \psi, \ \tau_3 = \varphi^2,$$

*we have*

$$(x_1, x_2)^{\tau_1} = (x_1 + 2x_2, x_2),$$
$$(x_1, x_2)^{\tau_2} = (x_1, x_1 + x_2),$$
$$(x_1, x_2)^{\tau_3} = (3x_1, x_2).$$

*The fixed-point subgroups* $N_1^{\tau_i}$ $(i = 1, 2, 3)$ *are*

$$N_1^{\tau_1} = \{(0,0), (1,0), (2,0), (3,0)\},$$
$$N_1^{\tau_2} = N_1^{\tau_3} = \{(0,0), (2,0), (0,1), (2,1)\}.$$

Let $\omega_0, \omega_1 \in Z^2(\mathbf{Z}_2 \times \mathbf{Z}_2, \mathbf{T})$ be non-trivial elements defined by

$$\omega_0(x,y) = \begin{cases} 1 & \text{if } x = y, \\ i & \text{if } (x,y) = ((1,0),(0,1)),\ ((0,1),(1,1)),\ ((1,1),(1,0)), \\ -i & \text{if } (x,y) = ((1,0),(1,1)),\ ((0,1),(1,0)),\ ((1,1),(0,1)), \end{cases}$$
$$\omega_1((a,b),(c,d)) = (-1)^{ad} \ (0 \le a,b,c,d \le 1).$$

Note that $\omega_0$ is the cocycle considered in §2 of Chapter 7. We denote by $\theta$ the surjective homomorphism from $\mathbf{Z}_4 \times \mathbf{Z}_2$ to $\mathbf{Z}_2 \times \mathbf{Z}_2$ given by $\theta(x_1, x_2) = (x_1, x_2)$, and define $\omega_{ij}^0, \omega_{ij}^1 \in Z^2(\mathbf{Z}_2 \times \mathbf{Z}_2 \times \mathbf{Z}_2, \mathbf{T})$, $1 \le i < j \le 3$ by

$$\omega_{ij}^k((x_1,x_2,x_3),(y_1,y_2,y_3)) = \omega_k((x_i,x_j),(y_i,y_j)).$$

On the other hand, we define $\omega^0, \omega^1 \in Z^2(\mathbf{Z}_4 \times \mathbf{Z}_2, \mathbf{T})$ by

$$\omega^k((x_1,x_2),(y_1,y_2)) = \omega_k(\theta(x_1,x_2), \theta(y_1,y_2)).$$

LEMMA 11.2. *(i) We have* $H^2(\mathbf{Z}_2 \times \mathbf{Z}_2 \times \mathbf{Z}_2, \mathbf{T}) \cong \mathbf{Z}_2 \times \mathbf{Z}_2 \times \mathbf{Z}_2$ *with generators* $[\omega_{ij}^0] = [\omega_{ij}^1]$ $((i,j) = (1,2), (1,3), (2,3))$.
*(ii) We have* $H^2(\mathbf{Z}_4 \times \mathbf{Z}_2, \mathbf{T}) \cong \mathbf{Z}_2$ *with the generator* $[\omega^0] = [\omega^1]$.
*(iii) The twisted group algebras of the above two groups relative to non-trivial cocycles are all isomorphic to* $M_2(\mathbf{C}) \oplus M_2(\mathbf{C})$.

PROOF. (i) See 11.4.16 of [**59**] for example. (ii) By simple computation, we can see that $\mathbf{Z}_4 \times \mathbf{Z}_2$ has only one non-trivial projective representation given by

$$\pi(1,0) = \begin{pmatrix} 1 & 0 \\ 0 & -1 \end{pmatrix}, \quad \pi(0,1) = \begin{pmatrix} 0 & 1 \\ 1 & 0 \end{pmatrix},$$

which corresponds to the above cocycle. (iii) This holds because $M_2(\mathbf{C}) \oplus M_2(\mathbf{C})$ is a unique semisimple algebra of dimension 8 without abelian simple components. □

PROPOSITION 11.3. *Let* $\mathcal{A}$ *be a Kac algebra of dimension* 16 *with* $G(\mathcal{A})$ *abelian of order* 8.
*(i) A Kac algebra* $\mathcal{A}$ *is classified by* $N \rtimes \mathbf{Z}_2$ *with the dual group* $N$ *of* $G(\mathcal{A})$ *and an element* $[(\eta, \zeta)]$ *in* $H^2((N, \mathbf{Z}_2), \mathbf{T})/\sim$ *satisfying*

$$\hat{\mathcal{A}} \cong \mathbf{C}(N) \oplus \mathbf{C}_{\eta_1}(N) \cong \mathbf{C}^8 \oplus M_2(\mathbf{C}) \oplus M_2(\mathbf{C}).$$

*The condition* $\mathbf{C}(N) \oplus \mathbf{C}_{\eta_1}(N) \cong \mathbf{C}^8 \oplus M_2(\mathbf{C}) \oplus M_2(\mathbf{C})$ *is equivalent to the requirement that the class of* $\eta_1$ *in* $H^2(N, \mathbf{T})$ *is not trivial, and consequently* $N$ *is isomorphic to*

$$\text{either } \mathbf{Z}_4 \times \mathbf{Z}_2 \quad \text{or } \mathbf{Z}_2 \times \mathbf{Z}_2 \times \mathbf{Z}_2.$$

*(ii) If* $\mathcal{A}$ *is not commutative, then the* $\mathbf{Z}_2$ *action on* $N$ *has exactly* 4 *fixed points, and*

$$\mathcal{A} \cong \ell^\infty(N) \rtimes_\zeta \mathbf{Z}_2 \cong \mathbf{C}^8 \oplus M_2(\mathbf{C}) \oplus M_2(\mathbf{C}).$$

PROOF. (i) Let $\mathcal{P} \supseteq \mathcal{Q}$ be a depth 2 inclusion corresponding to $\mathcal{A}$, and we set

$$\mathcal{P} \supseteq \mathcal{R} = \mathcal{P}^{G(\mathcal{A})} \supseteq \mathcal{Q}.$$

Then we have $[\mathcal{R} : \mathcal{Q}] = 2$ so that $\mathcal{Q} = \mathcal{R}^{\mathbf{Z}_2}$ with a $\mathbf{Z}_2$ action. Since $\mathcal{P} = \mathcal{R} \rtimes N$ with the dual group $N$ of $G(\mathcal{A})$, we get the first half. The second half follows from Lemma 11.2,(iii). (ii) If $\mathcal{A}$ is not commutative, the $\mathbf{Z}_2$ action is not trivial and we get the result from Lemma 11.1. □

For each of the cases described in Proposition 11.3,(i) we compute invariants in the rest of the section.

We begin with the case $N = \mathbf{Z}_2 \times \mathbf{Z}_2 \times \mathbf{Z}_2$, and in this case a non-trivial $\mathbf{Z}_2$ action on $N$ in essentially unique (see Lemma 11.1,(i)). We denote by $\chi_i$ $(i = 1, 2, 3)$ the generators of the dual group of $\mathbf{Z}_2 \times \mathbf{Z}_2 \times \mathbf{Z}_2$ defined by

$$\chi_i(x_1, x_2, x_3) = (-1)^{x_i}.$$

For $H = \mathbf{Z}_2$ we use the following notations for simplicity:

$$\eta = \eta_1, \ \zeta_n = \zeta_n(1,1), \ \xi(n) = \xi(1, n).$$

LEMMA 11.4. *For* $N = \mathbf{Z}_2 \times \mathbf{Z}_2 \times \mathbf{Z}_2$ *with the* $\mathbf{Z}_2$ *action* $\tau_0$, *we have*

$$H^2((N, \mathbf{Z}_2), \mathbf{T}) \cong \mathbf{Z}_2 \times \mathbf{Z}_2 \times \mathbf{Z}_2,$$

*six elements of which give rise to non-trivial Kac algebras. There are four isomorphic classes among them. The following are representatives of these classes:*
*(i)* $\eta = \omega_{23}^0$ *and* $\zeta = 1$
*In this case,* $G(\hat{\mathcal{A}}) \cong \mathbf{Z}_2 \times \mathbf{Z}_2 \times \mathbf{Z}_2$, *and* $\mathcal{A}$ *is the tensor product of the Kac-Paljutkin's 8-dimensional Kac algebra and the group algebra of* $\mathbf{Z}_2$. *In particular,* $\mathcal{A}$ *is self-dual.*
*(ii)* $\eta = \omega_{23}^0$ *and* $\zeta_n = \chi_1(n)$
*In this case, we have* $G(\hat{\mathcal{A}}) \cong \mathbf{Z}_4 \times \mathbf{Z}_2$.
*(iii)* $\eta = \omega_{12}^1 \omega_{13}^1$ *and* $\zeta = 1$
*In this case, we have* $G(\hat{\mathcal{A}}) \cong \mathbf{Z}_2 \times \mathbf{Z}_2 \times \mathbf{Z}_2$, *and* $\mathcal{A}$ *is self-dual.*
*(iv)* $\eta = \omega_{12}^1 \omega_{13}^1$ *and* $\zeta = \chi_1$
*In this case, we have* $G(\hat{\mathcal{A}}) \cong \mathbf{Z}_4 \times \mathbf{Z}_2$.

PROOF. By definition, $\tau_0$ leaves $[\omega_{23}^0]$ invariant and permutes $[\omega_{12}^0]$ and $[\omega_{13}^0]$. Since $[\eta_1][\eta_1^{\tau_0}] = 1$, we may assume that $\eta_1$ is one of the follows:

$$1, \ \omega_{23}^0, \ \omega_{12}^1 \omega_{13}^1, \ \omega_{23}^0 \omega_{12}^1 \omega_{13}^1.$$

Since $\eta_1 \eta_1^{\tau_0} = 1$ holds in these cases, the freedom coming from $\zeta$ is exactly the cardinality of $H^2(\mathbf{Z}_2, \hat{N})$ where the $\mathbf{Z}_2$ action on $\hat{N}$ is the flip of $\chi_2$ and $\chi_3$. Considering the corresponding extensions, we can see that $H^2(\mathbf{Z}_2, \hat{N}) \cong \mathbf{Z}_2$ and $\zeta_n(1,1) = \chi_1(n)$ gives the non-trivial element. Therefore, we conclude

$$H^2((N, \mathbf{Z}_2), \mathbf{T}) = \{[(1,1)], [(1,\chi_1)], [(\omega_{23}^0, 1)], [(\omega_{23}^0, \chi_1)], [(\omega_{12}^1\omega_{13}^1, 1)], [(\omega_{12}^1\omega_{13}^1, \chi_1)], [(\omega_{23}^0\omega_{12}^1\omega_{13}^1, 1)], [(\omega_{23}^0\omega_{12}^1\omega_{13}^1, \chi_1)]\},$$

and $\eta_1 \neq 1$ cases give rise to non-trivial Kac algebras. To prove $H^2((N, \mathbf{Z}_2), \mathbf{T}) \cong \mathbf{Z}_2 \times \mathbf{Z}_2 \times \mathbf{Z}_2$, it suffices to show that $((\omega_{23}^0)^2, 1)$ is a coboundary. But, $\xi$ defined as follows gives an appropriate coboundary:

$$\xi(x_1, x_2, x_3) = \begin{cases} 1 & \text{if } (x_2, x_3) \neq (1,1), \\ -1 & \text{if } (x_2, x_3) = (1,1). \end{cases}$$

By direct computation based on Lemma 11.1 we see that $[\omega_{23}^0]$ is transformed into $[\omega_{23}^0\omega_{12}^1\omega_{13}^1]$ by the centralizer $C_{Aut(N)}(\tau_0)$ while $[\omega_{12}^1\omega_{13}^1]$ is invariant under $C_{Aut(N)}(\tau_0)$. Thus, there are 4 isomorphic classes of Kac algebras of this type.

We compute $G(\hat{\mathcal{A}})$ in the above cases. In case (i), the first component of $N$ is involved neither in the $\mathbf{Z}_2$ action nor in the invariant $(\omega_{23}^0, 1)$. Therefore, the resulting Kac algebra is isomorphic to the tensor product of the group algebra of $\mathbf{Z}_2$ and the Kac-Paljutkin's 8-dimensional Kac algebra (see §2 of Chapter 7). This Kac algebra is self-dual and we have $G(\mathcal{A}) \cong \mathbf{Z}_2 \times \mathbf{Z}_2 \times \mathbf{Z}_2$. In case (ii), let $N_0$ be the fixed-point subgroup of $N$ under $\tau_0$:

$$N_0 = \{(0,0,0), (1,0,0), (0,1,1), (1,1,1)\}.$$

Note $\zeta_{(x_1,x_2,x_2)} = (-1)^{x_1}$. We define $\xi : N \longrightarrow \mathbf{T}$ by

$$\xi(x_1, x_2, x_3) = \begin{cases} 1 & \text{if } x_1 = 0 \\ i & \text{if } x_1 = 1, \end{cases}$$

and set

$$\begin{cases} \eta'(n_1, n_2) = \eta(n_1, n_2)\xi(n_1)\xi(n_2)\overline{\xi(n_1 + n_2)}, \\ \zeta'_n = \zeta_n \overline{\xi(n)\xi(n^{\tau_0})}. \end{cases}$$

Then, $(\eta, \zeta)$ is cohomologous to $(\eta', \zeta')$ and $\zeta'_n = 1$ for $n \in N_0$. Now $G(\hat{\mathcal{A}})$ is the extension determined by $\eta_{|N_0} \in Z^2(N_0, \hat{\mathbf{Z}}_2)$ and it is $\mathbf{Z}_4 \times \mathbf{Z}_2$. In the same way. we can obtain description for $G(\hat{\mathcal{A}})$ in (iii) and (iv). Note that $\mathcal{A}$ is self-dual in case (iii) because there are only two isomorphic classes of 16-dimensional Kac algebras with

$$G(\mathcal{A}) \cong G(\hat{\mathcal{A}}) \cong \mathbf{Z}_2 \times \mathbf{Z}_2 \times \mathbf{Z}_2,$$

and one of them is self-dual thanks to case (i). □

We next consider the case $N = \mathbf{Z}_4 \times \mathbf{Z}_2$, where we have to deal with three kinds of $\mathbf{Z}_2$ actions (see Lemma 11.1,(iii)). We denote by $\chi_i$, $i = 1, 2$ the generators of the dual group of $\mathbf{Z}_4 \times \mathbf{Z}_2$ defined by

$$\begin{cases} \chi_1(x_1, x_2) = i^{x_1}, \\ \chi_2(x_1, x_2) = (-1)^{x_2}. \end{cases}$$

LEMMA 11.5. *For $N = \mathbf{Z}_4 \times \mathbf{Z}_2$ with the $\mathbf{Z}_2$ action $\tau_1$, we have*

$$H^2((N, \mathbf{Z}_2), \mathbf{T}) \cong \mathbf{Z}_2 \times \mathbf{Z}_2,$$

*two elements of which give rise to non-isomorphic non-trivial Kac algebras. The following are representatives of these classes:*
*(i) $\eta = \omega^1$ and $\zeta = 1$*

*In this case, we have* $G(\hat{\mathcal{A}}) \cong \mathbf{Z}_4 \times \mathbf{Z}_2$.
*(ii)* $\eta = \omega^1$ *and* $\zeta = \chi_2$
*In this case, we have* $G(\hat{\mathcal{A}}) \cong \mathbf{Z}_4 \times \mathbf{Z}_2$.

PROOF. Thanks to Lemma 11.2, we may assume $\eta = 1$ or $\eta = \omega^1$. Since $\omega^1(\omega^1)^{\tau_1} = (\omega^1)^2 = 1$ holds, $\zeta \in Z^2(\mathbf{Z}_2, \hat{N})$, where $\mathbf{Z}_2$ acts on $\hat{N}$ by

$$\chi_1^{\tau_1} = \chi_1\chi_2, \quad \chi_2^{\tau_1} = \chi_2.$$

Considering the corresponding extensions, we can see that $H^2(\mathbf{Z}_2, \hat{N}) \cong \mathbf{Z}_2$ and $\zeta_n = \chi_2(n)$ represents a non-trivial element. Thus, we observe

$$H^2((N, \mathbf{Z}_2), \mathbf{T}) \cong \{[(1,1)], [(1,\chi_2)], [(\omega^1, 1)], [(\omega^1, \chi_2)]\} \cong \mathbf{Z}_2 \times \mathbf{Z}_2,$$

and $\{[(\omega^1, 1)], [(\omega^1, \chi_2)]\}$ give rise to non-trivial Kac algebras. Direct computation shows that $C_{Aut(N)}(\tau_1)$ acts on $H^2((N, \mathbf{Z}_2), \mathbf{T})$ trivially and these two Kac algebras are not isomorphic. Note that $\chi_2$ and $\omega^1$ are trivial on $N^{\tau_1}$. Thus, $G(\hat{\mathcal{A}})$ is a trivial extension of $N^{\tau_1}$ by $\hat{\mathbf{Z}}_2$ so that it is isomorphic to $\mathbf{Z}_4 \times \mathbf{Z}_2$. □

LEMMA 11.6. *For* $N = \mathbf{Z}_4 \times \mathbf{Z}_2$ *with the* $\mathbf{Z}_2$ *action* $\tau_2$, *we have*

$$H^2((N, \mathbf{Z}_2), \mathbf{T}) \cong \mathbf{Z}_2 \times \mathbf{Z}_2,$$

*two elements of which give rise to non-isomorphic non-trivial Kac algebras. The following are representatives of these classes:*
*(i)* $\eta = \omega^0$ *and* $\zeta = 1$
*In this case, we have* $G(\hat{\mathcal{A}}) \cong \mathbf{Z}_2 \times \mathbf{Z}_2 \times \mathbf{Z}_2$.
*(ii)* $\eta = \omega^0$ *and* $\zeta = \chi_1$
*In this case, we have* $G(\hat{\mathcal{A}}) \cong \mathbf{Z}_4 \times \mathbf{Z}_2$.

PROOF. Thanks to Lemma 11.2, we may assume $\eta = 1$ or $\eta = \omega^0$. Since $\omega^0(\omega^0)^{\tau_2} = 1$, we have $\zeta \in Z^2(\mathbf{Z}_2, \hat{N})$, where $\mathbf{Z}_2$ acts on $\hat{N}$ by

$$\chi_1^{\tau_2} = \chi_1, \quad \chi_2^{\tau_2} = \chi_1^2\chi_2.$$

Considering the corresponding extensions, we can see that $H^2(\mathbf{Z}_2, \hat{N}) \cong \mathbf{Z}_2$ and $\zeta_n = \chi_1(n)$ represents a non-trivial element. Thus, we get

$$H^2((N, \mathbf{Z}_2), \mathbf{T}) \cong \{[(1,1)], [(1,\chi_1)], [(\omega^1, 1)], [(\omega^1, \chi_1)]\} \cong \mathbf{Z}_2 \times \mathbf{Z}_2,$$

and $\{[(\omega^1, 1)], [(\omega^1, \chi_2)]\}$ give rise to non-trivial Kac algebras. Direct computation shows that $C_{Aut(N)}(\tau_2)$ acts on $H^2((N, \mathbf{Z}_2), \mathbf{T})$ trivially and these two Kac algebras are not isomorphic. Note that $N^{\tau_2} \cong \mathbf{Z}_2 \times \mathbf{Z}_2$ and $\omega^0$ restricted to $N^{\tau_2}$ equals to 1. Thus, in the first case, $G(\hat{\mathcal{A}})$ is the trivial extension of $N^{\tau_2}$ by $\hat{\mathbf{Z}}_2$, and isomorphic to $\mathbf{Z}_2 \times \mathbf{Z}_2 \times \mathbf{Z}_2$. In the second case, we define $\xi((a,b)) = e^{\frac{\pi i}{4}a}$, $0 \leq a \leq 3$, and set

$$\begin{cases} \eta'(n_1, n_2) = \eta(n_1, n_2)\xi(n_1)\xi(n_2)\overline{\xi(n_1 n_2)}, \\ \zeta'_n = \zeta_n \overline{\xi(n)\xi(n^{\tau_2})}. \end{cases}$$

Then, $\zeta'$ is trivial on $N^{\tau_2}$ and $\eta'$ corresponds to the exact sequence

$$1 \longrightarrow \hat{\mathbf{Z}}_2 \longrightarrow G(\hat{\mathcal{A}}) \longrightarrow N^{\tau_2} \longrightarrow 1.$$

Since $\eta'$ is symmetric on $N^{\tau_2}$ with $\eta'((2,0),(2,0)) = -1$, we conclude $G(\hat{\mathcal{A}}) \cong \mathbf{Z}_4 \times \mathbf{Z}_2$. □

LEMMA 11.7. *For $N = \mathbf{Z}_4 \times \mathbf{Z}_2$ with the $\mathbf{Z}_2$ action $\tau_3$, we have*

$$H^2((N, \mathbf{Z}_2), \mathbf{T}) \cong \mathbf{Z}_2 \times \mathbf{Z}_2 \times \mathbf{Z}_2,$$

*four elements of which give rise to non-trivial Kac algebras. There are three isomorphism classes of these Kac algebras, and the following are representatives of these classes:*
*(i) $\eta = \omega^1$ and $\zeta = 1$*
*In this case, we have $G(\hat{\mathcal{A}}) \cong \mathbf{Z}_2 \times \mathbf{Z}_2 \times \mathbf{Z}_2$.*
*(ii) $\eta = \omega^1$ and $\zeta = \chi_2$*
*In this case, we have $G(\hat{\mathcal{A}}) \cong \mathbf{Z}_4 \times \mathbf{Z}_2$.*
*(iii) $\eta = \omega^1$ and $\zeta = \chi_1^2\chi_2$*
*In this case, we have $G(\hat{\mathcal{A}}) \cong \mathbf{Z}_4 \times \mathbf{Z}_2$.*

PROOF. Thanks to Lemma 11.2, we may assume $\eta = 1$ or $\eta = \omega^1$. Since $\omega^1(\omega^1)^{\tau_3} = (\omega^1)^2 = 1$, we have $\zeta \in Z^2(\mathbf{Z}_2, \hat{N})$, where $\mathbf{Z}_2$ acts on $\hat{N}$ by

$$\chi_1^{\tau_3} = \chi_1^3, \; \chi_2^{\tau_3} = \chi_2.$$

Considering the corresponding extensions, we can see that $H^2(\mathbf{Z}_2, \hat{N}) \cong \mathbf{Z}_2 \times \mathbf{Z}_2$ and $\zeta = \chi_1(n)^2, \chi_2(n), \chi_1(n)^2\chi_2(n)$ represents non-trivial elements. Thus, we have

$$\begin{aligned} H^2((N, \mathbf{Z}_2), \mathbf{T}) &\cong \{[(1,1)], [(1, \chi_1^2)], [(1, \chi_2)], [(1, \chi_1^2\chi_2)], \\ &\qquad [(\omega^1, 1)], [(\omega^1, \chi_1^2)], [(\omega^1, \chi_2)], [(\omega^1, \chi_1^2\chi_2)]\} \\ &\cong \mathbf{Z}_2 \times \mathbf{Z}_2 \times \mathbf{Z}_2, \end{aligned}$$

and $\eta \neq 1$ cases give rise to non-trivial Kac algebras. To get the number of isomorphism classes of these Kac algebras, we have to examine the action of $C_{Aut(N)}(\tau_3) = Aut(N) = \langle \tau_1, \tau_2 \rangle$ on $H^2((N, \mathbf{Z}_2), \mathbf{T})$. Direct computation shows that $\tau_1$ acts on $H^2((N, \mathbf{Z}_2), \mathbf{T})$ trivially. We define $\xi$ by

$$\xi(n) = \begin{cases} 1 & \text{if } n \neq (0,1), (2,1), \\ i & \text{if } n = (0,1), (2,1). \end{cases}$$

Then, we have

$$\begin{cases} (\omega^1)^{\tau_2}(n_1, n_2) = \omega^1(n_1, n_2)\xi(n_1)\xi(n_2)\overline{\xi(n_1 n_2)}, \\ \xi(n)\xi(n^{\tau_2}) = \chi_1^2(n). \end{cases}$$

Since $\chi_1^{\tau_2} = \chi_1$, $\chi_2^{\tau_2} = \chi_1^2\chi_2$, we get $[(\omega^1, 1)^{\tau_2}] = [(\omega^1, \chi_1^2)]$, $[(\omega^1, \chi_2)^{\tau_2}] = [(\omega^1, \chi_2)]$, and $[(\omega^1, \chi_1^2\chi_2)^{\tau_2}] = [(\omega^1, \chi_1^2\chi_2)]$. Note that $N^{\tau_3} \cong \mathbf{Z}_2 \times \mathbf{Z}_2$ and $\omega^1$ restricted to $N^{\tau_2}$ equals to 1. Thus in case (i) $G(\hat{\mathcal{A}})$ is the trivial extension of $N^{\tau_3}$ by $\hat{\mathbf{Z}}_2$, and isomorphic to $\mathbf{Z}_2 \times \mathbf{Z}_2 \times \mathbf{Z}_2$. In case (ii) and (iii), a similar reasoning as in the proof of the previous lemma shows $G(\hat{\mathcal{A}}) \cong \mathbf{Z}_4 \times \mathbf{Z}_2$. □

## 2. Computation of the invariant $H^2((D_8, \mathbf{Z}_2), \mathbf{T})/\sim$

Here we determine all possible $\mathbf{Z}_2$ actions on $D_8$, the dihedral group of order 8, for each of which we compute invariants, i.e., $H^2((D_8, \mathbf{Z}_2), \mathbf{T})/\sim$. Computations here will be used in the classification parts in Proposition 11.15 and Proposition 11.16 in the next section.

It is routine to show the following:

LEMMA 11.8.
*(i) Let $D_8 = \mathbf{Z}_4 \rtimes \mathbf{Z}_2$ be the dihedral group of order 8, where the action of $\mathbf{Z}_2$ on $\mathbf{Z}_4$ is given by*

$$(0,1)(x,0)(0,1) = (4-x,0).$$

*Then, $Aut(D_8)$ is isomorphic to $D_8$ itself with generators*

$$(x_1,x_2)^{\varphi} = (x_1+x_2,x_2) \quad \text{and} \quad (x_1,x_2)^{\psi} = (4-x_1,x_2)$$

*satisfying*

$$\varphi^4 = \psi^2 = id, \ \psi\varphi\psi = \varphi^{-1}.$$

*(ii) The conjugacy classes of period two automorphisms of $D_8$ consist of*

$$\{\varphi^2\}, \quad \{\psi, \varphi^2\psi\}, \quad \{\varphi\psi, \varphi^3\psi\},$$

*and the centralizers of their representatives are as follows:*

$$C_{Aut(D_8)}(\varphi^2) = Aut(D_8),$$
$$C_{Aut(D_8)}(\psi) = \{id, \varphi^2, \psi, \varphi^2\psi\},$$
$$C_{Aut(D_8)}(\varphi\psi) = \{id, \varphi^2, \varphi\psi, \varphi^3\psi\}.$$

*(iii) We set*

$$\tau_1 = \varphi\psi, \ \tau_2 = \psi, \ \tau_3 = \varphi^2.$$

*Then, we have*

$$(x_1,x_2)^{\tau_1} = (4-x_1-x_2,x_2),$$
$$(x_1,x_2)^{\tau_2} = (4-x_1,x_2),$$
$$(x_1,x_2)^{\tau_3} = (x_1+2x_2,x_2),$$

*and the fixed-point subgroups $D_8^{\tau_i}$, $i = 1,2,3$ are*

$$D_8^{\tau_1} = \{(0,0),(2,0)\},$$
$$D_8^{\tau_2} = \{(0,0),(2,0),(0,1),(2,1)\},$$
$$D_8^{\tau_3} = \langle(1,0)\rangle.$$

We define $\omega \in Z^2(D_{4n}, \mathbf{T})$ by

$$\begin{cases} \omega((x,0),(y,0)) = \omega((x,0),(y,1)) = 1, \\ \omega((x,1),(y,0)) = \omega((x,1),(y,1)) = e^{\frac{\pi i}{n}y}, \end{cases}$$

where $x, y = 0,1,2,\cdots,2n-1$.

LEMMA 11.9.
*(i) We have $H^2(D_{4n}, \mathbf{T}) \cong \mathbf{Z}_2$ with the generator $[\omega]$.*
*(ii) The twisted group algebra $\mathbf{C}_\omega(D_{4n})$ is isomorphic to*

$$M_2(\mathbf{C}) \oplus M_2(\mathbf{C}) \oplus \cdots \oplus M_2(\mathbf{C}).$$

PROOF. Let $\pi$ be an irreducible projective, not genuine, representation of $D_{4n}$. Then, by changing phase factors if necessary, we may assume

$$\pi(1,0)^{2n} = \pi(0,1)^2 = 1 \quad \text{and} \quad \pi(0,1)\pi(1,0)\pi(0,1) = e^{\frac{\pi i}{n}}\pi(1,0)^{-1},$$

which shows there is a unique non-trivial class $[\omega]$ in $H^2(D_{4n}, \mathbf{T})$. (ii) We can see that the dimension of every irreducible projective, not genuine, representation of $D_{4n}$ is 2 by diagonalizing $\pi(1,0)$. Therefore, $\mathbf{C}_\omega(D_{4n})$ is the direct sum of $M_2(\mathbf{C})$'s. $\square$

The next result was shown in §6 of Chapter 7. (see especially Proposition 7.5 and Lemma 7.6)

LEMMA 11.10. *For $N = D_8$ with the $\mathbf{Z}_2$ action $\tau_1$ we have*

$$H^2((N, \mathbf{Z}_2), \mathbf{T}) \cong \mathbf{Z}_2.$$

*The corresponding Kac algebras are non-trivial in both cases. Without a cocycle the algebra structure is given by*

$$\begin{cases} \mathcal{A} \cong \mathbf{C}^4 \oplus M_2(\mathbf{C}) \oplus M_2(\mathbf{C}) \oplus M_2(\mathbf{C}) \\ \hat{\mathcal{A}} \cong \mathbf{C}^8 \oplus M_2(\mathbf{C}) \oplus M_2(\mathbf{C}), \end{cases}$$

*and the intrinsic groups are*

$$G(\mathcal{A}) \cong D_8, \quad G(\hat{\mathcal{A}}) \cong \mathbf{Z}_2 \times \mathbf{Z}_2.$$

*On the other hand, with a non-trivial cocycle we have*

$$\mathcal{A} \cong \hat{\mathcal{A}} \cong \mathbf{C}^4 \oplus M_2(\mathbf{C}) \oplus M_2(\mathbf{C}) \oplus M_2(\mathbf{C}),$$

*and the intrinsic groups are*

$$G(\mathcal{A}) \cong G(\hat{\mathcal{A}}) \cong \mathbf{Z}_2 \times \mathbf{Z}_2.$$

We define $\chi_1, \chi_2 \in Hom(D_8, \mathbf{T})$ by

$$\begin{cases} \chi_1((x_1, x_2)) = (-1)^{x_1}, \\ \chi_2((x_1, x_2)) = (-1)^{x_2}. \end{cases}$$

Note that $\tau_2$ and $\tau_3$ act on $Hom(D_8, \mathbf{T})$ trivially.

LEMMA 11.11. *For $N = D_8$ with the $\mathbf{Z}_2$ action $\tau_2$ we have*

$$H^2((N, \mathbf{Z}_2), \mathbf{T}) \cong \mathbf{Z}_2 \times \mathbf{Z}_2 \times \mathbf{Z}_2,$$

*and $\# \left(H^2((N, \mathbf{Z}_2), \mathbf{T})/\sim\right) = 6$. The underlying algebra of the corresponding Kac algebra $\mathcal{A}$ is always*

$$\mathcal{A} \cong \ell^\infty(N) \rtimes_\varsigma \mathbf{Z}_2 \cong \mathbf{C}^8 \oplus M_2(\mathbf{C}) \oplus M_2(\mathbf{C}).$$

*If $[\eta_1]$ is trivial, then the underlying algebra of $\hat{\mathcal{A}}$ is*

$$\hat{\mathcal{A}} \cong \mathbf{C}(D_8) \oplus \mathbf{C}(D_8) \cong \mathbf{C}^8 \oplus M_2(\mathbf{C}) \oplus M_2(\mathbf{C}).$$

*If $[\eta_1]$ is not trivial, then the underlying algebra of $\hat{\mathcal{A}}$ is*

$$\hat{\mathcal{A}} \cong \mathbf{C}(D_8) \oplus \mathbf{C}_{\eta_1}(D_8) \cong \mathbf{C}^4 \oplus M_2(\mathbf{C}) \oplus M_2(\mathbf{C}) \oplus M_2(\mathbf{C}).$$

*The representatives of $H^2((N, \mathbf{Z}_2), \mathbf{T})/\sim$ and the intrinsic groups are as follows:*
*(i)* $(\eta, \zeta) = (1, 1)$
*We have* $G(\mathcal{A}) \cong \mathbf{Z}_2 \times \mathbf{Z}_2 \times \mathbf{Z}_2$, $G(\hat{\mathcal{A}}) \cong \mathbf{Z}_2 \times \mathbf{Z}_2 \times \mathbf{Z}_2$.
*(ii)* $(\eta, \zeta) = (1, \chi_1)$
*We have* $G(\mathcal{A}) \cong \mathbf{Z}_4 \times \mathbf{Z}_2$, $G(\hat{\mathcal{A}}) \cong \mathbf{Z}_2 \times \mathbf{Z}_2 \times \mathbf{Z}_2$.
*(iii)* $(\eta, \zeta) = (1, \chi_2)$
*We have* $G(\mathcal{A}) \cong \mathbf{Z}_4 \times \mathbf{Z}_2$, $G(\hat{\mathcal{A}}) \cong \mathbf{Z}_4 \times \mathbf{Z}_2$.
*(iv)* $(\eta, \zeta) = (1, \chi_1\chi_2)$
*We have* $G(\mathcal{A}) \cong \mathbf{Z}_4 \times \mathbf{Z}_2$, $G(\hat{\mathcal{A}}) \cong \mathbf{Z}_4 \times \mathbf{Z}_2$.
*(v)* $(\eta, \zeta) = (\omega, 1)$
*We have* $G(\mathcal{A}) \cong \mathbf{Z}_2 \times \mathbf{Z}_2$, $G(\hat{\mathcal{A}}) \cong D_8$.
*(vi)* $(\eta, \zeta) = (\omega, \chi_1)$
*We have* $G(\mathcal{A}) \cong \mathbf{Z}_2 \times \mathbf{Z}_2$, $G(\hat{\mathcal{A}}) \cong D_8$.

PROOF. We may assume that either $\eta = 1$ or $\eta = \omega$. Since $\omega\omega^{\tau_2} = 1$, we have $\zeta \in Z^2(\mathbf{Z}_2, Hom(N, \mathbf{T}))$, where the action of $\mathbf{Z}_2$ on $Hom(N, \mathbf{T}))$ is trivial. Thus, we have $H^2(\mathbf{Z}_2, Hom(N, \mathbf{T})) \cong \mathbf{Z}_2 \times \mathbf{Z}_2$, and

$$\begin{aligned} H^2((N, \mathbf{Z}_2), \mathbf{T}) &= \{[(1,1)], [(1,\chi_1)], [(1,\chi_2)], [(1,\chi_1\chi_2)], \\ &\qquad [(\omega,1)], [(\omega,\chi_1)], [(\omega,\chi_2)], [(\omega,\chi_1\chi_2)]\}. \end{aligned}$$

To prove $H^2((N, \mathbf{Z}_2), \mathbf{T}) \cong \mathbf{Z}_2 \times \mathbf{Z}_2 \times \mathbf{Z}_2$, it suffices to show that $[(\omega^2, 1)]$ is trivial. Indeed, for

$$\xi(x_1, x_2) = i^{x_1}$$

we compute

$$\begin{cases} \omega(n_1, n_2)^2 = \xi(n_1)\xi(n_2)\overline{\xi(n_1 n_2)}, \\ \xi(n_1, n_2)\xi(n_1^{\tau_2}, n_2^{\tau_2}) = 1, \end{cases}$$

showing that $(\omega^2, 1)$ is coboundary. Note that $C_{Aut(D_8)}(\tau_2) = \langle \tau_2, \tau_3 \rangle$ thanks to Lemma 11.8. To obtain the equivalence relation in $H^2((N, \mathbf{Z}_2), \mathbf{T})$, it suffices to compute the action of $\tau_3$ on $H^2((N, \mathbf{Z}_2), \mathbf{T})$. We set

$$\xi'(x_1, x_2) = \begin{cases} 1 & \text{if } x_2 = 0, \\ i & \text{if } x_2 = 1. \end{cases}$$

Then, we get the following for $n, n_1, n_2 \in N$:

$$\begin{cases} \omega((n_1, n_2)^{\tau_3}) = \omega((n_1, n_2))\xi'(n_1)\xi'(n_2)\overline{\xi'(n_1 n_2)}, \\ \xi'(n)\xi'(n^{\tau_3}) = \chi_2(n). \end{cases}$$

Thus, we have

$$[(\omega, 1)] \sim [(\omega, \chi_2)] \quad \text{and} \quad [(\omega, \chi_1)] \sim [(\omega, \chi_1\chi_2)].$$

When $\eta = 1$, the cocycle $\zeta$ describes the exact sequence

$$1 \longrightarrow Hom(N, \mathbf{T}) \longrightarrow G(\mathcal{A}) \longrightarrow \mathbf{Z}_2 \longrightarrow 1.$$

On the other hand, when $\eta = \omega$, we have $G(\mathcal{A}) = Hom(N, \mathbf{T})$ and so we get the results for $G(\mathcal{A})$. Since $\chi_1|_{N^{\tau_2}}$ is trivial, in case (i) and (ii) $G(\hat{\mathcal{A}})$ is the trivial extension, and in case (v) and (vi) it is described by $\omega|_{N^{\tau_2}}$. It is routine to see that in the latter case we have $G(\hat{\mathcal{A}}) \cong D_8$. In case (iii) and (iv), we set

$$\xi''(x_1, x_2) = \begin{cases} 1 & \text{if } x_2 = 1, \\ i & \text{if } x_2 = 1. \end{cases}$$

Then, $\eta'(n_1, n_2) = \xi''(n_1)\xi''(n_2)\overline{\xi''(n_1 n_2)}$ describes the exact sequence

$$1 \longrightarrow \hat{\mathbf{Z}}_2 \longrightarrow G(\hat{\mathcal{A}}) \longrightarrow N^{\tau_2} \longrightarrow 1,$$

and we get $G(\hat{\mathcal{A}}) \cong \mathbf{Z}_4 \times \mathbf{Z}_2$. □

LEMMA 11.12. *For $N = D_8$ with the $\mathbf{Z}_2$ action $\tau_3$ we have*

$$H^2((N, \mathbf{Z}_2), \mathbf{T}) \cong \mathbf{Z}_2 \times \mathbf{Z}_2,$$

*and $\#\left(H^2((N, \mathbf{Z}_2), \mathbf{T})/\sim\right) = 4$. The underlying algebras of the corresponding Kac algebra $\mathcal{A}$ and its dual $\hat{\mathcal{A}}$ are*

$$\begin{cases} \mathcal{A} \cong \ell^\infty(N) \rtimes_\zeta \mathbf{Z}_2 \cong \mathbf{C}^8 \oplus M_2(\mathbf{C}) \oplus M_2(\mathbf{C}), \\ \hat{\mathcal{A}} \cong \mathbf{C}(D_8) \oplus \mathbf{C}(D_8) \cong \mathbf{C}^8 \oplus M_2(\mathbf{C}) \oplus M_2(\mathbf{C}). \end{cases}$$

*The representatives of* $H^2((N, \mathbf{Z}_2), \mathbf{T})/\sim$ *and the intrinsic groups are as follows:*
(*i*) $(\eta, \zeta) = (1, 1)$
*We have* $G(\mathcal{A}) \cong \mathbf{Z}_2 \times \mathbf{Z}_2 \times \mathbf{Z}_2$, $G(\hat{\mathcal{A}}) \cong \mathbf{Z}_4 \times \mathbf{Z}_2$.
(*ii*) $(\eta, \zeta) = (1, \chi_1)$
*We have* $G(\mathcal{A}) \cong \mathbf{Z}_4 \times \mathbf{Z}_2$, $G(\hat{\mathcal{A}}) \cong \mathbf{Z}_4 \times \mathbf{Z}_2$.
(*iii*) $(\eta, \zeta) = (1, \chi_2)$
*We have* $G(\mathcal{A}) \cong \mathbf{Z}_4 \times \mathbf{Z}_2$, $G(\hat{\mathcal{A}}) \cong \mathbf{Z}_4 \times \mathbf{Z}_2$.
(*iv*) $(\eta, \zeta) = (1, \chi_1\chi_2)$
*We have* $G(\mathcal{A}) \cong \mathbf{Z}_4 \times \mathbf{Z}_2$, $G(\hat{\mathcal{A}}) \cong \mathbf{Z}_4 \times \mathbf{Z}_2$.

PROOF. First we show $[\eta] = 1$. Suppose not, then we may assume $\eta = \omega$ and $\zeta$ satisfies

$$\zeta_n = \zeta_{n^{\tau_3}} \quad \text{and} \quad \omega(n_1, n_2)\omega(n_1^{\tau_3}, n_2^{\tau_3}) = \zeta_{n_1 n_2}\overline{\zeta_{n_1}\zeta_{n_2}}.$$

However, direct computation shows that there is no such $\zeta$ and we get $[\eta] = 1$. Thus, we get

$$H^2((N, \mathbf{Z}_2), \mathbf{T}) \cong H^2(\mathbf{Z}_2, Hom(N, \mathbf{T})) \cong H^2(\mathbf{Z}_2, \mathbf{Z}_2 \times \mathbf{Z}_2),$$

where the action of $\mathbf{Z}_2$ on $Hom(N, \mathbf{T})$ is trivial. Therefore, we get

$$H^2((N, \mathbf{Z}_2), \mathbf{T}) \cong \{[(1, 1)], [(1, \chi_1)], [(1, \chi_2)], [(1, \chi_1\chi_2)]\}.$$

Since $\eta = 1$, the cocycle $\zeta$ describes the extension

$$1 \longrightarrow Hom(N, \mathbf{T}) \longrightarrow G(\mathcal{A}) \longrightarrow \mathbf{Z}_2 \longrightarrow 1,$$

and we get the result for $G(\mathcal{A})$. Note that $N^{\tau_3} = \langle(1, 0)\rangle$ and $\chi_2|_{N^{\tau_3}}$ is trivial. Thus, in case (i) and (iii), $G(\hat{\mathcal{A}})$ is the trivial extension of $N^{\tau_3}$ by $\hat{\mathbf{Z}}_2$. We set

$$\begin{cases} \xi(x_1, x_2) = 1 & \text{for } x_1 \text{ even,} \\ \xi(x_1, x_2) = i & \text{for } x_1 \text{ odd.} \end{cases}$$

Then, in case (ii) and (iv) $\eta'(n_1, n_2) = \xi(n_1)\xi(n_2)\overline{\xi(n_1 n_2)}$ describes the exact sequence

$$1 \longrightarrow \hat{\mathbf{Z}}_2 \longrightarrow G(\hat{\mathcal{A}}) \longrightarrow N^{\tau_3} \longrightarrow 1.$$

Therefore, we get the statement for $G(\hat{\mathcal{A}})$. □

## 3. Classification

At first the difficult cases (see the beginning of the chapter) for classification are handled as Proposition 11.15 and Proposition 11.16. To prove them we begin with two lemmas.

LEMMA 11.13. *Let* $(\alpha, \beta)$ *be an outer near action of* $(\mathbf{Z}_2 \times \mathbf{Z}_2, \mathbf{Z}_2)$ *on* $\mathcal{R}$ *with an invariant* $(1, \zeta)$, *and we set*

$$\mathcal{P} = \mathcal{R} \rtimes_\alpha (\mathbf{Z}_2 \times \mathbf{Z}_2) \supseteq \mathcal{Q} = \mathcal{R}^\beta.$$

(*i*) *If the action of* $\mathbf{Z}_2$ *on* $\mathbf{Z}_2 \times \mathbf{Z}_2$ *is not trivial, then* $\mathcal{Q}$ *is the fixed-point subalgebra of* $\mathcal{P}$ *under a* $D_8$ *action.*
(*ii*) *If the action of* $\mathbf{Z}_2$ *on* $\mathbf{Z}_2 \times \mathbf{Z}_2$ *is trivial, then* $\mathcal{Q}$ *is the fixed-point subalgebra of* $\mathcal{P}$ *under an abelian group action.*

PROOF. In fact, $\mathcal{A}$ is co-commutative since $N$ is abelian with $\eta = 1$, and the intrinsic group $G(\mathcal{A})$ can be easily computed based on Theorem 6.1. (see the second remark after Theorem 6.2) □

LEMMA 11.14. *Let $\rho : \mathcal{Q} \hookrightarrow \mathcal{P}$ be the inclusion map for a depth 2 inclusion $\mathcal{P} \supseteq \mathcal{Q}$, and*

$$[\rho\bar{\rho}] = \oplus_{g\in G(\mathcal{A})}[\tau_g] \oplus (\oplus_{i\in I} d(\pi_i)[\pi_i])$$

*be the irreducible decomposition with $d(\pi_i) \geq 2$. For*

$$G_i = \{g \in G(\mathcal{A});\ [\tau_g\pi_i] = [\pi_i]\}$$

*we have the following:*
*(i) If $\#G_i = d(\pi_i)^2$, then $H^2(G_i, \mathbf{T})$ is not trivial.*
*(ii) We set $I_0 = \{i \in I;\ G_i = G_{\bar{i}} = G(\mathcal{A}),\ d(\pi_i)^2 = \#G(\mathcal{A})\}$. Then, we have*

$$\#I_0 \leq \#H^2(G_i, \mathbf{T}) - 1.$$

PROOF. (i) We take representatives $\{\tau_g\}_{g\in G(\mathcal{A})}$ satisfying $\tau_g \cdot \rho = \rho$, and take unitaries $\{u_g^i\}_{g\in G_i}$ satisfying $Ad(u_g^i) \cdot \tau_g \cdot \pi_i = \pi_i$. We set $\tau_g^i = Ad(u_g^i) \cdot \tau_g$, which is an action of $G_i$. Let $\omega^i \in Z^2(G_i, \mathbf{T})$ be the cocycle carried by $\{u_g^i\}$. We claim that $[\omega_i] \neq 1$. To show this by contradiction, let us assume that $\omega^i$ is a coboundary for some $i \in I$ and $\#G_i = d(\pi_i)^2$. Then, we may assume that $u_g^i = 1$ and $\pi_i(\mathcal{P})$ is the fixed-point algebra $\mathcal{P}^{(G_i,\tau)}$ by choosing an appropriate representative for $\pi_i$. Let $\rho_1 : \mathcal{P}^{(G_i,\tau)} \hookrightarrow \mathcal{P}$, and $\rho_2 : \mathcal{Q} \hookrightarrow \mathcal{P}^{(G_i,\tau)}$ be the inclusion maps. Then, there exists an isomorphism $\rho_0$ from $\mathcal{P}$ to $\mathcal{P}^{(G_i,\tau)}$ (considered as a $\mathcal{P}$-$\mathcal{P}^{(G_i,\tau)}$ sector) satisfying $[\pi_i] = [\rho_1\rho_0]$. On one hand, we have $[\pi_{\bar{i}}][\rho] = d(\pi_i)[\rho]$ by the Frobenius reciprocity. On the other hand, $[\pi_{\bar{i}}][\rho] = [\bar{\rho}_0\bar{\rho}_1\rho_1][\rho_2]$ contains the irreducible sector $[\bar{\rho}_0\rho_2]$. Thus, we get $[\rho] = [\bar{\rho}_0\rho_2]$ and $d(\rho) = d(\rho_0)d(\rho_2) = d(\rho_2)$, which is a contradiction.
(ii) Assume $i, j \in I_0$ and $[\omega^i] = [\omega^j]$. Then, $\overline{\pi}_j\pi_i$ contains an automorphism and there exists $g \in G(\mathcal{A})$ satisfying $[\pi_i] = [\pi_j\tau_g]$. However, since $G_{\bar{j}} = G(\mathcal{A})$, we get $[\pi_j\tau_g] = [\pi_j]$ and $i = j$. This means that $\{[\omega_i]\}_{i\in I_0}$ are distinct non-trivial classes in $H^2(G(\mathcal{A}), \mathbf{T})$ and we get the result. □

PROPOSITION 11.15. *Let $\mathcal{A}$ be a non-trivial Kac algebra of dimension 16. If $G(\mathcal{A})$ is a non-commutative group of order 8, then $G(\mathcal{A})$ must be isomorphic to $D_8$. Such a Kac algebra is classified by $D_8 \rtimes \mathbf{Z}_2$ and an element $[(\eta, \zeta)]$ in*

$$H^2((D_8, \mathbf{Z}_2), \mathbf{T})/\sim$$

*satisfying*

$$\mathcal{A} = \mathbf{C}(D_8) \oplus \mathbf{C}_{\eta_1}(D_8),$$
$$\hat{\mathcal{A}} = \ell^\infty(D_8) \rtimes_\zeta \mathbf{Z}_2 \cong \mathbf{C}^8 \oplus M_2(\mathbf{C}) \oplus M_2(\mathbf{C}),$$
$$G(\mathcal{A}) \cong D_8.$$

*Moreover, there are two isomorphism classes of such Kac algebras, and in both cases we have*

$$\mathcal{A} \cong \mathbf{C}^4 \oplus M_2(\mathbf{C}) \oplus M_2(\mathbf{C}) \oplus M_2(\mathbf{C}),$$
$$G(\hat{\mathcal{A}}) \cong \mathbf{Z}_2 \times \mathbf{Z}_2.$$

PROOF. Let $\mathcal{P} \supseteq \mathcal{Q}$ be a depth 2 inclusion of factors corresponding to $\mathcal{A}$, and $\rho : \mathcal{Q} \hookrightarrow \mathcal{P}$ be the inclusion map. Then, the irreducible decomposition of $\rho\bar{\rho}$ is

$$[\rho\bar{\rho}] = \oplus_{g\in G(\mathcal{A})}[\tau_g] \oplus 2[\pi_1] \oplus 2[\pi_2],$$

with $d(\pi_1) = d(\pi_2) = 2$. We set

$$G_i = \{g \in G(\mathcal{A});\ [\tau_g\pi_i] = [\pi_i]\}, \quad i = 1, 2.$$

Then, it is easy to see $\#G_i = 4$. Thanks to Lemma 11.14,(i), $H^2(G_i, \mathbf{T})$ is not trivial so that we must have $G_i \cong \mathbf{Z}_2 \times \mathbf{Z}_2$. Since the non-abelian group $G(\mathcal{A})$ of order 8 contains a subgroup isomorphic to $\mathbf{Z}_2 \times \mathbf{Z}_2$, we conclude $G(\mathcal{A}) \cong D_8$.

Let $\mathcal{L} = \mathcal{P}^{G(\mathcal{A})}$, and $\rho_1 : \mathcal{L} \hookrightarrow \mathcal{P}$ and $\rho_2 : \mathcal{Q} \hookrightarrow \mathcal{L}$ be the inclusion maps. Then, the irreducible decomposition of $\bar{\rho}_1\rho_1$ is

$$[\bar{\rho}_1\rho_1] = [\theta_g]_{g\in K} \oplus 2[\sigma]$$

with $K = Hom(D_8, \mathbf{T}) \cong \mathbf{Z}_2 \times \mathbf{Z}_2$. We choose representatives $\{\theta_g\}_{g\in K}$, which forms a group, such that $Ad(\lambda_g)|_{\mathcal{L}} = \theta_g$ with a unitary representation $\{\lambda_g\}_{g\in K}$ in $\mathcal{P}$. We set

$$\mathcal{Q} \subseteq \mathcal{L} \subseteq \mathcal{R} = \langle \mathcal{L}, \{\lambda_g\}_{g\in K}\rangle \ (\cong \mathcal{L} \rtimes_\theta K) \subseteq \mathcal{P}$$

Because of $[\mathcal{L} : \mathcal{Q}] = 2$ there exists a period two automorphism $\mu$ satisfying $\mathcal{Q} = \mathcal{L}^\mu$ and we have $[\mu\bar{\rho}_1\rho_1] = [\bar{\rho}_1\rho_1\mu]$ (since $\mathcal{P} \supseteq \mathcal{Q}$ is of depth 2). Thus, $[\mu]$ normalizes $\{[\theta_g]\}_{g\in K}$ and $[\mu\sigma] = [\sigma\mu]$. Therefore, $(\theta, \mu)$ is a near action (on $\mathcal{L}$) of $K \rtimes \mathbf{Z}_2$. We use the notation $[\theta_{g^\mu}] = [\mu\theta_g\mu]$ in the rest of the proof.

We denote by $(\eta, \zeta)$ the invariant of the near action $(\theta, \mu)$, and claim that $\eta$ is trivial. Since $[\mu\sigma] = [\sigma\mu]$ there exists a unitary $u$ satisfying $Ad(u) \cdot \mu \cdot \sigma = \sigma \cdot \mu$, and $u\mu(u)$ is a scalar (since $\mu$ is of period 2). After changing a phase of $u$, we may assume $u\mu(u) = 1$, i.e., $\{1, u\}$ is a $\mu$-cocycle and there exists a unitary $v$ satisfying $u = v^*\mu(v)$ thanks to Connes' theorem ([**4**]). Then, replacing $\sigma$ by $Ad(v) \cdot \sigma$, we may and do assume $\mu\cdot\sigma = \sigma\cdot\mu$ from the beginning. Since $[\theta_g\sigma] = [\sigma]$ $(g \in K)$, there exist unitaries $w_g$ satisfying $Ad(w_g) \cdot \theta_g \cdot \sigma = \sigma$. This shows that $\theta'_g = Ad(w_g) \cdot \theta_g$ also give rise to a $\mathbf{Z}_2 \times \mathbf{Z}_2$ action. Note

$$\mu \cdot \theta'_g \cdot \mu \cdot \sigma = \mu \cdot \theta'_g \cdot \sigma \cdot \mu = \mu \cdot \sigma \cdot \mu = \sigma,$$

implying $\mu\cdot\theta'_g\cdot\mu = \theta'_{g^\mu}$. We thus have seen that $(\theta', \mu)$ gives rise to a genuine action of the semi-direct product group $(\mathbf{Z}_2 \times \mathbf{Z}_2) \rtimes \mathbf{Z}_2$. We note $(\theta, \mu) = (Ad(w_g^*) \cdot \theta'_g, \mu)$ so that $(\theta, \mu)$ has trivial $\eta$ thanks to Lemma 2.4.

We have shown $\eta = 1$, and hence $\mathcal{Q}$ is the fixed-point subalgebra of $\mathcal{R}$ either under a $D_8$ action or under an abelian group action thanks to Lemma 11.13. Actually the former case occurs. Indeed, if the group were abelian, $G(\mathcal{A})$ would be abelian too as was seen in §1. Thus, we conclude $\mathcal{Q} = \mathcal{R}^{(\alpha, D_8)}$ with an outer action $\alpha : D_8 \longrightarrow Aut(\mathcal{R})$. On the other hand, since $[\mathcal{P} : \mathcal{R}] = 2$, there exists a period two automorphism $\beta$ on $\mathcal{R}$ such that $\mathcal{P} = \mathcal{R} \rtimes_\beta \mathbf{Z}_2$. Therefore, our Kac algebras are classified by the data described in the statement, and Lemma 11.11 (see cases (v) and (vi)) gives actual classification. □

PROPOSITION 11.16. *Let $\mathcal{A}$ be a non-trivial Kac algebra of dimension 16. If $\#G(\mathcal{A}) = 4$, then $G(\mathcal{A})$ is isomorphic to $\mathbf{Z}_2 \times \mathbf{Z}_2$. Such a Kac algebra is classified by $D_8 \rtimes \mathbf{Z}_2$ and an element $[(\eta, \zeta)]$ in $H^2((D_8, \mathbf{Z}_2), \mathbf{T})/\sim$ satisfying*

$$\mathcal{A} = \ell^\infty(D_8) \rtimes_\zeta \mathbf{Z}_2,$$
$$\hat{\mathcal{A}} = \mathbf{C}(D_8) \oplus \mathbf{C}_{\eta_1}(D_8) \cong \mathbf{C}^4 \oplus M_2(\mathbf{C}) \oplus M_2(\mathbf{C}) \oplus M_2(\mathbf{C}),$$
$$G(\mathcal{A}) \cong \mathbf{Z}_2 \times \mathbf{Z}_2.$$

*There are three isomorphism classes of such Kac algebras. In one case, $\mathcal{A}$ is self-dual. The other two Kac algebras are dual to those in the previous proposition and* *we have*

$$\mathcal{A} \cong \mathbf{C}^8 \oplus M_2(\mathbf{C}) \oplus M_2(\mathbf{C}),$$
$$G(\hat{\mathcal{A}}) \cong D_8.$$

PROOF. Let $\mathcal{P} \supseteq \mathcal{Q}$ be a depth 2 inclusion of factors corresponding to $\mathcal{A}$, and $\rho : \mathcal{Q} \hookrightarrow \mathcal{P}$ be the inclusion map. Then, the irreducible decomposition of $\rho\bar{\rho}$ is

$$[\rho\bar{\rho}] = \oplus_{g \in G(\mathcal{A})}[\tau_g] \oplus 2[\pi_1] \oplus 2[\pi_2] \oplus 2[\pi_3]$$

with $d(\pi_1) = d(\pi_2) = d(\pi_3) = 2$. Let

$$G_i = \{g \in G(\mathcal{A});\ [\tau_g \pi_i] = [\pi_i]\}, \quad i = 1, 2, 3.$$

Then, it is easy to see that $G_i = G(\mathcal{A})$ for some $i$ and we may assume $G_3 = G(\mathcal{A})$. Note that $G(\mathcal{A}) \cong \mathbf{Z}_4$ or $\mathbf{Z}_2 \times \mathbf{Z}_2$ and hence $G(\mathcal{A})$ is abelian. Since $\pi_3(\mathcal{P})$ is the fixed-point algebra of $\mathcal{P}$ under an abelian group action, so is $\overline{\pi}_3(\mathcal{P})$. This means that we also have $G_{\bar{3}} = G(\mathcal{A})$. Lemma 11.14,(i) implies that $H^2(G(\mathcal{A}), \mathbf{T})$ is not trivial, and so we must have $G(\mathcal{A}) \cong \mathbf{Z}_2 \times \mathbf{Z}_2$. Lemma 11.14,(ii) and $H^2(G(\mathcal{A}), \mathbf{T}) = \mathbf{Z}_2$ imply $\#G_1 = \#G_2 = 2$ (and $[\pi_3]$ is self-dual). Therefore, $\{\tau_g\}_{g \in G(\mathcal{A})} \cup \{\pi_3\}$ is closed under conjugation and irreducible decomposition of products. We emphasize that this system is canonically associated with the inclusion $\mathcal{P} \supseteq \mathcal{Q}$.

Let $\mathcal{L}$ be the unique intermediate subfactor of $\mathcal{P} \supseteq \mathcal{Q}$ such that

$$[\rho_1 \bar{\rho}_1] = \oplus_{g \in G(\mathcal{A})}[\tau_g] \oplus 2[\pi_3],$$

where $\rho_1 : \mathcal{L} \hookrightarrow \mathcal{P}$ is the inclusion map. There are 3 possibilities for the inclusion $\mathcal{P} \supseteq \mathcal{L}$ of depth 2:

$$\begin{cases} \text{(i)} & \text{crossed product by Kac-Paljutkin's 8-dimensional Kac algebra,} \\ \text{(ii)} & \text{crossed product by the quaternion group } Q_8, \\ \text{(iii)} & \text{crossed product by the dihedral group } D_8. \end{cases}$$

Firstly we show that (i) does not occur. Suppose that (i) holds. The same argument as in the proof of the previous proposition shows that we would have an intermediate subfactor $\mathcal{R}$ between $\mathcal{P} \supseteq \mathcal{L}$ and a near action $(\alpha, \beta)$ of $D_8 \rtimes \mathbf{Z}_2$ on $\mathcal{R}$ such that $\mathcal{P} = \mathcal{R} \rtimes_\beta \mathbf{Z}_2$ and $\mathcal{Q} = \mathcal{R}^\alpha$. Note that we would have a $\mathbf{Z}_2$-invariant subgroup $K \subseteq D_8$, which is isomorphic to $\mathbf{Z}_2 \times \mathbf{Z}_2$, satisfying $\mathcal{L} = \mathcal{R}^{(K,\alpha)}$. Thus, our Kac algebras would be classified by $D_8 \rtimes \mathbf{Z}_2$ and an element $[(\eta, \zeta)]$ in $H^2((D_8, \mathbf{Z}_2), \mathbf{T})/\sim$ satisfying $\mathcal{A} = \mathbf{C}(D_8) \oplus \mathbf{C}_{\eta_1}(D_8)$ and

$$\hat{\mathcal{A}} = \ell^\infty(D_8) \rtimes_\zeta \mathbf{Z}_2 \cong \mathbf{C}^4 \oplus M_2(\mathbf{C}) \oplus M_2(\mathbf{C}) \oplus M_2(\mathbf{C}), \quad G(\mathcal{A}) \cong \mathbf{Z}_2 \times \mathbf{Z}_2,$$

with the following additional property: there exists a $\mathbf{Z}_2$-invariant subgroup $K \subseteq D_8$ isomorphic to $\mathbf{Z}_2 \times \mathbf{Z}_2$ such that $(\eta, \zeta)$ restricted to $K \rtimes \mathbf{Z}_2$ gives rise to Kac-Paljutkin's 8-dimensional Kac algebra. However, Lemma 11.10 and Lemma 11.11 show that there is no such Kac algebra. Secondly, we show that case (ii) does not occur either. Suppose $\mathcal{P} = \mathcal{L} \rtimes_\theta Q_8$. Because of $[\mathcal{L} : \mathcal{Q}] = 2$ there would exist a period two automorphism $\mu$ satisfying $\mathcal{Q} = \mathcal{L}^\mu$. Thus, $\mathcal{P} \supseteq \mathcal{Q}$ would arise from the near action $(\theta, \mu)$. Then, $H^2(Q_8, \mathbf{T}) = \{1\}$ implies

$$\hat{\mathcal{A}} \cong \mathbf{C}(Q_8) \oplus \mathbf{C}(Q_8) \cong \mathbf{C}^8 \oplus M_2(\mathbf{C}) \oplus M_2(\mathbf{C}),$$

which is a contradiction.

Therefore, case (iii) occurs, and the actual classification follows from Lemma 11.10 (non-trivial cocycle case) and Lemma 11.11 (see cases (v) and (vi)). □

We are now ready to classify Kac algebras of dimension 16. At first we recall what was mentioned at the beginning of the chapter. In the theorem below case (vi) comes from Proposition 11.15 while cases (v) and (vii) come from Proposition 11.16. On the other hand, the situation described in Proposition 11.3 gives us the first four cases (i), (ii), (iii), and (iv), and the actual classification in these cases follows from the discussions in §1. More precisely, cases (i), (ii), (iii), (iv) come from

Lemma 11.4,(i),(iii)
Lemma 11.4,(ii),(iv)
Lemma 11.5,(i),(ii), Lemma 11.6,(ii), and Lemma 11.7,(ii),(iii)
Lemma 11.6,(i) and Lemma 11.7,(i)

respectively. In this way we get $2+2+5+2$ Kac algebras.

THEOREM 11.17. *Non-trivial Kac algebras of dimension* 16 *are classified as follows*:
*(i) There are two self-dual Kac algebras with algebra structure and the intrinsic groups*

$$\mathcal{A} \cong \hat{\mathcal{A}} \cong \mathbf{C}^8 \oplus M_2(\mathbf{C}) \oplus M_2(\mathbf{C}),$$
$$G(\mathcal{A}) \cong G(\hat{\mathcal{A}}) \cong \mathbf{Z}_2 \times \mathbf{Z}_2 \times \mathbf{Z}_2.$$

*(ii) There are two Kac algebras with algebra structure and the intrinsic groups*

$$\mathcal{A} \cong \hat{\mathcal{A}} \cong \mathbf{C}^8 \oplus M_2(\mathbf{C}) \oplus M_2(\mathbf{C}),$$
$$G(\mathcal{A}) \cong \mathbf{Z}_2 \times \mathbf{Z}_2 \times \mathbf{Z}_2, \quad G(\hat{\mathcal{A}}) \cong \mathbf{Z}_4 \times \mathbf{Z}_2.$$

*(iii) There are five Kac algebras with algebra structure and the intrinsic groups*

$$\mathcal{A} \cong \hat{\mathcal{A}} \cong \mathbf{C}^8 \oplus M_2(\mathbf{C}) \oplus M_2(\mathbf{C}),$$
$$G(\mathcal{A}) \cong G(\hat{\mathcal{A}}) \cong \mathbf{Z}_4 \times \mathbf{Z}_2.$$

*(iv) There are two Kac algebras with algebra structure and the intrinsic groups*

$$\mathcal{A} \cong \hat{\mathcal{A}} \cong \mathbf{C}^8 \oplus M_2(\mathbf{C}) \oplus M_2(\mathbf{C}),$$
$$G(\mathcal{A}) \cong \mathbf{Z}_4 \times \mathbf{Z}_2, \quad G(\hat{\mathcal{A}}) \cong \mathbf{Z}_2 \times \mathbf{Z}_2 \times \mathbf{Z}_2.$$

*(v) There are two Kac algebras with algebra structure and the intrinsic groups*

$$\mathcal{A} \cong \mathbf{C}^8 \oplus M_2(\mathbf{C}) \oplus M_2(\mathbf{C}),$$
$$\hat{\mathcal{A}} \cong \mathbf{C}^4 \oplus M_2(\mathbf{C}) \oplus M_2(\mathbf{C}) \oplus M_2(\mathbf{C}),$$
$$G(\mathcal{A}) \cong \mathbf{Z}_2 \times \mathbf{Z}_2, \quad G(\hat{\mathcal{A}}) \cong D_8.$$

*(vi) There are two Kac algebras with algebra structure and the intrinsic groups*

$$\mathcal{A} \cong \mathbf{C}^4 \oplus M_2(\mathbf{C}) \oplus M_2(\mathbf{C}) \oplus M_2(\mathbf{C}),$$
$$\hat{\mathcal{A}} \cong \mathbf{C}^8 \oplus M_2(\mathbf{C}) \oplus M_2(\mathbf{C}),$$
$$G(\mathcal{A}) \cong D_8, \quad G(\hat{\mathcal{A}}) \cong \mathbf{Z}_2 \times \mathbf{Z}_2.$$

*(vii) There is one self-dual Kac algebra with algebra structure and the intrinsic groups*

$$\mathcal{A} \cong \hat{\mathcal{A}} \cong \mathbf{C}^4 \oplus M_2(\mathbf{C}) \oplus M_2(\mathbf{C}) \oplus M_2(\mathbf{C}),$$
$$G(\mathcal{A}) \cong G(\hat{\mathcal{A}}) \cong \mathbf{Z}_2 \times \mathbf{Z}_2.$$

Our proof shows that every non-trivial Kac algebras of dimension 16 arises from one of the following semi-direct product groups:

$$(\mathbf{Z}_2 \times \mathbf{Z}_2 \times \mathbf{Z}_2) \rtimes \mathbf{Z}_2, \quad (\mathbf{Z}_4 \times \mathbf{Z}_2) \rtimes \mathbf{Z}_2, \quad D_8 \rtimes \mathbf{Z}_2.$$

It is also possible to express some of Kac algebras in the above list as (twisted) crossed products of the Kac-Paljutkin algebra by some $\mathbf{Z}_2$-actions, which will be discussed in §3 of Chapter 13 (see Theorem 13.17 for details). Namely, the two Kac algebras in Theorem 11.17,(vi) are (twisted) crossed products of the Kac-Paljutkin algebra. On the other hand, one Kac algebra in Theorem 11.17,(i) is the crossed product relative to the trivial $\mathbf{Z}_2$ action, i.e., the tensor product of the Kac-Paljutkin algebra and $\mathbf{C}(\mathbf{Z}_2)$.

CHAPTER 12

# Group extensions of general Kac algebras

When $(\alpha, \beta)$ is an outer action of a product group $G = G_1 \cdot G_2$ on a factor $\mathcal{R}$, the inclusion $\mathcal{R} \rtimes_\alpha G_1 \supseteq \mathcal{R}^{(G_2,\beta)}$ is irreducible and of depth 2. So far we have been investigating the resulting Kac algebra (mainly) in the semi-direct product case $G = N \rtimes H$. In this chapter we will deal with a more general case where $\mathcal{R} \supseteq \mathcal{R}^{(N,\alpha)}$ is replaced by an arbitrary irreducible inclusion of depth 2.

We assume

$$\begin{cases} \mathcal{R} \supseteq \rho_1(\mathcal{R}) \text{ is an irreducible depth 2 inclusion of factors,} \\ \text{an outer action } \alpha : G \longrightarrow Aut(\mathcal{R}) \text{ of a finite group } G \text{ is given.} \end{cases}$$

Throughout the chapter we require that the canonical endomorphism $\gamma = \rho_1\bar{\rho}_1$ satisfies

$$\begin{cases} [\gamma\alpha_g] = [\alpha_g\gamma] \text{ for every } g \in G, \\ [\alpha_g]\ (g \neq e) \text{ does not appear in } \gamma. \end{cases}$$

We claim that $\mathcal{R} \rtimes_\alpha G \supseteq \rho_1(\mathcal{R})$ is an irreducible inclusion of depth 2 again. In fact, let $\rho_2$ be an endomorphism of $\mathcal{R}$ whose image is the fixed-point algebra $\rho_2(\mathcal{R}) = \mathcal{R}^\alpha$, and we set $\sigma = \bar{\rho}_2\rho_1$. What we have to consider here is the inclusion $\mathcal{R} \supseteq \sigma(\mathcal{R})$. The irreducibility comes from the above second requirement. On the other hand, the Frobenius reciprocity shows

$$\dim(\sigma\bar{\sigma}, \sigma\bar{\sigma}) = \dim(\rho_2\bar{\rho}_2\gamma, \gamma\rho_2\bar{\rho}_2)$$

Hence, $\rho_2\bar{\rho}_2 = \oplus_{g\in G}[\alpha_g]$ and the above first requirement show

$$\dim(\sigma\bar{\sigma}, \sigma\bar{\sigma}) = \#G \times \dim(\gamma, \gamma) = \#G \times (d\rho_1)^2 = (d\sigma)^2,$$

meaning that $\mathcal{R} \supseteq \sigma(\mathcal{R})$ is of depth 2. (See [**61**] for precise information.)

In this chapter, the Kac algebra arising from the depth 2 inclusion $\mathcal{R} \supseteq \sigma(\mathcal{R})$ will be investigated.

## 1. Description of cocycles

Here, we deal with the Kac algebra corresponding to the inclusion $\mathcal{R} \supseteq \rho_1(\mathcal{R})$ and various cocycles on this Kac algebra. We set

$$\mathcal{H} = (\rho_1, \rho_1\bar{\rho}_1\rho_1),$$

which is the $(d\rho_1)^2$-dimensional Hilbert space on which our Kac algebra acts. Let $\mathcal{O}_\mathcal{H}$ be the Cuntz algebra generated by $\mathcal{H}$. For a unitary $v \in \mathcal{O}_\mathcal{H}$, we denote by $\lambda_v$ the endomorphism of $\mathcal{O}_\mathcal{H}$ defined by

$$\lambda(S) = vS \quad (S \in \mathcal{H}).$$

Let $F$ be the flip of $\mathcal{H} \otimes \mathcal{H}$. The shift $\lambda_F$ will be denoted by $\sigma_\mathcal{H}$. The restriction of $\gamma$ $(= \rho_1\bar{\rho}_1)$ to $\mathcal{O}_\mathcal{H}$ determines a multiplicative unitary $V$ through the relation

$\gamma|_{\mathcal{O}_{\mathcal{H}}} = \lambda_{VF}$. Let $\mathcal{A}$ and $\hat{\mathcal{A}}$ be the Kac algebra and the dual Kac algebra associated with $V$. Recall that we have

$$V \in \hat{\mathcal{A}} \otimes \mathcal{A} \quad \text{and} \quad (\gamma, \gamma) = \hat{\mathcal{A}}'.$$

Let $\Gamma$ and $\hat{\Gamma}$ be the coproducts of $\mathcal{A}$ and $\hat{\mathcal{A}}$ respectively. Then, $\mathcal{A}, \hat{\mathcal{A}}$ act standardly on $\mathcal{H}$, and let $J$ and $\hat{J}$ be the modular conjugation of $\mathcal{A}$ and $\hat{\mathcal{A}}$. For $x \in \mathcal{A}$, $\hat{x} \in \hat{\mathcal{A}}$, and $\hat{x}' \in \hat{\mathcal{A}}'$, we have

$$\Gamma(x) = V(x \otimes 1)V^*, \tag{12.1}$$

$$\hat{\Gamma}(\hat{x}) = V^*(1 \otimes \hat{x})V, \tag{12.2}$$

$$(1 \otimes U)F\hat{\Gamma}(U\hat{x}'U)F(1 \otimes U) = V(1 \otimes \hat{x}')V^*, \tag{12.3}$$

where $U = J\hat{J}$. An action $\pi : \mathcal{R} \longrightarrow \mathcal{R} \otimes \hat{\mathcal{A}}$ of $\hat{\mathcal{A}}$ on $\mathcal{R}$ is given by

$$\pi(x) = \sum_{i,j} S_i^* \gamma(x) S_j \otimes e_{ij},$$

where $\{S_i\}$ is an orthonormal basis of $\mathcal{H}$ and $e_{ij} = S_i S_j^*$.

Choose and fix a unitary $u(g) \in \mathcal{R}$ satisfying

$$\gamma \cdot \alpha_g = Ad(u(g)) \cdot \alpha_g \cdot \gamma \quad (\text{for each } g \in G),$$

and we set

$$\zeta'(g,h) = u(g)\alpha_g(u(h))u(gh)^* \in (\gamma, \gamma) = \hat{\mathcal{A}}' \subseteq B(\mathcal{H}).$$

Then, $\alpha_g^u = Ad(u(g)) \cdot \alpha_g$ is a cocycle action of $G$ with the cocycle $\zeta'$ preserving $\hat{\mathcal{A}}'$ globally:

$$\begin{cases} \alpha_g^u \cdot \alpha_h^u = Ad(\zeta'(g,h)) \cdot \alpha_{gh}^u, \\ \alpha_g^u(\zeta'(h,k))\zeta'(g,hk) = \zeta'(g,h)\zeta'(gh,k). \end{cases}$$

We set

$$\theta'_g = \alpha_g^u|_{\hat{\mathcal{A}}'} \in Aut(\hat{\mathcal{A}}').$$

Let $v(g)$ be the canonical implementation (see [**67**]) of $\theta'_g$ ($\theta'_g = Ad(v(g))$), and we set

$$\theta_g = Ad(v(g))|_{\hat{\mathcal{A}}} \in Aut(\hat{\mathcal{A}}).$$

Note that $\theta$ is a cocycle action of $G$ on $\hat{\mathcal{A}}$ with the cocycle $\zeta$ defined by

$$\zeta(g,h) = \hat{J}\zeta'(g,h)\hat{J} \ (\in \hat{\mathcal{A}}).$$

Since the canonical implementation of the inner automorphism $Ad(\zeta(g,h))$ is the unitary $\zeta'(g,h)\zeta(g,h)$, we have

$$v(g)v(h) = \zeta(g,h)\zeta'(g,h)v(gh).$$

In the discussions so far, the choice of $u(g)$ is of course not unique. In fact, one may replace $u(g)$ by $\xi'(g)u(g)$ with a unitary $\xi'(g)$ in $\hat{\mathcal{A}}'$. This amounts to change from $(\theta_g, \zeta(g,h))$ to

$$(Ad(\xi(g)) \cdot \theta_g, \ \xi(g)\theta_g(\xi(h))\zeta(g,h)\xi(gh)^*)$$

with $\xi(g) = \hat{J}\xi'(g)\hat{J} \in \mathcal{A}$. In fact, we compute

$$\begin{aligned} Ad(\xi(g)) \cdot \theta_g \cdot Ad(\xi(h)) \cdot \theta_h &= Ad\left(\xi(g)\theta_g(\xi(h))\right) \cdot \theta_g \cdot \theta_h \\ &= Ad\left(\xi(g)\theta_g(\xi(h))\zeta(g,h)\right) \cdot \theta_{gh}. \end{aligned}$$

Note that the above gives rise to an equivalent cocycle action.

Let $\{S_i\}$ be an orthonormal basis for $\mathcal{H}$. The definition of $\alpha_g^u$ means

$$\alpha_g^u \cdot \gamma = \gamma \cdot \alpha_g.$$

By making use of this, we compute

$$\begin{aligned}\alpha_g^u(S_j)\gamma\alpha_g(x) &= \alpha_g^u(S_j)\alpha_g^u\gamma(x) = \alpha_g^u(S_j\gamma(x))\\ &= \alpha_g^u(\gamma^2(x)S_j) = \alpha_g^u\gamma^2(x)\alpha_g^u(S_j) = \gamma\alpha_g\gamma(x)\alpha_g^u(S_j).\end{aligned}$$

Since

$$\gamma \cdot \alpha_g \cdot \gamma = \gamma \cdot Ad(u(g)^*) \cdot \alpha_g^u \cdot \gamma = Ad(\gamma(u(g)^*)) \cdot \gamma^2 \cdot \alpha_g,$$

from the above we have

$$\gamma(u(g))\alpha_g^u(S_j)\gamma\alpha_g(x) = \gamma^2\alpha_g(x)\gamma(u(g))\alpha_g^u(S_j).$$

Therefore, by hitting $S_i^*$ from the left we get

$$\begin{aligned}S_i^*\gamma(u(g))\alpha_g^u(S_j)\gamma\alpha_g(x) &= S_i^*\gamma^2\alpha_g(x)\gamma(u(g))\alpha_g^u(S_j)\\ &= \gamma\alpha_g(x)S_i^*\gamma(u(g))\alpha_g^u(S_j),\end{aligned}$$

showing $S_i^*\gamma(u(g))\alpha_g^u(S_j) \in (\gamma, \gamma) = \hat{\mathcal{A}}'$. For each $g \in G$ we set

$$w(g) = \sum_j \gamma(u(g))\alpha_g^u(S_j)S_j^* \quad \left( = \sum_{ij} S_iS_i^*\gamma(u(g))\alpha_g^u(S_j)S_j^* \right).$$

The operator $w(g)$ is obviously a unitary in $B(\mathcal{H}) \otimes B(\mathcal{H})$ not depending upon the choice of a basis $\{S_i\}$. The preceding computation actually means $w(g) \in B(\mathcal{H}) \otimes \hat{\mathcal{A}}'$.

LEMMA 12.1. *(i) The unitary $w(g)$ in $B(\mathcal{H}) \otimes \hat{\mathcal{A}}'$ satisfies*

$$\alpha_g^u(S) = \gamma(u(g)^*)w(g)S \quad \textit{for each } S \in \mathcal{H}.$$

*(ii) The unitary $w_0(g) = w(g)v(g)^*$ (with the canonical implementation $v(g)$ of $\alpha_g^u$) belongs to $\hat{\mathcal{A}} \otimes \hat{\mathcal{A}}'$.*

PROOF. It remains to show (ii), and we claim $\alpha_g^u(x)w(g) = w(g)x$ for $x \in \hat{\mathcal{A}}'$. To show the claim we may and do assume that $x = u$ is a unitary. Since $\{u^*S_i\}$ is also a basis for $\mathcal{H}$, we have

$$w(g) = \sum_j \gamma(u(g))\alpha_g^u(u^*S_j)S_j^*u,$$

and the claim follows since $\alpha_g^u(u) \in (\gamma, \gamma) = \hat{\mathcal{A}}'$ and $\gamma(u(g))$ commutes with $\alpha_g^u(u)$. Therefore, we conclude

$$xw(g) = w(g)\left(\alpha_g^u\right)^{-1}(x) = w(g)v(g)^*xv(g) \ (x \in \hat{\mathcal{A}}'),$$

and (ii) is proved. □

We define a unitary $W_g$ $(g \in G)$ by

$$W_g = \sum_{i,j} \alpha_g(S_i^*v(g))u(g)^*S_j \otimes e_{ij}.$$

Lemma 10.1,(i) applied to $S_i^*v(g) \in \mathcal{H}^*$ shows

$$\begin{aligned}\alpha_g(S_i^*v(g))u(g)^* &= u(g)^*\alpha_g^u(S_i^*v(g))\\ &= u(g)^*S_i^*v(g)w(g)^*\gamma(u(g)) = u(g)^*S_i^*w(g)_0^*\gamma(u(g)).\end{aligned}$$

Thus, $W_g$ can be also written as

$$W_g = \sum_{ij} u(g)^* S_i^* w_0(g)^* \gamma(u(g)) S_j \otimes e_{ij},$$

and this means $W_g \in \mathcal{R} \otimes \hat{\mathcal{A}}$ because $w_0(g)^*\gamma(u(g))$ commutes with an element in $(\gamma, \gamma) = \hat{\mathcal{A}}'$.

LEMMA 12.2. *(i)* $Ad(W_g) \cdot \pi = (\alpha_g \otimes \theta_g) \cdot \pi \cdot \alpha_g^{-1}$.
*(ii)* $(W_g \otimes 1)(\pi \otimes id)(W_g)(id \otimes \hat{\Gamma})(W_g^*)$ *belongs to* $\mathbf{C} \otimes \hat{\mathcal{A}} \otimes \hat{\mathcal{A}}$.
*(iii)* $(\alpha_g \otimes \theta_g)(W_h) W_g = (1 \otimes \zeta(g,h)) W_{gh}$.

PROOF. (i) For $x \in \mathcal{R}$ we compute

$$\begin{aligned}(\alpha_g \otimes \theta_g) \cdot \pi \cdot \alpha_g^{-1}(x) &= \sum_{ij} \alpha_g(S_i^* \gamma(\alpha_g^{-1}(x)) S_j) \otimes v(g) e_{ij} v(g)^* \\ &= \sum_{ij} \alpha_g(S_i^*) u(g)^* \alpha_g^u \gamma \alpha_g^{-1}(x) u(g) \alpha_g(S_j) \otimes v(g) S_i S_j^* v(g)^* \\ &= \sum_{ij} \alpha_g(S_i^*) u(g)^* \gamma(x) u(g) \alpha_g(S_j) \otimes v(g) S_i S_j^* v(g)^* \\ &= \sum_{ij} \alpha_g(S_i^* v(g)) u(g)^* \gamma(x) u(g) \alpha_g(v(g)^* S_j) \otimes S_i S_j^* \\ &= \sum_{ij} \alpha_g(S_i^* v(g)) u(g)^* \gamma(x) u(g) \alpha_g(v(g)^* S_j) \otimes e_{ij} = W_g \pi(x) W_g^*.\end{aligned}$$

(ii) By setting $\pi_g = Ad(W_g) \cdot \pi$, we generally compute

$$\begin{aligned}(\pi_g \otimes id) \cdot \pi_g &= Ad\left((W_g \otimes 1)(\pi \otimes id)(W_g)\right) \cdot (\pi \otimes id) \cdot \pi \\ &= Ad\left((W_g \otimes 1)(\pi \otimes id)(W_g)\right) \cdot (id \otimes \hat{\Gamma}) \cdot \pi \\ &= Ad\left((W_g \otimes 1)(\pi \otimes id)(W_g)(id \otimes \hat{\Gamma})(W_g^*)\right) \cdot (id \otimes \hat{\Gamma}) \cdot \pi_g.\end{aligned}$$

For $x$ in the fixed-point algebra $\mathcal{R}^\pi = \{x \in \mathcal{R};\ \pi(x) = x \otimes 1\}$ we have $\pi_g(\alpha_g(x)) = (\alpha_g \otimes \theta_g) \cdot \pi(x) = (\alpha_g \otimes \theta_g)(x \otimes 1) = \alpha_g(x) \otimes 1$ thanks to (i) so that we have

$$(\pi_g \otimes id) \cdot \pi_g(\alpha_g(x)) = (id \otimes \hat{\Gamma}) \cdot \pi_g(\alpha_g(x)) = \alpha_g(x) \otimes 1 \otimes 1.$$

Thus, we get $(W_g \otimes 1)(\pi \otimes id)(W_g)(id \otimes \hat{\Gamma})(W_g^*) \in (\alpha_g(\mathcal{R}^\pi)' \cap \mathcal{R}) \otimes \hat{\mathcal{A}} \otimes \hat{\mathcal{A}} = \mathbf{C} \otimes \hat{\mathcal{A}} \otimes \hat{\mathcal{A}}$ since the inclusion $\mathcal{R} \supseteq \mathcal{R}^\pi$ is irreducible.
(iii) Note

$$\begin{aligned}(\alpha_g \otimes \theta_g)(W_h) W_g &= \sum_{i,j,k,l} \alpha_g\left(\alpha_h\left(S_i^* v(h)\right) u(h)^* S_j\right) \alpha_g\left(S_k^* v(g)\right) u(g)^* S_l \\ &\qquad\qquad \otimes v(g) e_{ij} v(g) e_{kl} \\ &= \sum_{i,j,k,l} \alpha_g\left(\alpha_h\left(S_i^* v(h)\right) u(h)^* S_j S_k^* v(g)\right) u(g)^* S_l \\ &\qquad\qquad \otimes v(g) S_i S_j^* v(g)^* e_{kl}.\end{aligned}$$

We can replace $S_i, S_j$ by $v(g)^*S_i, v(g)^*S_j$ in the above, and we get

$$\begin{aligned}
&(\alpha_g \otimes \theta_g)(W_h)W_g \\
&= \sum_{i,j,k,l} \alpha_g \left(\alpha_h \left(S_i^* v(g)v(h)\right) u(h)^* v(g)^* S_j S_k^* v(g)\right) u(g)^* S_l \otimes e_{ij}e_{kl} \\
&= \sum_{i,j,l} \alpha_g \left(\alpha_h \left(S_i^* v(g)v(h)\right) u(h)^* v(g)^* S_j S_j^* v(g)\right) u(g)^* S_l \otimes e_{il} \\
&= \sum_{i,l} \alpha_g \left(\alpha_h \left(S_i^* v(g)v(h)\right) u(h)^*\right) u(g)^* S_l \otimes e_{il} \\
&= \sum_{i,l} \alpha_{gh} \left(S_i^* v(g)v(h)\right) \alpha_g \left(u(h)^*\right) u(g)^* S_l \otimes e_{il} \\
&= \sum_{i,l} \alpha_{gh}(S_i^* \zeta(g,h)\zeta'(g,h)v(gh))u(gh)^* \zeta'(g,h)^* S_l \otimes e_{il} \\
&= \sum_{i,l} u(gh)^* \alpha^u_{gh} \left(S_i^* \zeta(g,h)\zeta'(g,h)v(gh)\right) \zeta'(g,h)^* S_l \otimes e_{il}.
\end{aligned}$$

Lemma 10.1,(i) applied to $v(gh)^*\zeta(g,h)^*\zeta'(g,h)^*S_i \in \mathcal{H}$ shows

$$\begin{aligned}
(\alpha_g \otimes \theta_g)(W_h)W_g &= \sum_{i,l} u(gh)^* S_i^* \zeta(g,h)\zeta'(g,h) w_0(gh)^* \gamma(u(gh))\zeta'(g,h)^* S_l \otimes e_{il} \\
&= \sum_{i,l} u(gh)^* S_i^* \zeta(g,h) w_0(gh)^* \gamma(u(gh)) S_l \otimes S_i S_l^*.
\end{aligned}$$

Here, the second equality follows from $w_0(gh) \in \hat{\mathcal{A}} \otimes \hat{\mathcal{A}}'$ (Lemma 10.1,(ii)) and $\zeta'(g,h) \in (\gamma,\gamma) = \hat{\mathcal{A}}'$. Replacing $S_i$ by $\zeta(g,h)S_i$ in the above, we get the desired result. □

Let $\eta_g$ be the unitary in $\hat{\mathcal{A}} \otimes \hat{\mathcal{A}}$ defined by

$$1 \otimes \eta_g = (W_g \otimes 1)(\pi \otimes id)(W_g)(id \otimes \hat{\Gamma})(W_g^*)$$

(see Lemma 12.2,(ii)).

PROPOSITION 12.3. *(i)* $Ad(\eta_g) \cdot \hat{\Gamma} = (\theta_g \otimes \theta_g) \cdot \hat{\Gamma} \cdot \theta_g^{-1}$.
*(ii)* $(\eta_g \otimes 1)(\hat{\Gamma} \otimes id)(\eta_g) = (1 \otimes \eta_g)(id \otimes \hat{\Gamma})(\eta_g)$.
*(iii)* $\eta_{gh} = (\zeta(g,h)^* \otimes \zeta(g,h)^*)(\theta_g \otimes \theta_g)(\eta_h)\eta_g\hat{\Gamma}(\zeta(g,h))$.

PROOF. (i) From the co-associativity of the action $\pi$ we have

$$(\alpha_g \otimes \theta_g \otimes \theta_g) \cdot (\pi \otimes id) \cdot \pi \cdot \alpha_g^{-1} = (\alpha_g \otimes \theta_g \otimes \theta_g) \cdot (id \otimes \hat{\Gamma}) \cdot \pi \cdot \alpha_g^{-1}.$$

Repeated use of Lemma 10.2,(i) shows that the left hand side is equal to

$$\begin{aligned}
&((Ad(W_g) \cdot \pi) \otimes id) \cdot (\alpha_g \otimes \theta_g) \cdot \pi \cdot \alpha_g^{-1} \\
&= Ad((W_g \otimes 1)(\pi \otimes id)(W_g)) \cdot (\pi \otimes id) \cdot \pi \\
&= Ad((W_g \otimes 1)(\pi \otimes id)(W_g)) \cdot (id \otimes \hat{\Gamma}) \cdot \pi = Ad(1 \otimes \eta_g) \cdot (id \otimes \hat{\Gamma}) \cdot \pi_g.
\end{aligned}$$

On the other hand, the right hand side is $\left(id \otimes \left((\theta_g \otimes \theta_g) \cdot \hat{\Gamma} \cdot \theta_g^{-1}\right)\right) \cdot \pi_g$ by Lemma 10.2,(i) again. Since $\pi$ is a faithful action of $\hat{\mathcal{A}}$, we get the result.

(ii) Using the definition of $\eta_g$, we compute

$$\begin{aligned}
&(1 \otimes \eta_g \otimes 1)\left(1 \otimes (\hat{\Gamma} \otimes id)(\eta_g)\right)\\
&= (W_g \otimes 1 \otimes 1)\,((\pi \otimes id)(W_g) \otimes 1)\left((id \otimes \hat{\Gamma})(W_g^*) \otimes 1\right)\\
&\qquad \times \left(id \otimes \hat{\Gamma} \otimes id\right)\left((W_g \otimes 1)(\pi \otimes id)(W_g)(id \otimes \hat{\Gamma})(W_g^*)\right)\\
&= (W_g \otimes 1 \otimes 1)\,((\pi \otimes id)(W_g) \otimes 1)\left(((id \otimes \hat{\Gamma}) \cdot \pi) \otimes id\right)(W_g)\\
&\qquad \times \left(id \otimes ((\hat{\Gamma} \otimes id) \cdot \hat{\Gamma})\right)(W_g^*)\\
&= (W_g \otimes 1 \otimes 1)\,(\pi \otimes id \otimes id)\,((W_g \otimes 1)(\pi \otimes id)(W_g))\\
&\qquad \times \left(id \otimes ((id \otimes \hat{\Gamma}) \cdot \hat{\Gamma})\right)(W_g^*)\\
&= (W_g \otimes 1 \otimes 1)\,(\pi \otimes id \otimes id)\left((1 \otimes \eta_g)(id \otimes \hat{\Gamma})(W_g)\right)\\
&\qquad \times \left(id \otimes ((id \otimes \hat{\Gamma}) \cdot \hat{\Gamma})\right)(W_g^*)\\
&= (1 \otimes 1 \otimes \eta_g)(id \otimes id \otimes \hat{\Gamma})((W_g \otimes 1)(\pi \otimes id)(W_g)(id \otimes \hat{\Gamma})(W_g^*))\\
&= (1 \otimes 1 \otimes \eta_g)\left(1 \otimes (id \otimes \hat{\Gamma})(\eta_g)\right).
\end{aligned}$$

(iii) Using Lemma 10.2,(iii), we compute

$$\begin{aligned}
1 \otimes \eta_{gh} &= (W_{gh} \otimes 1)(\pi \otimes id)(W_{gh})(id \otimes \hat{\Gamma})(W_{gh}^*)\\
&= (1 \otimes \zeta(g,h)^* \otimes 1)\,((\alpha_g \otimes \theta_g)(W_h) \otimes 1)\,(W_g \otimes 1)\\
&\qquad \times (1 \otimes 1 \otimes \zeta(g,h)^*)(\pi \cdot \alpha_g \otimes \theta_g)(W_h)(\pi \otimes id)(W_g)\\
&\qquad \times (id \otimes \hat{\Gamma})(W_g^*)(\alpha_g \otimes \hat{\Gamma} \cdot \theta_g)(W_h^*)(id \otimes \hat{\Gamma})(1 \otimes \zeta(g,h))\\
&= (1 \otimes \zeta(g,h)^* \otimes \zeta(g,h)^*)((\alpha_g \otimes \theta_g)(W_h) \otimes 1)(((\alpha_g \otimes \theta_g) \cdot \pi) \otimes \theta_g)(W_h)\\
&\qquad \times (W_g \otimes 1)(\pi \otimes id)(W_g)(id \otimes \hat{\Gamma})(W_g^*)\\
&\qquad \times (\alpha_g \otimes \hat{\Gamma} \cdot \theta_g)(W_h^*)(id \otimes \hat{\Gamma})(1 \otimes \zeta(g,h))\\
&= (1 \otimes \zeta(g,h)^* \otimes \zeta(g,h)^*)(\alpha_g \otimes \theta_g \otimes \theta_g)((W_h \otimes 1)(\pi \otimes id)(W_h))(1 \otimes \eta_g)\\
&\qquad \times (\alpha_g \otimes \hat{\Gamma} \cdot \theta_g)(W_h^*)(id \otimes \hat{\Gamma})(1 \otimes \zeta(g,h))\\
&= (1 \otimes \zeta(g,h)^* \otimes \zeta(g,h)^*)\\
&\qquad \times (\alpha_g \otimes \theta_g \otimes \theta_g)((W_h \otimes 1)(\pi \otimes id)(W_h)(id \otimes \hat{\Gamma})(W_h^*))\\
&\qquad \times (1 \otimes \eta_g)(id \otimes \hat{\Gamma})(1 \otimes \zeta(g,h))\\
&= (1 \otimes \zeta(g,h)^* \otimes \zeta(g,h)^*)(\alpha_g \otimes \theta_g \otimes \theta_g)(1 \otimes \eta_h)(1 \otimes \eta_g)\\
&\qquad \times (id \otimes \hat{\Gamma})(1 \otimes \zeta(g,h))\\
&= (1 \otimes (\zeta(g,h)^* \otimes \zeta(g,h)^*)(\theta_g \otimes \theta_g)(\eta_h)\eta_g\hat{\Gamma}(\zeta(g,h))).
\end{aligned}$$

□

**Remark.** The "coboundary" operation is given as follows: Let $\xi(g) \in \hat{\mathcal{A}}$ be a unitary. Then the perturbation by $\xi(g)$ changes $\theta_g$, $\zeta(g,h)$, and $\eta_g$ to

$$\begin{cases}
\tilde{\theta}_g = Ad(\xi(g)) \cdot \theta_g,\\
\tilde{\zeta}(g,h) = \xi(g)\theta_g(\xi(h))\zeta(g,h)\xi(gh)^*,\\
\tilde{\eta}_g = (\xi(g) \otimes \xi(g))\eta_g\hat{\Gamma}(\xi(g)^*).
\end{cases}$$

Let $\hat{\epsilon}$ be the counit of $\hat{\mathcal{A}}$. Applying $id \otimes \hat{\epsilon} \otimes id$ to the both side of (ii), we get

$$((id \otimes \hat{\epsilon})(\eta_g) \otimes 1)\eta_g = (1 \otimes (\hat{\epsilon} \otimes id)(\eta_g))\eta_g.$$

This shows that $(id \otimes \hat{\epsilon})(\eta_g) = (\hat{\epsilon} \otimes id)(\eta_g)$ is a scalar. By taking a appropriate coboundary stated as above, we may and do assume

$$(id \otimes \hat{\epsilon})(\eta_g) = (\hat{\epsilon} \otimes id)(\eta_g) = 1.$$

To keep this condition, $\xi_g$ should satisfy $\hat{\epsilon}(\xi_g) = 1$ in what follows.

We now compare $\eta_g$ with $w_0(g)$.

PROPOSITION 12.4. *With the counit $\hat{\epsilon}'$ of $\hat{\mathcal{A}}'$ we have*

$$\eta_g = ((id \otimes \hat{\epsilon}')(w_0(g)^*) \otimes 1) F(1 \otimes U)w_0(g)^*(1 \otimes U)F\hat{\Gamma}\left((id \otimes \hat{\epsilon}')(w_0(g))\right).$$

PROOF. By definition, we have

$$\begin{aligned}
(1 \otimes \eta_g) &= (W_g \otimes 1)(\pi \otimes id)(W_g)(id \otimes \hat{\Gamma})(W_g^*) \\
&= \left(\sum_{ij}(u(g)^* S_i^* w_0(g)^* \gamma(u(g))S_j) \otimes e_{ij} \otimes 1\right) \\
&\quad \times \left(\sum_{klmn}(S_k^* \gamma(u(g)^* S_m^* w_0(g)^* \gamma(u(g))S_n)S_l) \otimes e_{kl} \otimes e_{mn}\right) \\
&\quad \times \left(\sum_{pq}(S_p^* \gamma(u(g)^*) w_0(g) S_q u(g)) \otimes \hat{\Gamma}(e_{pq})\right) \\
&= \sum_{ilmnpqr} [u(g)^* S_i^* w_0(g)^* \gamma(S_m^* w_0(g)^*)\gamma^2(u(g))\gamma(S_n)S_l \\
&\quad \times S_p^* \gamma(u(g)^*) w_0(g) S_q u(g)] \otimes S_i S_m S_n^* S_l^* V^* S_r S_p S_q^* S_r^* V \\
&= \sum_{ilmnpqr} [u(g)^* S_i^* w_0(g)^* S_m^* F V^* \gamma(w_0(g)^*)\sigma_{\mathcal{H}} \cdot \gamma(u(g)) V S_l S_n \\
&\quad \times S_p^* \gamma(u(g)^*) w_0(g) S_q u(g)] \otimes S_i S_m S_n^* S_l^* V^* S_r S_p S_q^* S_r^* V.
\end{aligned}$$

Here, the third equality comes from (12.2) while we have used $\gamma^2 = \sigma_{\mathcal{H}} \cdot \gamma$ and $\gamma|_{\mathcal{O}_{\mathcal{H}}} = \lambda_{VF}$ to get the fourth. Since $S_n^* S_l^* V^* S_r S_p$ is a scalar, we can move it to the first tensor component, and so we get

$$\begin{aligned}
(1 \otimes \eta_g) &= \sum_{imqr} S_i^* w_0(g)^* S_m^* F V^* \gamma(w_0(g)^*)\sigma_{\mathcal{H}} \cdot \gamma(u(g)) S_r \gamma(u(g)^*) w_0(g) S_q \\
&\qquad \otimes S_i S_m S_q^* S_r^* V \\
&= \sum_{imqr} S_i^* w_0(g)^* S_m^* F V^* \gamma(w_0(g)^*) S_r w_0(g) S_q \otimes S_i S_m S_q^* S_r^* V \\
&= \sum_{imqrst} S_i^* w_0(g)^* S_m^* F V^* \gamma(w_0(g)^*) S_r w_0(g) S_q \otimes S_i S_m S_q^* S_r^* V S_s S_t S_t^* S_s^*
\end{aligned}$$

$$= \sum_{imqrst} S_i^* w_0(g)^* S_m^* F V^* \gamma(w_0(g)^*) S_r w_0(g) S_q S_q^* S_r^* V S_s S_t \otimes S_i S_m S_t^* S_s^*$$

$$= \sum_{imst} S_i^* w_0(g)^* S_m^* F V^* \gamma(w_0(g)^*) \sigma_{\mathcal{H}}(w_0(g)) V S_s S_t \otimes S_i S_m S_t^* S_s^*$$

$$= \sum_{imst} S_m^* S_i^* F \sigma_{\mathcal{H}}(w_0(g)^*) F V^* \gamma(w_0(g)^*) \sigma_{\mathcal{H}}(w_0(g)) V S_s S_t \otimes S_i S_m S_t^* S_s^*.$$

Since $w_0(g) \in B(\mathcal{H}) \otimes B(\mathcal{H})$, we have

$$\gamma(w_0(g)) = VF\sigma_{\mathcal{H}}(VF)(w_0(g))\sigma_{\mathcal{H}}(FV^*)FV^*,$$

and so

$$(1 \otimes \eta_g) = \sum_{imst} S_m^* S_i^* F \sigma_{\mathcal{H}}(w_0(g)^*) \sigma_{\mathcal{H}}(VF)(w_0(g)) \sigma_{\mathcal{H}}(FV^*) F V^* \sigma_{\mathcal{H}}(w_0(g)) V S_s S_t \otimes S_i S_m S_t^* S_s^*.$$

Thus, $\eta_g \otimes 1 \in B(\mathcal{H}) \otimes B(\mathcal{H}) \otimes B(\mathcal{H})$ can be written as

$$(F\otimes 1)(1\otimes w_0(g)^*)(1\otimes VF)(w_0(g)\otimes 1)(1\otimes FV^*)(F\otimes 1)(V^*\otimes 1)(1\otimes w_0(g))(V\otimes 1).$$

Using (12.2) and (12.3), we get

$$\begin{aligned}
&(\eta_g \otimes 1) \\
&= (F \otimes 1)(1 \otimes w_0(g)^*)(1 \otimes 1 \otimes U)(1 \otimes F)(id \otimes \hat{\Gamma})((1 \otimes U)w_0(g)^*(1 \otimes U)) \\
&\quad \times (1 \otimes F)(1 \otimes 1 \otimes U)(F \otimes 1)(\hat{\Gamma} \otimes id)(w_0(g)).
\end{aligned}$$

Applying $(id \otimes id \otimes \hat{\epsilon}')$ to this identity, we obtain the desired result. □

**Remark.** Let $\mu_g = F(1 \otimes U)w_0(g)^*(1 \otimes U)F$. Then, from the proposition we have

$$\eta_g = ((\hat{\epsilon} \otimes id)(\mu_g) \otimes 1)\, \mu_g \hat{\Gamma}\left((\hat{\epsilon} \otimes id)(\mu_g^*)\right).$$

Thus, $(\hat{\epsilon} \otimes id)(\eta_g) = (id \otimes \hat{\epsilon})(\eta_g) = 1$ implies $(id \otimes \hat{\epsilon})(\mu_g) = 1$, which is equivalent to $(\hat{\epsilon} \otimes id)(w_0(g)) = 1$.

## 2. Multiplicative unitary

To obtain detailed description for the multiplicative unitary associated with the inclusion $\mathcal{R} \rtimes_\alpha G \supseteq \rho_1(\mathcal{R})$ (i.e., $\mathcal{R} \supseteq \sigma(\mathcal{R})$ with $\sigma = \bar{\rho}_2\rho_1$) of depth 2, we need to express $w_0(g)$ in terms of $\eta_g$. To do so, we collect some auxiliary results. Let $m : \hat{\mathcal{A}} \otimes \hat{\mathcal{A}} \longrightarrow \hat{\mathcal{A}}$ be the multiplication map and $\hat{\kappa}$ be the antipode of $\hat{\mathcal{A}}$. Using $m \cdot (id \otimes \hat{\kappa}) \cdot \hat{\Gamma} = \hat{\epsilon}$, from the identity in the preceding remark we have

$$\begin{aligned}
m \cdot (id \otimes \hat{\kappa})(\eta_g) &= (\hat{\epsilon} \otimes id)(\mu_g)\,(m \cdot (id \otimes \hat{\kappa})(\mu_g))\,(\hat{\epsilon} \otimes \hat{\epsilon})(\mu_g^*) \\
&= (\hat{\epsilon} \otimes id)(\mu_g)\,(m \cdot (id \otimes \hat{\kappa})(\mu_g)).
\end{aligned}$$

Our first goal is to show $m \cdot (id \otimes \hat{\kappa})(\mu_g) = 1$.

LEMMA 12.5. *Let $h$ and $\hat{h}$ be the normalized Haar measures of $\mathcal{A}$ and $\hat{\mathcal{A}}$ respectively, and $\Omega_h \in \mathcal{H}$, $\Omega_{\hat{h}} \in \mathcal{H}$ be the GNS cyclic vectors of $h, \hat{h}$. Then, for $\hat{x}' \in \hat{\mathcal{A}}'$ and $\xi, \eta \in \mathcal{H}$, we have*

$$\langle V^*(1 \otimes \hat{x}')(\xi \otimes \Omega_{\hat{h}}), (\eta \otimes \Omega_h)\rangle = \tfrac{1}{\sqrt{n}}\langle \hat{J}\hat{x}'^* \hat{J}\xi, \eta\rangle$$

*with* $n = dim\mathcal{A} = d(\rho_1)^2$.

PROOF. The natural pairing $(x, \hat{y})$ of $x \in \mathcal{A}$ and $\hat{y} \in \hat{\mathcal{A}}$ is known to be given by

$$(x, \hat{y}) = \sqrt{n}\ \langle x\Omega_h, \hat{y}^*\Omega_{\hat{h}}\rangle,$$

where $\langle\cdot,\cdot\rangle$ means the inner product in $\mathcal{H}$. We take $\hat{a}' \in \hat{\mathcal{A}}'$, $b \in \mathcal{A}$ satisfying $\xi = \hat{a}'\Omega_{\hat{h}}$, $\eta = b\Omega_h$, and we compute

$$\begin{aligned}
&\langle V^*(1\otimes\hat{x}')(\xi\otimes\Omega_{\hat{h}}), (\eta\otimes\Omega_h)\rangle \\
&= \langle(\hat{a}'\otimes\hat{x}')(\Omega_{\hat{h}}\otimes\Omega_{\hat{h}}), V(b\otimes 1)(\Omega_h\otimes\Omega_h)\rangle \\
&= \langle(\hat{a}'\otimes\hat{x}')(\Omega_{\hat{h}}\otimes\Omega_{\hat{h}}), \Gamma(b)(\Omega_h\otimes\Omega_h)\rangle \\
&= \langle(\hat{J}\otimes\hat{J})\Gamma(b)(\Omega_h\otimes\Omega_h), (\hat{J}\hat{a}'\hat{J}\otimes\hat{J}\hat{x}'\hat{J})(\Omega_{\hat{h}}\otimes\Omega_{\hat{h}})\rangle \\
&= \langle(\kappa\otimes\kappa)\cdot\Gamma(b^*)(\Omega_h\otimes\Omega_h), (\hat{J}\hat{a}'\hat{J}\otimes\hat{J}\hat{x}'\hat{J})(\Omega_{\hat{h}}\otimes\Omega_{\hat{h}})\rangle \\
&= \langle F\Gamma(\kappa(b^*))F(\Omega_h\otimes\Omega_h), (\hat{J}\hat{a}'\hat{J}\otimes\hat{J}\hat{x}'\hat{J})(\Omega_{\hat{h}}\otimes\Omega_{\hat{h}})\rangle \\
&= \langle\Gamma(\kappa(b^*))(\Omega_h\otimes\Omega_h), (\hat{J}\hat{x}'\hat{J}\otimes\hat{J}\hat{a}'\hat{J})(\Omega_{\hat{h}}\otimes\Omega_{\hat{h}})\rangle \\
&= \left(\tfrac{1}{n}(\Gamma(\kappa(b^*)), \left(\hat{J}\hat{x}'^*\hat{J}\otimes\hat{J}\hat{a}'^*\hat{J}\right)\right) = \tfrac{1}{n}\left(\kappa(b^*), \hat{J}\hat{x}'^*\hat{a}'^*\hat{J}\right) \\
&= \tfrac{1}{\sqrt{n}}\langle\hat{J}b\hat{J}\Omega_h, \hat{J}\hat{a}'\hat{x}'\Omega_{\hat{h}}\rangle = \tfrac{1}{\sqrt{n}}\langle\hat{a}'\hat{x}'\Omega_{\hat{h}}, b\Omega_h\rangle \\
&= \tfrac{1}{\sqrt{n}}\langle\hat{J}\hat{x}'^*\hat{J}\hat{a}'\Omega_{\hat{h}}, b\Omega_h\rangle = \tfrac{1}{\sqrt{n}}\langle\hat{J}\hat{x}'^*\hat{J}\xi, \eta\rangle,
\end{aligned}$$

completing the proof. □

Let $R_{\rho_1} \in (id, \overline{\rho}_1\rho_1)$ and $\overline{R}_{\rho_1} \in (id, \rho_1\overline{\rho}_1)$ be isometries satisfying

$$R^*_{\rho_1}\overline{\rho}_1(\overline{R}_{\rho_1}) = \overline{R}^*_{\rho_1}\rho_1(R_{\rho_1}) = \frac{1}{d(\rho_1)}.$$

Then, we can identify $\Omega_h$ with $\overline{R}_{\rho_1}$ and $\Omega_{\hat{h}}$ with $\rho_1(R_{\rho_1})$.

LEMMA 12.6. *(i)* $v(g)\Omega_h = \Omega_h$ *and* $v(g)\Omega_{\hat{h}} = \Omega_{\hat{h}}$.
*(ii)* $m\cdot(id\otimes\hat{\kappa})((1\otimes U)w_0(g)(1\otimes U)) = 1$, *or equivalently,*

$$m\cdot(id\otimes\hat{\kappa})(\mu_g) = 1.$$

PROOF. (i) The second equality is obvious. Note that $\Omega_h = \sqrt{n}\ \hat{e}\Omega_{\hat{h}}$, where $\hat{e} \in \hat{\mathcal{A}}\cap\hat{\mathcal{A}}'$ is the integral of both $\hat{\mathcal{A}}$ and $\hat{\mathcal{A}}'$. Thus, to see the first it suffices to show $\alpha^u_g(\hat{e}) = \hat{e}$. It is plain to see that the isometry $u(g)\alpha_g(\overline{R}_{\rho_1})$ belongs to the one-dimensional space $(id, \gamma)$ so that $u(g)\alpha_g(\overline{R}_{\rho_1}) = \omega_g\overline{R}_{\rho_1}$ with some $\omega_g \in \mathbf{T}$. On the other hand, we compute

$$\begin{aligned}
\omega_g &= \overline{R}^*_{\rho_1}u(g)\alpha_g(\overline{R}_{\rho_1}) = \overline{R}^*_{\rho_1}\alpha^u_g(\overline{R}_{\rho_1})u(g) \\
&= \overline{R}^*_{\rho_1}\gamma(u(g)^*)w_0(g)v(g)\overline{R}_{\rho_1}u(g) \quad \text{(by Lemma 10.1)} \\
&= u(g)^*\overline{R}^*_{\rho_1}w_0(g)v(g)\overline{R}_{\rho_1}u(g) = u(g)^*(\hat{\epsilon}\otimes id)(w_0(g))\overline{R}^*_{\rho_1}v(g)\overline{R}_{\rho_1}u(g) \\
&= u(g)^*\overline{R}^*_{\rho_1}v(g)\overline{R}_{\rho_1}u(g) \quad \text{(see the remarks in the previous section)} \\
&= nu(g)^*\langle\alpha^u_g(\hat{e})\Omega_{\hat{h}}, \hat{e}\Omega_{\hat{h}}\rangle u(g) = n\langle\alpha^u_g(\hat{e})\Omega_{\hat{h}}, \hat{e}\Omega_{\hat{h}}\rangle.
\end{aligned}$$

This computation means that the right hand side is not zero, and so $\alpha^u_g(\hat{e}) = \hat{e}$ and $\omega_g = 1$, i.e., $u(g)\alpha_g(\overline{R}_{\rho_1}) = \overline{R}_{\rho_1}$.
(ii) First, we show

$$v(g) = d(\rho_1)\sigma_{\mathcal{H}}(\overline{R}_{\rho_1})^*V^*w(g)\sigma_{\mathcal{H}}(\rho_1(R_{\rho_1})).$$

Let $\Lambda_{\hat{h}'} : \hat{\mathcal{A}}' \longrightarrow \mathcal{H}$ be the natural map defined by identification of $\mathcal{H}$ and the GNS Hilbert space with the GNS cyclic vector $\Omega_{\hat{h}} = \rho_1(R_{\rho_1})$. Then, using the fact that $(\gamma, \gamma) = \hat{\mathcal{A}}'$, we get the following formula for $\Lambda_{\hat{h}'}^{-1}$:

$$\Lambda_{\hat{h}'}^{-1}(S) = d(\rho_1)\gamma(\overline{R}_{\rho_1}^*)S, \quad S \in \mathcal{H}.$$

Therefore, for $S \in \mathcal{H}$ we have

$$\begin{aligned} v(g)S &= \Lambda_{\hat{h}'}(\alpha_g^u(\Lambda_{\hat{h}'}^{-1}(S))) = d(\rho_1)\alpha_g^u(\gamma(\overline{R}_{\rho_1}^*)S)\rho_1(R_{\rho_1}) \\ &= d(\rho_1)\gamma(\alpha_g(\overline{R}_{\rho_1}^*))\alpha_g^u(S)\rho_1(R_{\rho_1}) = d(\rho_1)\gamma(\alpha_g(\overline{R}_{\rho_1}^*)u(g)^*)w(g)S\rho_1(R_{\rho_1}). \end{aligned}$$

Thus, using $u(g)\alpha_g(\overline{R}_{\rho_1}) = \overline{R}_{\rho_1}$, we get

$$v(g)S = d(\rho_1)\gamma(\overline{R}_{\rho_1}^*)w(g)S\rho_1(R_{\rho_1}) = d(\rho_1)\sigma_{\mathcal{H}}(\overline{R}_{\rho_1}^*)V^*w(g)S\rho_1(R_{\rho_1}).$$

Taking an orthonormal basis $\{S_i\} \subset \mathcal{H}$, we get

$$\begin{aligned} v(g) &= \sum_i v(g)S_iS_i^* = \sum_i d(\rho_1)\sigma_{\mathcal{H}}(\overline{R}_{\rho_1}^*)V^*w(g)S_i\rho_1(R_{\rho_1})S_i^* \\ &= d(\rho_1)\sigma_{\mathcal{H}}(\overline{R}_{\rho_1}^*)V^*w(g)\sigma_{\mathcal{H}}(\rho_1(R_{\rho_1})), \end{aligned}$$

which implies

$$d(\rho_1)\sigma_{\mathcal{H}}(\overline{R}_{\rho_1}^*)V^*w_0(g)\sigma_{\mathcal{H}}(\rho_1(R_{\rho_1})) = 1.$$

Using Lemma 12.5, we get the statement. □

COROLLARY 12.7. *(i)* $\nu_g = m \cdot (id \otimes \hat{\kappa})(\eta_g) = (id \otimes \hat{\epsilon}')(w_0(g)^*)$ *is a unitary.*
*(ii) We have*

$$\begin{aligned} w_0(g) &= (1 \otimes U)F\hat{\Gamma}(\nu_g^*)\eta_g^*(\nu_g \otimes 1)F(1 \otimes U) \\ &= V(1 \otimes U\nu_g^*U)V^*(1 \otimes U)F\eta_g^*(\nu_g \otimes 1)F(1 \otimes U). \end{aligned}$$

LEMMA 12.8.
*(i)* $\eta_g = (\hat{\kappa}(\nu_g) \otimes (\hat{\kappa}(\nu_g)))F(\hat{\kappa} \otimes \hat{\kappa})(\eta_g^*)F\hat{\Gamma}(\hat{\kappa}(\nu_g^*)) = F(\hat{\kappa} \otimes \hat{\kappa})((\nu_g^* \otimes \nu_g^*)\eta_g\hat{\Gamma}(\nu_g))^*F.$
*(ii)* $(v(g) \otimes v(g))V(v(g)^* \otimes v(g)^*) = w_0(g)^*V\eta_g^*.$

PROOF. (i) The identity $(\eta_g \otimes 1)(\hat{\Gamma} \otimes id)(\eta_g) = (1 \otimes \eta_g)(id \otimes \hat{\Gamma})(\eta_g)$ implies

$$\begin{aligned} &(F(\hat{\kappa} \otimes \hat{\kappa})(\eta_g^*)F \otimes 1 \otimes 1)(\hat{\Gamma} \otimes \hat{\Gamma})((\hat{\kappa} \otimes id)(\eta_g^*)) \\ &\quad = (F \otimes 1 \otimes 1)(\hat{\kappa} \otimes \hat{\kappa} \otimes id \otimes id)((id \otimes (id \otimes \hat{\Gamma}) \cdot \hat{\Gamma})(\eta_g^*)(1 \otimes (id \otimes \hat{\Gamma})(\eta_g^*))) \\ &\qquad \times (F \otimes 1 \otimes 1) \\ &\quad = (F \otimes 1 \otimes 1) \\ &\qquad \times (\hat{\kappa} \otimes \hat{\kappa} \otimes id \otimes id)((id \otimes (\hat{\Gamma} \otimes id) \cdot \hat{\Gamma})(\eta_g^*)(1 \otimes (\hat{\Gamma} \otimes id)(\eta_g^*)(\eta_g^* \otimes 1))) \\ &\qquad \times (F \otimes 1 \otimes 1)(1 \otimes 1 \otimes \eta_g) \\ &\quad = (F \otimes 1 \otimes 1) \\ &\qquad \times (\hat{\kappa} \otimes \hat{\kappa} \otimes id \otimes id)((id \otimes \hat{\Gamma} \otimes id)((id \otimes \hat{\Gamma})(\eta_g^*)(1 \otimes \eta_g^*))(1 \otimes \eta_g^* \otimes 1)) \\ &\qquad \times (F \otimes 1 \otimes 1)(1 \otimes 1 \otimes \eta_g). \end{aligned}$$

Let $m : \hat{\mathcal{A}} \otimes \hat{\mathcal{A}} \longrightarrow \hat{\mathcal{A}}$ be the multiplication map. We apply $m_{13} \cdot m_{24}$ to the both sides. The above left hand side gives us

$$F(\hat{\kappa} \otimes \hat{\kappa})(\eta_g^*)F\hat{\Gamma}(\phi(\eta_g^*)) = (F(\hat{\kappa} \otimes \hat{\kappa})(\eta_g^*)F)\hat{\Gamma}(\nu_g^*)$$

with $\phi = m \cdot (\hat{\kappa} \otimes id)$. On the other hand, the right hand side gives us

$$\begin{aligned}
&\phi_{13} \cdot \phi_{23}(((id \otimes \hat{\Gamma} \otimes id)((id \otimes \hat{\Gamma})(\eta_g^*)(1 \otimes \eta_g^*))(1 \otimes \eta_g^* \otimes 1)))\eta_g \\
&\quad = \phi_{13}(((id \otimes \hat{\epsilon} \otimes id)((id \otimes \hat{\Gamma})(\eta_g^*)(1 \otimes \eta_g^*))(1 \otimes \hat{\kappa}(\nu_g^*) \otimes 1)))\eta_g \\
&\quad = \phi_{13}((\eta_g^*)_{13}(1 \otimes \hat{\kappa}(\nu_g^*) \otimes 1))\eta_g \\
&\quad = (\hat{\kappa}(\nu_g^*) \otimes \hat{\kappa}(\nu_g^*))\eta_g.
\end{aligned}$$

(ii) Let $\hat{x} \in \hat{\mathcal{A}}$ and $\hat{x}' \in \hat{\mathcal{A}}'$. Then, $(v(g) \otimes v(g))V^*(v(g)^* \otimes v(g)^*)\hat{x}'\Omega_{\hat{h}} \otimes \hat{x}\Omega_{\hat{h}}$ is

$$\begin{aligned}
&(v(g) \otimes v(g))V^* v(g)^* \hat{x}'\Omega_{\hat{h}} \otimes \theta_g^{-1}(\hat{x})\Omega_{\hat{h}} \\
&\quad = (v(g) \otimes v(g))\hat{\Gamma}(\theta_g^{-1}(\hat{x}))v(g)^*\hat{x}'\Omega_{\hat{h}} \otimes \Omega_{\hat{h}} \\
&\quad = (\theta_g \otimes \theta_g)(\hat{\Gamma}(\theta_g^{-1}(\hat{x})))\hat{x}'\Omega_{\hat{h}} \otimes \Omega_{\hat{h}} \\
&\quad = \eta_g \hat{\Gamma}(\hat{x})\eta_g^* \hat{x}'\Omega_{\hat{h}} \otimes \Omega_{\hat{h}} \\
&\quad = (\hat{x}' \otimes 1)\eta_g \hat{\Gamma}(\hat{x})\eta_g^* \Omega_{\hat{h}} \otimes \Omega_{\hat{h}}.
\end{aligned}$$

On the other hand, $\eta_g V^* w_0(g)\hat{x}'\Omega_{\hat{h}} \otimes \hat{x}\Omega_{\hat{h}}$ is

$$\begin{aligned}
\eta_g V^*(\hat{x}' \otimes \hat{x})w_0(g)\Omega_{\hat{h}} \otimes \Omega_{\hat{h}} &= \eta_g \hat{\Gamma}(\hat{x})V^*(\hat{x}' \otimes 1)w_0(g)\Omega_{\hat{h}} \otimes \Omega_{\hat{h}} \\
&= (\hat{x}' \otimes 1)\eta_g \hat{\Gamma}(\hat{x})V^* w_0(g)\Omega_{\hat{h}} \otimes \Omega_{\hat{h}}.
\end{aligned}$$

Therefore, it suffices to show $V^* w_0(g)\Omega_{\hat{h}} \otimes \Omega_{\hat{h}} = \eta_g^* \Omega_{\hat{h}} \otimes \Omega_{\hat{h}}$. We take $\{a_i\}, \{b_i\} \subset \hat{\mathcal{A}}$ satisfying $\eta_g = \sum a_i \otimes b_i$. Thanks to Corollary 12.7, we have

$$\begin{aligned}
&V^* w_0(g)\Omega_{\hat{h}} \otimes \Omega_{\hat{h}} = (1 \otimes U\nu_g^* U)V^*(1 \otimes U)F\eta_g^*(\nu_g \otimes 1)F(1 \otimes U)\Omega_{\hat{h}} \otimes \Omega_{\hat{h}} \\
&\quad = \sum_i (1 \otimes U\nu_g^* U)V^*(b_i^*(\Omega_{\hat{h}} \otimes U a_i^* \nu_g \Omega_{\hat{h}}) \\
&\quad = \sum_i (1 \otimes U\nu_g^* U)V^*(b_i^* \Omega_{\hat{h}} \otimes \hat{\kappa}(a_i^* \nu_g)\Omega_{\hat{h}}) \\
&\quad = \sum_i (1 \otimes U\nu_g^* U)\hat{\Gamma}(\hat{\kappa}(a_i^* \nu_g))(b_i^* \Omega_{\hat{h}} \otimes \Omega_{\hat{h}}) \\
&\quad = \sum_i \hat{\Gamma}(\hat{\kappa}(\nu_g))F(\hat{\kappa} \otimes \hat{\kappa}) \cdot \hat{\Gamma}(a_i^*)(1 \otimes b_i^*)F(\Omega_{\hat{h}} \otimes U\nu_g^* \Omega_{\hat{h}}) \\
&\quad = \sum_i \hat{\Gamma}(\hat{\kappa}(\nu_g))F(\hat{\kappa} \otimes \hat{\kappa})(\hat{\Gamma}(a_i)(1 \otimes \hat{\kappa}(b_i)))^* F(\Omega_{\hat{h}} \otimes \hat{\kappa}(\nu_g^*)\Omega_{\hat{h}}) \\
&\quad = \hat{\Gamma}(\hat{\kappa}(\nu_g))F(\hat{\kappa} \otimes \hat{\kappa})((id \otimes m \cdot (id \otimes \hat{\kappa}))(\hat{\Gamma} \otimes id)(\eta_g)))^* F(\Omega_{\hat{h}} \otimes \hat{\kappa}(\nu_g^*)\Omega_{\hat{h}}).
\end{aligned}$$

Using $(\hat{\Gamma} \otimes id)(\eta_g) = (\eta_g^* \otimes 1)(1 \otimes \eta_g)(id \otimes \hat{\Gamma})(\eta_g)$, we get

$$(id \otimes m \cdot (id \otimes \hat{\kappa}))(\hat{\Gamma} \otimes id)(\eta_g^*)) = \eta_g^*(1 \otimes \nu_g).$$

Thus, we obtain

$$V^* w_0(g)\Omega_{\hat{h}} \otimes \Omega_{\hat{h}} = \hat{\Gamma}(\hat{\kappa}(\nu_g))F(\hat{\kappa} \otimes \hat{\kappa})(\eta_g)F(\hat{\kappa}(\nu_g^*) \otimes \hat{\kappa}(\nu_g^*))(\Omega_{\hat{h}} \otimes \Omega_{\hat{h}}) = \eta_g^*(\Omega_{\hat{h}} \otimes \Omega_{\hat{h}}),$$

which proves the statement. □

We are now ready to describe the multiplicative unitary $\overline{V}$ associated with $\sigma = \overline{\rho}_2 \rho_1$ (see the beginning of the chapter). Recall $\rho_2(\mathcal{R}) = \mathcal{R}^\alpha$, the fixed-point algebra, and take isometries $T_g \in (\alpha_g, \rho_2\overline{\rho}_2), g \in G$ so that

$$\rho_2\overline{\rho}_2(x) = \sum_g T_g\alpha_g(x)T_g^*.$$

Note that $T_g$'s can be taken in such a way that

$$\alpha_g(T_h) = T_{gh}\ (g, h \in H).$$

Take a unitary representation $\{\lambda_g\}_{g\in G}$ in $(\overline{\rho}_2\rho_2, \overline{\rho}_2\rho_2)$ satisfying

$$Ad(\lambda_g)\cdot\overline{\rho}_2 = \overline{\rho}_2\cdot\alpha_g, \quad \lambda_g\rho_2(T_h) = T_{hg^{-1}} \quad \text{and} \quad \lambda_g R_{\rho_2} = R_{\rho_2}.$$

For $S \in \mathcal{H}$, we set

$$W(S,g) = \overline{\rho}_2(\gamma(T_g)u(g)\alpha_g(S))\lambda_g = \overline{\rho}_2(\gamma(T_g)u(g))\lambda_g\overline{\rho}_2(S),$$

and we write $W(i,g) = W(S_i,g)$ for simplicity, where $\{S_i\}$ is a fixed orthonormal basis of $\mathcal{H}$.

LEMMA 12.9. *$\{W(i,g)\}_{i,g}$ is an orthonormal basis of $(\sigma, \sigma\overline{\sigma}\sigma)$.*

PROOF. For $x \in \mathcal{R}$ we have

$$\begin{aligned}
\sigma\cdot\overline{\sigma}\cdot\sigma(x) &= \overline{\rho}_2\cdot\gamma\cdot\rho_2\cdot\overline{\rho}_2\cdot\rho_1(x) = \overline{\rho}_2\cdot\gamma\left(\sum_g T_g\alpha_g(\rho_1(x))T_g^*\right)\\
&= \sum_g \overline{\rho}_2(\gamma(T_g))\overline{\rho}_2\alpha_g^u\cdot\gamma\cdot\rho_1(x)\overline{\rho}_2(\gamma(T_g)^*)\\
&= \sum_g \overline{\rho}_2(\gamma(T_g)u(g))\lambda_g\overline{\rho}_2\cdot\sigma_{\mathcal{H}}\cdot\rho_1(x)\lambda_g^*\overline{\rho}_2(u(g)^*\gamma(T_g)^*)\\
&= \sum_{ig} \overline{\rho}_2(\gamma(T_g)u(g))\lambda_g\overline{\rho}_2(S_i)\sigma(x)\overline{\rho}_2(S_i^*)\lambda_g^*\overline{\rho}_2(u(g)^*\gamma(T_g)^*),
\end{aligned}$$

which proves the statement. □

Let $\{e_{g,h}\}_{g,h\in G}$ the natural system of matrix units of $B(\ell^2(G))$. Define the unitary $\rho_g^\zeta \in B(\mathcal{H}\otimes\ell^2(G))$ by

$$\rho_g^\zeta = \sum_{h\in G}\theta_h^{'-1}(\zeta'(h,g)^*)v(g)\otimes e_{h,hg}.$$

Direct computations show

$$\begin{cases} \rho_g^\zeta\rho_h^\zeta = (\zeta(g,h)\otimes 1)\rho_{gh}^\zeta \\ \rho_g^\zeta(\hat{x}\otimes 1)\rho_g^{\zeta^*} = \theta_g(\hat{x})\otimes 1 \end{cases}$$

for $\hat{x} \in \hat{\mathcal{A}}$. Note that the $C^*$-algebra generated by $\rho_g^\zeta$ $(g \in G)$ and $\hat{\mathcal{A}}\otimes\mathbf{C}$ is the twisted crossed product $\hat{\mathcal{A}} \rtimes_{\theta,\zeta} G$ (see [**76**]).

In the rest of the chapter, we will identify $(\sigma, \sigma\overline{\sigma}\sigma)$ (see Lemma 12.9) with $\mathcal{H}\otimes\ell^2(G)$ in the obvious way.

THEOREM 12.10. *The multiplicative unitary $\overline{V}$ corresponding to the inclusion $\mathcal{R} \supseteq \sigma(\mathcal{R})$ of depth 2 is given by*

$$\begin{aligned}
\overline{V} &= \sum_g [F_{23}(((id\otimes\theta_g^{'-1})(w_0(g))(\theta_g\otimes id)(V))\otimes 1)F_{23}\,(\rho_g^\zeta\otimes 1)]\otimes e_{g,g}\\
&= \sum_g [(1\otimes 1\otimes v(g)^*)F_{23}(V\eta_g^*\otimes 1)F_{23}\,(\rho_g^\zeta\otimes v(g))]\otimes e_{g,g}.
\end{aligned}$$

PROOF. By using the definition of $W(i,g)$, we compute

$$
\begin{aligned}
&\sigma\cdot\overline{\sigma}(W(i,g))W(j,h)\\
&= \overline{\rho}_2\cdot\gamma\cdot\rho_2\left(\overline{\rho}_2\left(\gamma(T_g)u(g)\alpha_g(S_i)\right)\lambda_g\right)\overline{\rho}_2\left(\gamma(T_h)u(h)\alpha_h(S_j)\right)\lambda_h\\
&= \overline{\rho}_2\left(\gamma\left(\rho_2\cdot\overline{\rho}_2\left(\gamma(T_g)u(g)\alpha_g(S_i)\right)\rho_2(\lambda_g)T_h\right)u(h)\alpha_h(S_j)\right)\lambda_h\\
&= \overline{\rho}_2\left(\gamma\left(\rho_2\cdot\overline{\rho}_2\left(\gamma(T_g)u(g)\alpha_g(S_i)\right)T_{hg^{-1}}\right)u(h)\alpha_h(S_j)\right)\lambda_h\\
&\qquad\qquad\qquad\text{(because of } \rho_2(\lambda_g)T_h = T_{hg^{-1}})\\
&= \overline{\rho}_2\left(\gamma\left(T_{hg^{-1}}\alpha_{hg^{-1}}\left(\gamma(T_g)u(g)\alpha_g(S_i)\right)\right)u(h)\alpha_h(S_j)\right)\lambda_h\\
&\qquad\qquad\qquad\text{(because of } T_{hg^{-1}}\in(\sigma_{hg^{-1}},\rho_2\overline{\rho}_2))\\
&= \overline{\rho}_2\Big(\gamma\left(T_{hg^{-1}}\right)\alpha^u_{hg^{-1}}\cdot\gamma\left(\gamma(T_g)u(g)\alpha_g(S_i)\right)u(h)\alpha_h(S_j)\Big)\lambda_h\\
&\qquad\qquad\qquad\text{(because of } \gamma\cdot\alpha_{hg^{-1}} = \alpha^u_{hg^{-1}}\cdot\gamma)\\
&= \overline{\rho}_2\Big(\gamma(T_{hg^{-1}})u(hg^{-1})\\
&\qquad\times\alpha_{hg^{-1}}\left(\sigma_{\mathcal{H}}\cdot\gamma(T_g)\gamma(u(g)\alpha_g(S_i))\right)u(hg^{-1})^*u(h)\alpha_h(S_j)\Big)\lambda_h\\
&= \sum_k\overline{\rho}_2(\gamma(T_{hg^{-1}})u(hg^{-1})\\
&\qquad\times\alpha_{hg^{-1}}\left(S_k\gamma(T_g)S_k^*\gamma(u(g)\alpha_g(S_i))\right)u(hg^{-1})^*u(h)\alpha_h(S_j))\lambda_h\\
&= \sum_k\overline{\rho}_2\left(\gamma(T_{hg^{-1}})u(hg^{-1})\alpha_{hg^{-1}}(S_k)\right)\lambda_{hg^{-1}}\\
&\qquad\times\lambda_{gh^{-1}}\overline{\rho}_2\left(\alpha_{hg^{-1}}\left(\gamma(T_g)S_k^*\gamma\left(u(g)\alpha_g(S_i)\right)\right)u(hg^{-1})^*u(h)\alpha_h(S_j)\right)\lambda_h\\
&= \sum_k W(k,hg^{-1})\overline{\rho}_2(\gamma(T_g)S_k^*\gamma\left(u(g)\alpha_g(S_i)\right)\\
&\qquad\times\alpha_{gh^{-1}}\left(u(hg^{-1})^*u(h)\alpha_h(S_j)\right))\lambda^*_{hg^{-1}}\lambda_h\\
&= \sum_k W(k,hg^{-1})\overline{\rho}_2\left(\gamma(T_g)S_k^*\gamma\left(u(g)\alpha_g(S_i)\right)\alpha_{gh^{-1}}\left(u(hg^{-1})^*u(h)\right)\alpha_g(S_j)\right)\lambda_g.
\end{aligned}
$$

The above last expression suggests that we need to compute

$$S_k^*\gamma\left(u(g)\alpha_g(S_i)\right)\alpha_{gh^{-1}}\left(u(hg^{-1})^*u(h)\right)\alpha_g(S_j),$$

which is equal to

$$
\begin{aligned}
&S_k^*\gamma(u(g))\alpha^u_g(\gamma(S_i))\alpha_g\left(\alpha_{h^{-1}}\left(u(hg^{-1})^*u(h)\right)S_j\right)\\
&\quad= S_k^*\gamma(u(g))u(g)\alpha_g(\gamma(S_i))u(g)^*\alpha_g\left(\alpha_{h^{-1}}\left(u(hg^{-1})^*u(h)\right)S_j\right)\\
&\quad= S_k^*\gamma(u(g))u(g)\alpha_g\left(VFS_i\alpha_{h^{-1}}\left(\alpha_{hg^{-1}}(u(g)^*)u(hg^{-1})^*u(h)\right)S_j\right)\\
&\quad= S_k^*\gamma(u(g))u(g)\alpha_g\left(VFS_i\alpha_{h^{-1}}\left(u(h)^*\zeta'(hg^{-1},g)^*u(h)\right)S_j\right)\\
&\quad= S_k^*\gamma(u(g))u(g)\alpha_g(VFS_i\theta_h^{'-1}(\zeta'(hg^{-1},g)^*)S_j).
\end{aligned}
$$

We note that the flip $F$ exchanges $S_i$ and $\theta_h^{'-1}(\zeta'(hg^{-1},g)^*)S_j$ in $\mathcal{H}$ so that the above quantity is equal to

$$
\begin{aligned}
&S_k^*\gamma(u(g))u(g)\alpha_g\left(V\theta_h^{'-1}(\zeta'(hg^{-1},g)^*)S_jS_i\right)\\
&\quad= u(g)\alpha_g\left(\left(\alpha^u_g\right)^{-1}\left(S_k^*\gamma(u(g))\right)\right)\alpha_g\left(V\theta_h^{'-1}(\zeta'(hg^{-1},g)^*)S_jS_i\right)\\
&\quad= u(g)\alpha_g\left(\left(\alpha^u_g\right)^{-1}\left(S_k^*\gamma(u(g))\right)V\theta_h^{'-1}(\zeta'(hg^{-1},g)^*)S_jS_i\right).
\end{aligned}
$$

Substituting this computation back to the previous one, we see

$$\begin{aligned}&\sigma\cdot\overline{\sigma}(W(i,g))W(j,h)\\&\quad=\sum_k W(k,hg^{-1})W\left(\left(\alpha^u_g\right)^{-1}\left(S^*_k\gamma(u(g))\right)V\theta'^{-1}_h\left(\zeta'(hg^{-1},g)^*\right)S_jS_i,g\right).\end{aligned}$$

Note that $V\theta'^{-1}_h\left(\zeta'(hg^{-1},g)^*\right)S_jS_i$ appearing in the second $W$ sits in $V\mathcal{H}^2\subseteq\mathcal{H}^2$, and for $S,T\in\mathcal{H}$ we compute

$$\begin{aligned}\left(\alpha^u_g\right)^{-1}&(S^*_k\gamma(u(g)))ST=\left(\alpha^u_g\right)^{-1}\left(S^*_k\gamma(u(g))\alpha^u_g(S)\alpha^u_g(T)\right)\\&=\left(\alpha^u_g\right)^{-1}\left(S^*_k\gamma(u(g))\gamma(u(g)^*)w(g)S\gamma(u(g)^*)w(g)T\right)\\&\qquad\qquad\qquad\qquad\text{(by Lemma 10.1,(i))}\\&=\left(\alpha^u_g\right)^{-1}\left(S^*_kw_0(g)v(g)S\gamma(u(g)^*)w_0(g)v(g)T\right)\\&=\left(\alpha^u_g\right)^{-1}\left(\gamma(u(g)^*)w_0(g)S^*_kw_0(g)v(g)Sv(g)T\right)\\&=\left(\alpha^u_g\right)^{-1}\left(\gamma(u(g)^*)w(g)v(g)^*S^*_kw_0(g)v(g)Sv(g)T\right)\\&=v(g)^*S^*_kw_0(g)v(g)Sv(g)T\\&=v(g)^*S^*_kw_0(g)v(g)\sigma_{\mathcal{H}}(v(g))ST\\&=v(g)^*S^*_kw_0(g)\sigma_{\mathcal{H}}(v(g))v(g)ST\\&=S^*_k(id\otimes\theta'^{-1}_g)(w_0(g))v(g)ST.\end{aligned}$$

Here, the fourth equality follows from $S^*_kw_0(g)v(g)S\in\hat{\mathcal{A}}'=(\gamma,\gamma)$ and $w_0(g)\in\hat{\mathcal{A}}\otimes\hat{\mathcal{A}}'$ (see Lemma 10.1) while the sixth follows from Lemma 10.1,(i) applied for $v(g)^*S^*_kw_0(g)v(g)Sv(g)T\in\mathcal{H}$. Therefore, we have

$$\begin{aligned}&\sigma\cdot\overline{\sigma}(W(i,g))W(j,h)\\&=\sum_k W(k,hg^{-1})W\left(S^*_k(id\otimes\theta'^{-1}_g)(w_0(g))v(g)V\theta'^{-1}_h(\zeta'(hg^{-1},g)^*)S_jS_i,g\right)\\&=\sum_k W(k,hg^{-1})\\&\qquad\times W\left(S^*_k(id\otimes\theta'^{-1}_g)(w_0(g))(\theta_g\otimes id)(V)\theta'_g\cdot\theta'^{-1}_h(\zeta'(hg^{-1},g)^*)v(g)S_jS_i,g\right)\\&=\sum_k W(k,hg^{-1})\\&\qquad\times W(S^*_k(id\otimes\theta'^{-1}_g)(w_0(g))(\theta_g\otimes id)(V)\theta'^{-1}_{hg^{-1}}(\zeta'(hg^{-1},g)^*)v(g)S_jS_i,g).\end{aligned}$$

Here, the last equality follows from $\theta'_g\cdot\theta'^{-1}_h=\theta'^{-1}_{hg^{-1}}\cdot Ad(\zeta'(hg^{-1},g))$, which is a consequence of

$$\theta'_{hg^{-1}}\cdot\theta'_g\cdot\theta'^{-1}_h=Ad(\zeta'(hg^{-1},g))\cdot\theta'_h\cdot\theta'^{-1}_h=Ad(\zeta'(hg^{-1},g)).$$

Since $\sigma\cdot\overline{\sigma}(W(i,g))W(j,h)=\overline{V}W(j,h)W(i,g)$, we have

$$\overline{V}=\sum_{i,j,g,h}\sigma\cdot\overline{\sigma}(W(i,g))W(j,h)W(i,g)^*W(j,h)^*.$$

Therefore, by changing the variable $h$ to $hg$ in the final result of the preceding computation, we conclude

$$\begin{aligned}\overline{V} = &\sum_{i,j,k,g,h} W(k,h)\\ &\times W\left(S_k^*(id\otimes\theta_g^{'-1})(w_0(g))(\theta_g\otimes id)(V)\theta_h^{'-1}(\zeta'(h,g)^*)v(g)S_jS_i, g\right)\\ &\times W(i,g)^*W(j,hg)^*\\ = &\sum_{i,j,k,g,h} W(k,h)W\left(S_k^*(id\otimes\theta_g^{'-1})(w_0(g))(\theta_g\otimes id)(V)S_jS_i,g\right)\\ &\times W(i,g)^*W\left(v(g)^*\theta_h^{'-1}(\zeta'(h,g))S_j,hg\right)^*.\end{aligned}$$

It is easy to see that the above is exactly the first expression in the theorem while the second follows from Lemma 12.8. □

Let $\mathcal{B}$ and $\hat{\mathcal{B}}$ be the Kac algebra and the dual Kac algebra associated with $\overline{V}$. We determine the structure of $\hat{\mathcal{B}}$. Let $\hat{\overline{\Gamma}}$, $\hat{\overline{\kappa}}$, $\hat{\overline{\epsilon}}$, and $\hat{\overline{h}}$ be the comultiplication, the antipode, the counit, and the normalized Haar measure of $\hat{\mathcal{B}}$. For vectors $\xi,\eta\in\mathcal{H}\otimes\ell^2(G)$, $\omega_{\xi,\eta}$ denotes the linear functional of $B(\mathcal{H}\otimes\ell^2(G))$ defined by $\omega_{\xi,\eta}(x)=\langle x\xi,\eta\rangle$.

LEMMA 12.11. *(i)* $\sqrt{n}\,\left(id\otimes\omega_{\Omega_{\hat{h}}\otimes\delta_g,\Omega_h\otimes\delta_g}\right)\left(\overline{V}\right)=(\nu_g^*\otimes 1)\rho_g^{\zeta}$.
*(ii)* $\sqrt{n}\,\left(id\otimes\omega_{\Omega_h\otimes\delta_g,\Omega_{\hat{h}}\otimes\delta_g}\right)\left(\overline{V}^*\right)=\rho_g^{\zeta *}$.

PROOF. (i) The first expression of $\overline{V}$ in the above theorem implies

$$\begin{aligned}&\sqrt{n}\,\left(id\otimes\omega_{\Omega_{\hat{h}}\otimes\delta_g,\Omega_h\otimes\delta_g}\right)\left(\overline{V}\right)\\ &\quad=\sqrt{n}\,\left((id\otimes\omega_{\Omega_{\hat{h}},\Omega_h})\left((id\otimes\theta_g^{'-1})(w_0(g))(\theta_g\otimes id)(V)\right)\otimes 1\right)\rho_g^{\zeta}.\end{aligned}$$

Thus, (i) follows from $V(\xi\otimes\Omega_{\hat{h}})=\xi\otimes\Omega_{\hat{h}}$ and $w_0(g)^*(\xi\otimes\Omega_h)=\nu_g\xi\otimes\Omega_h$.
(ii) The second expression of $\overline{V}$ implies

$$\begin{aligned}&\sqrt{n}\,\left(id\otimes\omega_{\Omega_h\otimes\delta_g,\Omega_{\hat{h}}\otimes\delta_g}\right)\left(\overline{V}^*\right)\\ &\quad=\sqrt{n}\,\rho_g^{\zeta *}((id\otimes\omega_{\Omega_{\hat{h}},\Omega_h})((1\otimes v(g)^*)\eta_gV^*(1\otimes v(g)))\otimes 1),\end{aligned}$$

which proves (ii) in the same way as above. □

THEOREM 12.12. *The dual Kac algebra* $\hat{\mathcal{B}}$ *is generated by* $\hat{\mathcal{A}}\otimes\mathbf{C}$ *and* $\{\rho_g^{\zeta}\}_{g\in G}$*, i.e.,* $\hat{\mathcal{B}}=\hat{\mathcal{A}}\rtimes_{\theta,\zeta}G$*, a twisted crossed product. We identify* $\hat{\mathcal{A}}$ *with* $\hat{\mathcal{A}}\otimes\mathbf{C}\subseteq\hat{\mathcal{B}}$ *in what follows, and the Kac algebra structure of* $\hat{\mathcal{B}}$ *is given as follows:*
*(i) For* $\hat{x}\in\hat{\mathcal{A}}$ *and* $g\in G$*, we have*

$$\hat{\overline{\Gamma}}(\hat{x})=\hat{\Gamma}(\hat{x}),\quad \hat{\overline{\Gamma}}\left(\rho_g^{\zeta}\right)=\eta_g^*\left(\rho_g^{\zeta}\otimes\rho_g^{\zeta}\right).$$

*(ii) For* $\hat{x}\in\hat{\mathcal{A}}$ *and* $g\in G$*, we have*

$$\hat{\overline{\kappa}}(\hat{x})=\hat{\kappa}(\hat{x}),\quad \hat{\overline{\kappa}}\left(\rho_g^{\zeta}\right)=\rho_g^{\zeta *}\nu_g^*.$$

*(iii) For* $\hat{x}\in\hat{\mathcal{A}}$ *and* $g\in G$*, we have*

$$\hat{\overline{\epsilon}}(\hat{x})=\hat{\epsilon}(\hat{x}),\quad \hat{\overline{\epsilon}}\left(\rho_g^{\zeta}\right)=1.$$

*(iv) For $\hat{x} \in \hat{\mathcal{A}}$ and $g \in G$, we have*

$$\hat{\bar{h}}\left(\hat{x}\rho_g^{\zeta}\right) = \hat{h}\left(\hat{x}\right)\delta_{g,e}.$$

PROOF. We have

$$\hat{\mathcal{A}} \otimes \mathbf{C} = \left\{(id \otimes \omega_{\xi\otimes\delta_e,\eta\otimes\delta_e})\left(\overline{V}\right);\ \xi,\eta \in \mathcal{H}\right\}$$

from the definition of $\hat{\mathcal{A}}$, and Lemma 12.11 shows $(\hat{\mathcal{A}}\otimes\mathbf{C})\cup\{\rho_g^{\zeta}\}_{g\in G} \subseteq \hat{\mathcal{B}}$. Therefore, by comparing dimensions we conclude

$$\hat{\mathcal{B}} = \langle \hat{\mathcal{A}} \otimes \mathbf{C}, \rho_g^{\zeta}\ (g \in G)\rangle'' = \hat{\mathcal{A}} \rtimes_{\theta,\zeta} G.$$

(i) For $\hat{x} \in \hat{\mathcal{A}}$, we compute

$$\begin{aligned}
\hat{\bar{\Gamma}}\left(\hat{x}\right) &= \overline{V}^*\left(1 \otimes 1 \otimes \hat{x} \otimes 1\right)\overline{V} \\
&= \sum_g [\left(\rho_g^{\zeta^*} \otimes v(g)^*\right) F_{23}\left((\eta_g V^*(1 \otimes v(g)\hat{x}v(g)^*)V\eta_g^*) \otimes 1\right) \\
&\qquad \times F_{23}\left(\rho_g^{\zeta} \otimes v(g)\right)] \otimes e_{g,g} \\
&= \sum_g [\left(\rho_g^{\zeta^*} \otimes v(g)^*\right) F_{23}\left((\eta_g\hat{\bar{\Gamma}}(\theta_g(\hat{x}))\eta_g^*) \otimes 1\right) F_{23}\left(\rho_g^{\zeta} \otimes v(g)\right)] \otimes e_{g,g} \\
&= \sum_g [\left(\rho_g^{\zeta^*} \otimes v(g)^*\right) F_{23}\left((\theta_g \otimes \theta_g)\cdot(\hat{\bar{\Gamma}}(\hat{x})) \otimes 1\right) F_{23}\left(\rho_g^{\zeta} \otimes v(g)\right)] \otimes e_{g,g} \\
&= \sum_g [F_{23}(\hat{\bar{\Gamma}}(\hat{x}) \otimes 1)F_{23}] \otimes e_{g,g},
\end{aligned}$$

which proves the first formula. On the other hand, $\hat{\bar{\Gamma}}(\rho_g^{\zeta})$ is

$$\begin{aligned}
&\overline{V}^*\left(1 \otimes 1 \otimes \rho_g^{\zeta}\right)\overline{V} \\
&= \sum_h [\left(\rho_h^{\zeta^*} \otimes v(h)^*\right) F_{23}((\eta_h V^*(1 \otimes v(h)\theta_h'^{-1}(\zeta'(h,g)^*)v(g)v(hg)^*)V\eta_{hg}^*) \otimes 1) \\
&\qquad \times F_{23}\left(\rho_{hg}^{\zeta} \otimes v(hg)\right)]e_{h,hg} \\
&= \sum_h [\left(\rho_h^{\zeta^*} \otimes v(h)^*\right) F_{23}((\eta_h V^*(1 \otimes \zeta'(h,g)^*v(h)v(g)v(hg)^*)V\eta_{hg}^*) \otimes 1) \\
&\qquad \times F_{23}\left(\rho_{hg}^{\zeta} \otimes v(hg)\right)]e_{h,hg} \\
&= \sum_h [\left(\rho_h^{\zeta^*} \otimes v(h)^*\right) F_{23}((\eta_h V^*(1 \otimes \zeta(h,g))V\eta_{hg}^*) \otimes 1) \\
&\qquad \times F_{23}\left(\rho_{hg}^{\zeta} \otimes v(hg)\right)] \otimes e_{h,hg} \\
&= \sum_h [\left(\rho_h^{\zeta^*} \otimes v(h)^*\right) F_{23}((\eta_h\hat{\Gamma}(\zeta(h,g))\eta_{hg}^*) \otimes 1)F_{23}\left(\rho_{hg}^{\zeta} \otimes v(hg)\right)] \otimes e_{h,hg}
\end{aligned}$$

$$= \sum_h [\left(\rho_h^{\zeta *} \otimes v(h)^*\right) F_{23}(((\theta_g \otimes \theta_g)(\eta_g^*)(\zeta(h,g) \otimes \zeta(h,g))) \otimes 1)$$
$$\times F_{23}\left(\rho_{hg}^{\zeta} \otimes v(hg)\right)] \otimes e_{h,hg}$$
$$= \sum_h [\left(\rho_h^{\zeta *} \otimes v(h)^*\right) F_{23}(((\theta_g \otimes \theta_g)(\eta_g^*)) \otimes 1) F_{23}\left(\rho_h^{\zeta}\rho_g^{\zeta} \otimes \zeta'(h,g)^* v(h) v(g)\right)]$$
$$\otimes e_{h,hg}$$
$$= \sum_h [F_{23}(\eta_g^* \otimes 1) F_{23}\left(\rho_g^{\zeta} \otimes \theta_h'^{-1}(\zeta'(h,g)^*) v(g)\right)] \otimes e_{h,hg}$$
$$= F_{23}(\eta_g^* \otimes 1 \otimes 1) F_{23}\left(\rho_g^{\zeta} \otimes \rho_g^{\zeta}\right).$$

Thus, (i) has been proved.
(ii) It is obvious from Lemma 12.11.
(iii) Note that $\overline{R}_\sigma = \overline{\rho}_2(\overline{R}_{\rho_1})R_{\rho_2}$ holds. To express it in terms of the basis, we compute

$$\overline{\rho}_2(\gamma(T_g^*))\overline{R}_\sigma = \overline{\rho}_2(\overline{R}_{\rho_1}T_g^*)R_{\rho_2} = \overline{\rho}_2(\overline{R}_{\rho_1})\lambda_g\overline{\rho}_2(T_e^*)\lambda_g^* R_{\rho_2} = \overline{\rho}_2(\overline{R}_{\rho_1})\lambda_g$$
$$= \overline{\rho}_2(\overline{R}_{\rho_2})R_{\rho_2} = \overline{\rho}_2(\overline{R}_{\rho_2})R_{\rho_2} = \frac{1}{d(\rho_2)}\overline{\rho}_2(\overline{R}_{\rho_1})\lambda_g.$$

Thus, we get

$$\overline{R}_\sigma = \frac{1}{d(\rho_2)}\sum_g \overline{\rho}_2(\gamma(T_g^*)\overline{R}_{\rho_1})\lambda_g = \frac{1}{d(\rho_2)}\sum_g W(\overline{R}_{\rho_1}, g) = \frac{1}{d(\rho_2)}\sum_g \Omega_h \otimes \delta_g.$$

Since $\hat{\overline{\epsilon}}(X)\overline{R}_\sigma = X\overline{R}_\sigma$ for $X \in \mathcal{B}$, the statement is obvious for $\hat{x} \in \hat{\mathcal{A}}$. For $g \in G$, we have

$$\hat{\overline{\epsilon}}(\rho_g^{\zeta})\overline{R}_\sigma = \frac{1}{d(\rho_2)}\sum_{h,k} \theta_h'^{-1}(\zeta'(h,g))v(g)\Omega_h \otimes e_{h,hg}\delta_k$$
$$= \frac{1}{d(\rho_2)}\sum_h \hat{\epsilon}'(\zeta'(h,g))\Omega_h \otimes \delta_h.$$

This shows $\hat{\epsilon}'(\zeta'(h,g)) = 1$ and $\hat{\overline{\epsilon}}(\rho_g^{\zeta}) = 1$.
(iv) Since $R_\sigma = \overline{\rho}_1(\overline{R}_{\rho_2})R_{\rho_1}$, the GNS cyclic vector for the Haar measure $\Omega_{\hat{\overline{h}}}$ is identified with

$$\sigma(R_\sigma) = \overline{\rho}_2(\gamma(\overline{R}_{\rho_2})\rho_1(R_{\rho_1})) = W(\rho_1(R_{\rho_1}), e) = \Omega_{\hat{h}} \otimes \delta_e,$$

which proves the statement. □

**Remark.** Assume that the action $\alpha$ preserves the subfactor $\rho_1(\mathcal{R})$ globally invariant, i.e., $\alpha_g \in Aut(\mathcal{R}, \rho_1(\mathcal{R}))$. Then, the cocycle we have introduced is trivial and we get $\hat{\mathcal{B}} = \hat{\mathcal{A}} \rtimes_\alpha G$, the ordinary crossed product. This is exactly the situation analyzed by De Cannière ([**11**]), and full details are left to the reader.

COROLLARY 12.13. *(i)* $\hat{\kappa}(\nu_g) = \nu_g$ *for* $g \in G$.
*(ii)* $\hat{\epsilon}(\zeta(g,h)) = 1$ *for* $g, h \in G$.

PROOF. Since the antipode is involutive, (i) follows from (iii) of the above theorem. On the other hand, (ii) follows from $\hat{\epsilon}'(\zeta'(g,h)) = 1$. □

This result is used to show the next useful fact on normalization of cocycles (see the beginning of the next chapter). We say that $\eta_g$ is **normalized** when the following conditions are satisfied:

$$\begin{cases} (\hat{\epsilon} \otimes id)(\eta_g) = (id \otimes \hat{\epsilon})(\eta_g) = 1, \\ m \cdot (id \otimes \hat{\kappa})(\eta_g) = 1. \end{cases}$$

COROLLARY 12.14. *(i) There always exists a normalized cocycle $\tilde{\eta}_g$ equivalent to $\eta_g$.*
*(ii) If $\eta_g$ is normalized, then $F(\hat{\kappa} \otimes \hat{\kappa})(\eta_g^*)F = \eta_g$.*

PROOF. (i) Note that the $*$-algebra generated by $\nu_g$ is in the fixed-point set of $\hat{\kappa}$. Thus, we can take a unitary $\xi_g \in \hat{\mathcal{A}}$ satisfying $\xi_g^2 = \nu_g^*$, $\hat{\kappa}(\xi_g) = \xi_g$, and $\hat{\epsilon}(\xi_g) = 1$, and $(\xi_g \otimes \xi_g)\eta_g \hat{\Gamma}(\xi_g^*)$ does the job. (ii) It follows from Lemma 12.8. □

**Remark.** Assume that a cocycle $\eta_g$ is already normalized. Then, the perturbed one (via $\xi_g$) remains normalized as long as $\hat{\epsilon}(\xi_g) = 1$ and $\hat{\kappa}(\xi_g) = \xi_g^*$.

CHAPTER 13

# 2-cocycles of Kac algebras

Here discussions on 2-cocycles of Kac algebra are given. Recall that in the group algebra case they are closely related to ergodic actions (with full multiplicities) as was studied in [**73**]. We also closely examine Kac algebras of the form $\mathcal{A} \rtimes_{\theta,\varsigma} \mathbf{Z}_2$ appeared in the previous chapter (with $G = \mathbf{Z}_2$) for the following $\mathcal{A}$:

the Kac-Paljutkin algebra and the group rings $\mathbf{C}(D_8)$, $\mathbf{C}(D_{12})$, $\mathbf{C}(\mathfrak{A}_4)$.

## 1. Preliminaries

**1.1. Definition of 2-cocycles.** Let $\mathcal{A}$ be a finite-dimensional Kac algebra with coproduct $\Gamma$, antipode $\kappa$, counit $\epsilon$, and normalized Haar measure $h$. Following [**73**], we introduce the notion of 2-cocycles of $\mathcal{A}$.

DEFINITION 13.1.
*(i) A unitary $\omega \in \mathcal{A}\otimes\mathcal{A}$ is called a 2-cocycle of $\mathcal{A}$ if it satisfies the following cocycle relation:*

$$(\Gamma \otimes id)(\omega)(\omega \otimes 1) = (id \otimes \Gamma)(\omega)(1 \otimes \omega).$$

*The set of 2-cocycles of $\mathcal{A}$ is denoted by $Z^2(\mathcal{A}, \mathbf{T})$.*
*(ii) 2-cocycles $\omega_1, \omega_2 \in Z^2(\mathcal{A}, \mathbf{T})$ are equivalent if*

$$\omega_2 = \Gamma(\xi^*)\omega_1(\xi \otimes \xi) \quad \textit{for some unitary } \xi \in \mathcal{A}.$$

*A cocycle equivalent to 1 is called a coboundary, and the set of equivalence classes of 2-cocycles is denoted by $H^2(\mathcal{A}, \mathbf{T})$.*
*(iii) We say that $\omega \in Z^2(\mathcal{A}, \mathbf{T})$ is normalized if it satisfies*

$$(\epsilon \otimes id)(\omega) = (id \otimes \epsilon)(\omega) = 1 \quad \textit{and} \quad m \cdot (\kappa \otimes id)(\omega) = 1,$$

*where $m : \mathcal{A} \otimes \mathcal{A} \longrightarrow \mathcal{A}$ means the multiplication map.*

Normalization of cocycles here is essentially the same as in [**73**] though they look different. The next result is essentially obtained in [**73**] (for group algebras of compact groups), but we present a proof here for the sake of completeness.

LEMMA 13.2. *Let $\mathcal{A}$ be a finite-dimensional Kac algebra.*
*(i) Every 2-cocycle of $\mathcal{A}$ is equivalent to a normalized one.*
*(ii) When a 2-cocycle $\omega$ is normalized, we have*

$$F(\kappa \otimes \kappa)(\omega)^* F = \omega.$$

*(iii) If normalized 2-cocycles $\omega_1$ and $\omega_2$ are equivalent with $\omega_2 = \Gamma(\xi^*)\omega_1(\xi \otimes \xi)$, then $\xi$ satisfies*

$$\epsilon(\xi) = 1 \quad \textit{and} \quad \kappa(\xi) = \xi^*.$$

PROOF. (i) It is easy to see that we may assume $(\epsilon \otimes id)(\omega) = (id \otimes \epsilon)(\omega) = 1$. The same argument as in the proof of Lemma 12.7 shows that $\xi = m \cdot (\kappa \otimes id)(\omega)$ satisfies

$$\Gamma(\xi^*)F(\kappa \otimes \kappa)(\omega)^*F = \omega(\xi^* \otimes \xi^*). \tag{13.1}$$

Note that $\xi$ is not zero because of $\epsilon(\xi) = (\epsilon \otimes \epsilon)(\omega) = 1$. First, we show that $\xi$ is a unitary. From (13.1) we get

$$\Gamma(\xi^*\xi) = \omega(\xi^*\xi \otimes \xi^*\xi)\omega^*. \tag{13.2}$$

Taking the norm of the both sides of (13.2), we get $||\xi|| = 1$. Applying $(h \otimes h)$ to the same, we obtain $h(\xi^*\xi) = 1$, which shows that $\xi$ is a unitary. Applying $m \cdot (\kappa \otimes id)$ to the both sides of

$$F(\kappa \otimes \kappa)(\omega)^*F = \Gamma(\xi)\omega(\xi^* \otimes \xi^*),$$

we obtain $\kappa(\xi) = \xi$. Thus, the argument in the proof of Corollary 12.14 shows the result. We note that (ii) follows from (13.1) while (iii) is straight-forward. □

**1.2. Twisted dual algebras.** Let $\omega$ be a normalized cocycle of $\mathcal{A}$. Now, we introduce the notion of $\omega$-**twisted dual algebra** $\hat{\mathcal{A}}_\omega$ by generalizing arguments in [**73**]. We equip the dual space $\mathcal{A}^*$ with a product $\circ_\omega$ and a $*$-operation $+$ as follows:

$$\begin{cases} \langle x, \varphi \circ_\omega \psi \rangle = \langle \Gamma(x)\omega, \varphi \otimes \psi \rangle \text{ for } x \in \mathcal{A} \text{ and } \varphi, \psi \in \mathcal{A}^*, \\ \langle x, \varphi^+ \rangle = \overline{\langle \kappa(x)^*, \varphi \rangle} \text{ for } x \in \mathcal{A} \text{ and } \varphi \in \mathcal{A}^*. \end{cases}$$

With these operations $\mathcal{A}^*$ becomes a finite-dimensional C$^*$-algebra, which we denote by $\hat{\mathcal{A}}_\omega$. Note that it is the ordinary dual Hopf algebra $\hat{\mathcal{A}}$ when $\omega$ is trivial.

LEMMA 13.3. *(i) The C$^*$-algebra structure of $\hat{\mathcal{A}}_\omega$ depends only on the class $[\omega] \in H^2(\mathcal{A}, \mathbf{T})$.*
*(ii) A cocycle $\omega$ is a coboundary if and only if $\hat{\mathcal{A}}_\omega$ has an abelian simple component.*

PROOF. (i) is easy. (ii) The "only if" part follows from (i). Let $\varphi_0 \in \mathcal{A}^*$ be a central minimal projection corresponding to an abelian simple component, and $v \in \mathcal{A}^{**} = \mathcal{A}$ be the corresponding 1-dimensional representation of $\hat{\mathcal{A}}_\omega$. Then, since

$$\varphi_0 \circ_\omega \varphi = \varphi \circ_\omega \varphi_0 = \langle v, \varphi \rangle \varphi_0 \quad (\text{for } \varphi \in \mathcal{A}^*),$$

for $x \in \mathcal{A}$ we have

$$\langle \Gamma(x)\omega, \varphi_0 \otimes \varphi \rangle = \langle \Gamma(x)\omega, \varphi \otimes \varphi_0 \rangle = \langle v, \varphi \rangle \langle x, \varphi_0 \rangle,$$

which means

$$(\varphi_0 \otimes id)(\Gamma(x)\omega) = (id \otimes \varphi_0)(\Gamma(x)\omega) = \varphi_0(x)v.$$

Applying $\Gamma$ to the above, we have

$$\begin{aligned} \varphi_0(x)\Gamma(v) &= (\varphi_0 \otimes id \otimes id)((id \otimes \Gamma) \cdot \Gamma(x)(id \otimes \Gamma)(\omega)) \\ &= (\varphi_0 \otimes id \otimes id)((\Gamma \otimes id)(\Gamma(x)\omega)(\omega \otimes 1)(1 \otimes \omega^*)) \\ &= (v \otimes (\varphi_0 \otimes id)(\Gamma(x)\omega))\omega^* \\ &= \varphi_0(x)(v \otimes v)\omega^*. \end{aligned}$$

Since $\varphi_0$ is not zero, we get $\Gamma(v) = (v \otimes v)\omega^*$, which implies $\Gamma(vv^*) = vv^* \otimes vv^*$. Thus, $vv^*$ is a positive group-like, and hence $vv^* = 1$. Therefore, $\omega = \Gamma(v^*)(v \otimes v)$ is a coboundary. □

**1.3. Deformation of coproducts.** Let $\mathcal{A}$ be a finite-dimensional Kac algebra with a normalized 2-cocycle $\omega$, which gives rise to a new coproduct $\Gamma^\omega$. Namely, we set

$$\Gamma^\omega = Ad(\omega^*) \cdot \Gamma$$

(see [**43, 44, 73**]). Thanks to the normalization of $\omega$, we can define a new Kac algebra $\mathcal{A}^\omega$ with the coproduct $\Gamma^\omega$ and the other operations unchanged. We denote by $\hat{\mathcal{A}}^\omega$ its dual Kac algebra.

PROPOSITION 13.4. *Let $\mathcal{A}$ and $\omega$ be as above.*
*(i) The multiplicative unitary associated with $\mathcal{A}^\omega$ is given by*

$$\omega^* VF(1 \otimes U)\omega(1 \otimes U)F,$$

*where $V$ is the multiplicative unitary associated with $\mathcal{A}$ and $U = J\hat{J}$.*
*(ii) We have $G(\hat{\mathcal{A}}^\omega) \cong G(\hat{\mathcal{A}})$.*

PROOF. (i) This can be shown by a similar argument as in the proof of Lemma 12.6,(ii). (ii) is easy. □

**1.4. Twisted crossed products.** Let $\mathcal{A}$ be a finite-dimensional Kac algebra and $H$ a finite group. We consider a triple $(\theta, \{\eta_h\}_{h\in H}, \{\zeta(h,k)\}_{h,k\in H})$ consisting of
(i) a cocycle action $\theta$ of $H$ on $\mathcal{A}$ with a cocycle $\zeta$ satisfying $\epsilon \cdot \theta_h = \epsilon$:

$$\begin{cases} \theta_h \cdot \theta_k = Ad(\zeta(h,k)) \cdot \theta_{hk}, \\ \zeta(h,k)\zeta(hk,l) = \theta_h(\zeta(k,l))\zeta(h,kl), \\ \zeta(e,h) = \zeta(h,e) = 1, \end{cases}$$

and (ii) a 2-cocycle $\eta_h^*$ of $\mathcal{A}$ satisfying

$$\begin{cases} \eta_h \Gamma(\cdot)\eta_h^* = (\theta_h \otimes \theta_h) \cdot \Gamma \cdot \theta_h^{-1}, \\ \eta_{hk} = (\zeta(h,k)^* \otimes \zeta(h,k)^*)(\theta_h \otimes \theta_h)(\eta_k)\eta_h\Gamma(\zeta(h,k)) \end{cases}$$

(see Proposition 12.3). We say that two such triples $(\theta, \zeta, \eta)$ and $(\tilde{\theta}, \tilde{\eta}, \tilde{\zeta})$ are **equivalent** if there exists $\{\xi(h)\}_{h\in H} \subset U(\mathcal{A})$ satisfying

$$\begin{cases} \tilde{\theta}_h = Ad(\xi(h)) \cdot \theta_h, \\ \tilde{\eta}_h = (\xi(h) \otimes \xi(h))\eta_h\Gamma(\xi(h)^*), \\ \tilde{\zeta}(h,k) = \xi(h)\theta_h(\xi(k))\zeta(h,k)\xi(hk)^*. \end{cases}$$

We say that two triples are **weakly equivalent** if there exist a Kac algebra automorphism $\alpha$ of $\mathcal{A}$ and a group automorphism $\tau \in Aut(H)$ such that

$$(\alpha \cdot \theta_{\tau(\cdot)} \cdot \alpha^{-1}, (\alpha \otimes \alpha)(\eta_{\tau(\cdot)}), \alpha(\zeta(\tau(\cdot), \tau(\cdot)))) \text{ is equivalent to } (\tilde{\theta}, \tilde{\eta}, \tilde{\zeta}).$$

We may and do assume that $\eta_h^*$ is a normalized 2-cocycle passing to an equivalent triple if necessary. To keep this condition, we always assume

$$\xi(e) = 1, \quad \epsilon(\xi(h)) = 1, \quad \kappa(\xi(h)) = \xi(h)^*.$$

Note that $\theta_h$ can be regarded as a Kac algebra isomorphism between $\mathcal{A}$ and $\mathcal{A}^{\eta_h^*}$, and consequently that it commutes with $\kappa$. Since $\eta_{hk}^*$ and $\theta_h(\eta_k^*)\eta_h^*$ are normalized 2-cocycles, we get

$$\epsilon(\zeta(h,k)) = 1, \quad \kappa(\zeta(h,k)) = \zeta(h,k)^*.$$

The twisted crossed product $\mathcal{B} = \mathcal{A} \rtimes_{\theta,\zeta} H$ with the natural implementing unitaries $\left\{\lambda_h^\zeta\right\}_{h\in H}$ (see [**76**]) turns out to be a Kac algebra equipped with the following operations (that extend those of $\mathcal{A}$):

$$\Gamma(\lambda_k^\zeta) = \eta_k^*(\lambda_k^\zeta \otimes \lambda_k^\zeta), \quad \kappa(\lambda_k^\zeta) = \lambda_k^{\zeta^*},$$
$$\epsilon(\lambda_k^\zeta) = 1, \quad h(x\lambda_k^\zeta) = h(x)\delta_{k,e}, \quad (\text{for } k \in H \text{ and } x \in \mathcal{A}).$$

It is routine to check that the above Kac algebra structure depends only on the equivalence class $[(\theta, \eta, \zeta)]$. The underlying algebra structure of the dual Kac algebra $\hat{\mathcal{B}}$ of $\mathcal{B} = \mathcal{A} \rtimes_{\theta,\zeta} H$ and the intrinsic groups $G(\mathcal{B}), G(\hat{\mathcal{B}})$ are described by

PROPOSITION 13.5. *Let $\hat{\mathcal{B}}$ be the dual Kac algebra of $\mathcal{B} = \mathcal{A} \rtimes_{\theta,\zeta} H$.*
*(i) We have*

$$\hat{\mathcal{B}} \cong \oplus_{h\in H} \hat{\mathcal{A}}_{\eta_h^*}.$$

*(ii) We set*

$$H_0 = \{h \in H;\ \eta_h^* \in B^2(\mathcal{A}, \mathbf{T})\},$$

*and take a representative $(\theta, \eta, \zeta)$ satisfying $\eta_h = 1$ for all $h \in H_0$. Then, $\zeta$ restricted to $H_0$ describes the following exact sequence:*

$$1 \longrightarrow G(\mathcal{A}) \longrightarrow G(\mathcal{B}) \longrightarrow H_0 \longrightarrow 1.$$

*(iii) We identify $G(\hat{\mathcal{A}})$ with the set of 1-dimensional representations of $\mathcal{A}$, where the product of $G(\hat{\mathcal{A}})$ corresponds to the tensor product of representations. We set*

$$G(\hat{\mathcal{A}})_0 = \{\chi \in Hom(\mathcal{A}, \mathbf{C});\ \chi \cdot \theta_h = \chi \text{ for } h \in H,\ \chi(\zeta(\cdot,\cdot)) \in B^2(H, \mathbf{T})\},$$

*and take a representative $(\theta, \eta, \zeta)$ satisfying $\chi(\zeta(h,k)) = 1$ for $\chi \in G(\hat{\mathcal{A}})_0$, $h, k \in H$. Then, $(\chi_1 \otimes \chi_2)(\eta_\cdot)$ gives an element of $Z^2(G(\hat{\mathcal{A}})_0, Hom(H, \mathbf{T}))$, which describes*

$$1 \longrightarrow Hom(H, \mathbf{T}) \longrightarrow G(\hat{\mathcal{B}}) \longrightarrow G(\hat{\mathcal{A}})_0 \longrightarrow 1.$$

PROOF. (i) For $\varphi \in \mathcal{A}^*$ and $h \in H$, we define $(\varphi)_h \in \mathcal{B}^*$ by

$$(\varphi)_h(x\lambda_k^\zeta) = \varphi(x)\delta_{h,k}.$$

Clearly, $\mathcal{B}^*$ is spanned by $\{(\varphi)_h\}_{\varphi\in\mathcal{A}^*,\, h\in H}$. Using the definition of the product of the dual Kac algebra, we get

$$\begin{aligned}
\langle x\lambda_l^\zeta, (\varphi)_h \circ (\psi)_k\rangle &= \langle \Gamma(x)\eta_l^* \left(\lambda_l^\zeta \otimes \lambda_l^\zeta\right), (\varphi)_h \otimes (\psi)_k\rangle \\
&= \delta_{h,k}\delta_{h,l}\langle \Gamma(x)\eta_l^*, \varphi \otimes \psi\rangle \\
&= \delta_{h,k}\langle x\lambda_l^\zeta, (\varphi \circ_{\eta_h^*} \psi)_h\rangle.
\end{aligned}$$

Thus, we get $(\varphi)_h \circ (\psi)_k = \delta_{h,k}(\varphi \circ_{\eta_h^*} \psi)_h$. In a similar way, we can show $((\varphi)_h)^+ = (\varphi^+)_h$, which proves the statement.
(ii) By seeking 1-dimensional representations of $\hat{\mathcal{B}}$ based on Lemma 1.3, we can show that $G(\mathcal{B})$ is generated by $G(\mathcal{A})$ and $\{\lambda_h^\zeta\}_{h\in H_0}$. When $\eta_h = \eta_k = \eta_{hk} = 1$, we have

$$1 = (\zeta(h,k)^* \otimes \zeta(h,k)^*)\Gamma(\zeta(h,k)),$$

which means $\zeta(h,k) \in G(\mathcal{A})$. Thus, we get the result.
(iii) can be proved in the same way as in the case of $\mathcal{A} = \ell^\infty(N)$. □

When $G$ is a finite group, we use the notations $Z^2(\hat{G}, \mathbf{T})$, $H^2(\hat{G}, \mathbf{T})$, and $\ell^\infty_\omega(G)$ for $Z^2(\mathbf{C}(G), \mathbf{T})$, $H^2(\mathbf{C}(G), \mathbf{T})$, and $\widehat{\mathbf{C}(G)}_\omega$. We call $\omega \in Z^2(\mathbf{C}(G), \mathbf{T})$ a 2-cocycle of $\hat{G}$. Note that $(\widehat{\mathbf{C}(G)}_\omega)^* = \ell^\infty(G)$ and $\ell^\infty_\omega(G) = \ell^\infty(G)$ as linear spaces.

The next result comes from Goldman's theorem and Theorem 12.8. (Note that the requirement $[\gamma\alpha_g] = [\alpha_g\gamma]$ at the beginning of Chapter 12 is automatic since the group involved here is $\mathbf{Z}_2$.)

COROLLARY 13.6. *Let $\mathcal{B}$ be a Kac algebra with* $\dim \mathcal{B} = 2 \times \#G(\mathcal{B})$ *and we set* $G = G(\mathcal{B})$. *Then, there exists a triple $(\theta, \eta, \zeta)$ for $\mathcal{A} = \mathbf{C}(G)$ and $H = \mathbf{Z}_2$ with $[\eta_1^*]$ non-trivial such that $\mathcal{B} \cong \mathbf{C}(G) \rtimes_{\theta,\zeta} \mathbf{Z}_2$. Such $\mathcal{A}$ is classified by the weak equivalence classes of $(\theta, \eta, \zeta)$. The dual Hopf algebra $\hat{\mathcal{B}}$ is as follows:*

$$\hat{\mathcal{B}} \cong \ell^\infty(G) \oplus \ell^\infty_{\eta_1^*}(G).$$

*The intrinsic group $G(\hat{\mathcal{B}})$ is given by the following exact sequence:*

$$1 \longrightarrow \hat{\mathbf{Z}}_2 \longrightarrow G(\hat{\mathcal{B}}) \longrightarrow Hom(G, \mathbf{T})_0 \longrightarrow 1,$$

*where $Hom(G, \mathbf{T})_0$ is the fixed-point subgroup of $Hom(G, \mathbf{T})$ under $\theta$.*

## 2. Ergodic actions and 2-cocycles

To compute concrete examples of 2-cocycles, we summarize basic facts on the relationship between ergodic actions of finite groups and 2-cocycles of group algebras, which have been established in [**73**] in the case of compact groups.

DEFINITION 13.7. *Assume that $\mathcal{A}$ is a finite-dimensional von Neumann algebra equipped with an action $\alpha : G \longrightarrow Aut(\mathcal{A})$ of a finite group $G$.*
*(i) We say that $\alpha$ is* **ergodic** *if the fixed-point algebra $\mathcal{A}^\alpha$ is trivial.*
*(ii) We say that $\alpha$ has* **full multiplicity** *if* $\dim \mathcal{A} = \#G$.

Note that if $\mathcal{A}$ possesses an ergodic action of $G$ then $\mathcal{A}$ is a direct sum of full matrix algebras with the same rank because $G$ acts ergodically on the center of $\mathcal{A}$.

It is easy to show that the left translation of $G$ on $\ell^\infty(G)$ gives a full multiplicity ergodic action of $G$ on $\ell^\infty_\omega(G)$.

THEOREM 13.8. (*A. Wassermann* [**73**]) *The above construction gives rise to a one-to-one correspondence between $H^2(\hat{G}, \mathbf{T})$ and the conjugacy classes of ergodic actions of $G$ with full multiplicity.*

Let $\beta$ be an action of a subgroup $H$ ($\subseteq G$) on $\mathcal{A}$. We denote by $\rho$ the right translation of $G$ on $\ell^\infty(G)$. Then, the induced action $Ind_N^G\beta$ is (by definition) the restriction of the left translation of $G$ to $(\ell^\infty(G) \otimes \mathcal{A})^H$, where the $H$ action is given by $\rho \otimes \beta$. When $\beta$ is an ergodic action with full multiplicity, so is $Ind_N^G\beta$. It is easy to show that every ergodic action of a finite group with full multiplicity is conjugate to an induced action of an ergodic action of the isotropy subgroup with full multiplicity on a simple component (or more precisely, a minimal central projection). Since ergodic actions act transitively on the minimal central projections, the above subgroup is uniquely determined up to conjugacy. This shows the following:

LEMMA 13.9.
*(i)* $\#H^2(\hat{D}_8, \mathbf{T}) = 3$ *for the dihedral group $D_8$ of order* 8.
*(ii)* $\#H^2(\hat{Q}_8, \mathbf{T}) = 1$ *for the quaternion group $Q_8$.*
*(iii)* $\#H^2(\hat{D}_{10}, \mathbf{T}) = 1$ *for the dihedral group $D_{10}$ of order* 10.

*(iv)* $\#H^2(\hat{D}_{12}, \mathbf{T}) = 2$ *for the dihedral group* $D_{12}$ *of order* 12.
*(v)* $\#H^2(\hat{\mathfrak{A}}_4, \mathbf{T}) = 2$ *for the alternating group* $\mathfrak{A}_4$.

PROOF. (i) and (ii): Dimension counting argument shows that every non-trivial full multiplicity ergodic action of a group of order 8 is induced by a full multiplicity ergodic action of a subgroup of order 4 on $M_2(\mathbf{C})$. Since $H^2(\mathbf{Z}_4, \mathbf{T})$ is trivial, this subgroup should be isomorphic to $\mathbf{Z}_2 \times \mathbf{Z}_2$. Since $D_8$ has two mutually non-conjugate subgroups isomorphic to $\mathbf{Z}_2 \times \mathbf{Z}_2$, $H^2(\hat{D}_8, \mathbf{T})$ has two non-trivial elements. (ii) follows from that fact that $Q_8$ does not have a subgroup isomorphic to $\mathbf{Z}_2 \times \mathbf{Z}_2$. (iii): Dimension counting argument shows that there is no non-trivial ergodic action of a group of order 10. (iv) and (v) follow from a similar argument as above. □

Let $G$ be a group with a subgroup $H$, and $\omega$ be a 2-cocycle of $\hat{H}$. Since $\mathbf{C}(H)$ is a Hopf-subalgebra of $\mathbf{C}(G)$, we can regard a 2-cocycle $\omega$ of $\hat{H}$ as a 2-cocycle of $\hat{G}$ as well. Thus, we have two natural full multiplicity ergodic actions of $G$ on $\ell^\infty_\omega(G)$ and $(\ell^\infty(G) \otimes \ell^\infty_\omega(H))^H$.

LEMMA 13.10. *The above two actions are conjugate.*

PROOF. We define $\pi : \ell^\infty_\omega(G) \longrightarrow (\ell^\infty(G) \otimes \ell^\infty_\omega(H))^H$ by

$$\pi(f) = \sum_{h \in H} \rho_h(f) \otimes \delta_h, \quad f \in \ell^\infty(G).$$

Then, it is routine to show that $\pi$ is a $*$-isomorphism intertwining the two actions. □

This immediately implies the following:

COROLLARY 13.11. *Let* $G$ *be a finite group with a subgroup* $H$. *When* $\omega \in Z^2(\hat{H}, \mathbf{T})$ *is not a coboundary,* $\omega \in Z^2(\hat{G}, \mathbf{T})$ *is not a coboundary either.*

Let $\omega$ be a 2-cocycle of $\hat{G}$. We define a "bicharacter" $\beta_\omega$ for $\omega$ by $\beta_\omega = F\omega^* F\omega$. We define a map $\Lambda : \hat{\mathcal{A}}^\omega \longrightarrow \mathcal{A}^\omega$ by $\Lambda(\varphi) = (\varphi \otimes id)(\beta_\omega)$ with $\mathcal{A} = \mathbf{C}(G)$. For finite groups Theorem 12 in [**73**] shows

THEOREM 13.12. *The following two conditions are equivalent:*
*(i)* $\ell^\infty_\omega(G)$ *is a full matrix algebra.*
*(ii)* $\Lambda$ *is a* $*$*-isomorphism between* $\hat{\mathcal{A}}^\omega$ *and* $\mathcal{A}^\omega$.
*(In fact,* $\Lambda$ *is a co-algebra anti-isomorphism as well in this case.)*

Note that an ergodic action of $G$ on a full matrix algebra simply means an irreducible projective representation. Therefore, a group possesses a cocycle as above if and only if it admits an irreducible projective representation of dimension equal to the square root of the order of the group. Such a group is called a central type factor group (see [**23, 24, 64**]). It is easy to see that the order of a non-commutative group in this class is larger than or equal to 16 for it is a square number. Indeed, $G = D_8 \times \mathbf{Z}_2$ is one of smallest examples. We construct an irreducible projective representation of $G$ of dimension 4. We use the following presentation of $D_8$:

$$D_8 = \langle a, b |\ a^4 = b^2 = 1,\ bab = a^{-1} \rangle.$$

Let $c$ be a generator of $\mathbf{Z}_2$. Then, the following formula gives a desired projective representation:

$$\begin{cases} \pi(a) &= \begin{pmatrix} e^{\frac{\pi i}{4}} & 0 \\ 0 & e^{-\frac{\pi i}{4}} \end{pmatrix} \otimes \begin{pmatrix} 1 & 0 \\ 0 & -1 \end{pmatrix}, \\ \pi(b) &= \begin{pmatrix} 0 & 1 \\ 1 & 0 \end{pmatrix} \otimes \begin{pmatrix} 1 & 0 \\ 0 & 1 \end{pmatrix}, \\ \pi(c) &= \begin{pmatrix} 1 & 0 \\ 0 & 1 \end{pmatrix} \otimes \begin{pmatrix} 0 & 1 \\ 1 & 0 \end{pmatrix}. \end{cases}$$

Let $\omega$ be the cocycle of $\hat{G}$ corresponding to the ergodic action $Ad(\pi)$. Then, the Kac algebra $\mathbf{C}(G)^{\omega}$ and its dual have intrinsic groups isomorphic to $\mathbf{Z}_2 \times \mathbf{Z}_2 \times \mathbf{Z}_2$ thanks to Proposition 13.4. There are actually two such Kac algebras and they are self-dual (Chapter 12). In the next section, we will give a sketch of a proof of the following:

PROPOSITION 13.13. *Let $G$ be as above. Then, $\mathbf{C}(G)^{\omega}$ is isomorphic to the one obtained in (iii) of Lemma* 11.4.

**Problem.** Let $G$ be a central type factor group and $\pi$ be a projective irreducible representation of dimension equal to the square root of the order of $G$. On one hand, $\pi$ carries a 2-cocycle of $G$ in the usual sense. On the other hand, the ergodic action $Ad(\pi)$ with full multiplicity determines a 2-cocycle of $\hat{G}$. Describe the explicit relationship between these 2-cocycles (and also $\mathbf{C}(G)^{\omega}$).

## 3. Examples I Kac-Paljutkin algebra

Throughout this section, $\mathcal{A}$ denotes the self-dual Kac algebra of dimension 8 considered by Kac-Paljutkin ([**38**], and see also §2 of Chapter 7). The underlying algebra of $\mathcal{A}$ is

$$\mathbf{C} \oplus \mathbf{C} \oplus \mathbf{C} \oplus \mathbf{C} \oplus M_2(\mathbf{C}).$$

We denote by $A_\sigma$ the simple component of rank 2, and by $z(\sigma)$ the unit of $\mathcal{A}_\sigma$. Let $z = z(e)$, $z(a)$, $z(b)$, $z(c)$ be minimal central projections corresponding to the first 4 abelian simple components and $\{e(\sigma)_{ij}\}_{ij}$ be a system of matrix units of $\mathcal{A}_\sigma$. Note that $\mathcal{A}$ is generated by four group-like elements $u(e) = 1$, $u(a)$, $u(b)$, $u(c)$ and one 2-dimensional unitary co-representation $\{u(\sigma)_{ij}\}$ defined as follows:

$$\begin{cases} u(e) = 1 \oplus 1 \oplus 1 \oplus 1 \oplus \begin{pmatrix} 1 & 0 \\ 0 & 1 \end{pmatrix} & u(a) = 1 \oplus 1 \oplus 1 \oplus 1 \oplus \begin{pmatrix} -1 & 0 \\ 0 & -1 \end{pmatrix} \\ u(b) = 1 \oplus 1 \oplus -1 \oplus -1 \oplus \begin{pmatrix} -1 & 0 \\ 0 & 1 \end{pmatrix} & u(c) = 1 \oplus 1 \oplus -1 \oplus -1 \oplus \begin{pmatrix} 1 & 0 \\ 0 & -1 \end{pmatrix} \\ u(\sigma)_{11} = 1 \oplus -1 \oplus -1 \oplus 1 \oplus \begin{pmatrix} 0 & 0 \\ 0 & 0 \end{pmatrix} & u(\sigma)_{12} = 0 \oplus 0 \oplus 0 \oplus 0 \oplus \begin{pmatrix} 0 & e^{\frac{\pi i}{4}} \\ e^{-\frac{\pi i}{4}} & 0 \end{pmatrix} \\ u(\sigma)_{21} = 0 \oplus 0 \oplus 0 \oplus 0 \oplus \begin{pmatrix} 0 & e^{-\frac{\pi i}{4}} \\ e^{\frac{\pi i}{4}} & 0 \end{pmatrix} & u(\sigma)_{22} = 1 \oplus -1 \oplus 1 \oplus -1 \oplus \begin{pmatrix} 0 & 0 \\ 0 & 0 \end{pmatrix}. \end{cases}$$

The coproduct $\Gamma$ is given by

$$\begin{cases} \Gamma(u(g)) = u(g) \otimes u(g) \text{ for } g = e, a, b, c, \\ \Gamma(u(\sigma)_{ij}) = \sum_{k=1}^{2} u(\sigma)_{ik} \otimes u(\sigma)_{kj}. \end{cases}$$

The antipode $\kappa$ acts trivially on the abelian simple components and restriction of $\kappa$ to $\mathcal{A}_\sigma$ is the transposition, and $\{z(e), z(a), z(b), z(c), e(\sigma)_{ij}\}$ are expressed as follows:

$$\begin{cases} z(e) = \frac{1}{8}[u(e) + u(a) + u(b) + u(c) + 2u(\sigma)_{11} + 2u(\sigma)_{22}], \\ z(a) = \frac{1}{8}[u(e) + u(a) + u(b) + u(c) - 2u(\sigma)_{11} - 2u(\sigma)_{22}], \\ z(b) = \frac{1}{8}[u(e) + u(a) - u(b) - u(c) - 2u(\sigma)_{11} + 2u(\sigma)_{22}], \\ z(c) = \frac{1}{8}[u(e) + u(a) - u(b) - u(c) + 2u(\sigma)_{11} - 2u(\sigma)_{22}], \\ e(\sigma)_{11} = \frac{1}{4}[u(e) - u(a) - u(b) + u(c)], \\ e(\sigma)_{12} = \frac{1}{2}[e^{-\frac{\pi i}{4}} u(\sigma)_{12} + e^{\frac{\pi i}{4}} u(\sigma)_{21}], \\ e(\sigma)_{21} = \frac{1}{2}[e^{\frac{\pi i}{4}} u(\sigma)_{12} + e^{-\frac{\pi i}{4}} u(\sigma)_{21}], \\ e(\sigma)_{22} = \frac{1}{4}[u(e) - u(a) + u(b) - u(c)]. \end{cases}$$

Note that our choice of bases and parametrization immediately show that $\mathcal{A}$ is self-dual ([**29**]). We introduce the group structure on $G_0 = \{e, a, b, c\} \cong \mathbf{Z}_2 \times \mathbf{Z}_2$ with the unit $e$. This is compatible with the group structure of both $G(\mathcal{A})$ and $G(\hat{\mathcal{A}})$ regarded as the set of 1-dimensional representations of $\mathcal{A}$.

We define two actions $\mu$ and $\nu$ of $G_0$ on $\mathcal{A}$, which do not preserve the co-algebra structure, by the following relations:

$$\begin{cases} \mu_g(x) \otimes z(g) = \Gamma(x)(1 \otimes z(g)), \\ z(g) \otimes \nu_g(x) = \Gamma(x)(z(g) \otimes 1). \end{cases}$$

We also introduce a $*$-homomorphism $\pi : \mathcal{A} \longrightarrow \mathcal{A}_\sigma \otimes \mathcal{A}_\sigma$ by

$$\pi(x) = \Gamma(x)(z(\sigma) \otimes z(\sigma)).$$

LEMMA 13.14.
*(i)* $\mu_e = \nu_e = id$, $\mu_a = \nu_a$ *and* $\mu_g(z(h)) = \nu_g(z(h)) = z(g+h)$ *for* $g, h \in G_0$.
*(ii)* $\mu_b|_{\mathcal{A}_\sigma} = \nu_c|_{\mathcal{A}_\sigma}$, $\mu_c|_{\mathcal{A}_\sigma} = \nu_b|_{\mathcal{A}_\sigma}$ *and* $\mu_g|_{\mathcal{A}_\sigma} = Ad(\lambda_g)$ *for* $g \in G_0$, *where*

$$\begin{cases} \lambda_e = \begin{pmatrix} 1 & 0 \\ 0 & 1 \end{pmatrix}, & \lambda_a = \begin{pmatrix} 1 & 0 \\ 0 & -1 \end{pmatrix}, \\ \lambda_b = \begin{pmatrix} 0 & e^{\frac{\pi i}{4}} \\ e^{-\frac{\pi i}{4}} & 0 \end{pmatrix}, & \lambda_c = \begin{pmatrix} 0 & e^{-\frac{\pi i}{4}} \\ e^{\frac{\pi i}{4}} & 0 \end{pmatrix}. \end{cases}$$

*(iii)* $\pi|_{\mathcal{A}_\sigma} = 0$ *and*

$$\begin{cases} \pi(z(e)) = \frac{1}{2}\begin{pmatrix} 1 & 0 & 0 & 1 \\ 0 & 0 & 0 & 0 \\ 0 & 0 & 0 & 0 \\ 1 & 0 & 0 & 1 \end{pmatrix}, & \pi(z(a)) = \frac{1}{2}\begin{pmatrix} 1 & 0 & 0 & -1 \\ 0 & 0 & 0 & 0 \\ 0 & 0 & 0 & 0 \\ -1 & 0 & 0 & 1 \end{pmatrix}, \\ \pi(z(b)) = \frac{1}{2}\begin{pmatrix} 0 & 0 & 0 & 0 \\ 0 & 1 & -i & 0 \\ 0 & i & 1 & 0 \\ 0 & 0 & 0 & 0 \end{pmatrix}, & \pi(z(c)) = \frac{1}{2}\begin{pmatrix} 0 & 0 & 0 & 0 \\ 0 & 1 & i & 0 \\ 0 & -i & 1 & 0 \\ 0 & 0 & 0 & 0 \end{pmatrix}. \end{cases}$$

*Here, our convention is*

$$\begin{pmatrix} a & b \\ c & d \end{pmatrix} \otimes A = \begin{pmatrix} aA & bA \\ cA & dA \end{pmatrix}.$$

PROOF. It is clear that $\mu_e = \nu_e = id$ because $z(e)$ is the integral of $\mathcal{A}$. Direct computation shows

$$\begin{cases} \mu_a(u(g)) = \nu_g(u(g)) = u(g) \text{ for } g \in G_0, \\ \mu_a(u(\sigma)_{ij}) = \nu_a(u(\sigma)_{ij}) = -u(\sigma)_{ij} \text{ for } 1 \le i,j \le 2, \\ \mu_b(u(g)) = \mu_c(u(g)) = \nu_b(u(g)) = \nu_c(u(g)) = \begin{cases} u(g) & \text{when } g = e, a, \\ -u(g) & \text{when } g = b, c, \end{cases} \\ \mu_b(u(\sigma)_{ij}) = (-1)^j u(\sigma)_{ij} \text{ for } 1 \le i,j \le 2, \\ \mu_c(u(\sigma)_{ij}) = (-1)^{j+1} u(\sigma)_{ij} \text{ for } 1 \le i,j \le 2, \\ \nu_b(u(\sigma)_{ij}) = (-1)^i u(\sigma)_{ij} \text{ for } 1 \le i,j \le 2, \\ \nu_c(u(\sigma)_{ij}) = (-1)^{i+1} u(\sigma)_{ij} \text{ for } 1 \le i,j \le 2. \end{cases}$$

It is easy to show the statements from these equations. □

THEOREM 13.15. *Every 2-cocycle of Kac-Paljutkin's 8-dimensional Kac algebra $\mathcal{A}$ is a coboundary.*

PROOF. Let $\omega$ be a normalized 2-cocycle of $\mathcal{A}$. The normalization condition implies that $\omega$ is of the following form:

$$\omega = \sum_{g,h \in G_0} \omega_0(g,h)(z(g) \otimes z(h)) + \sum_g v(g) \otimes z(g) + \sum_g z(g) \otimes \overline{v(g)} + W.$$

Here, $\omega(g,h) \in \mathbf{T}$, $v(g) \in \mathcal{A}_\sigma$, $W \in \mathcal{A}_\sigma \otimes \mathcal{A}_\sigma$ are unitaries, and $\overline{v(g)}$ is the complex conjugate matrix of $v(g)$. Let $W = \sum_{i,j,k,l} W_{ik,jl} e(\sigma)_{i,j} \otimes e(\sigma)_{k,l}$. Then, we have

$$\begin{cases} \overline{\omega_0(g,h)} = \omega_0(h,g), \\ \omega_0(e,g) = \omega_0(g,e) = 1, \\ v(e) = 1, \\ W_{ik,jl} = \overline{W_{ki,lj}}, \\ \sum_{j=1}^2 W_{ik,jj} = \delta_{i,k}. \end{cases}$$

When we perturb $\omega$ to a normalized cocycle by $\xi$, then $\xi$ should satisfy

$$\xi = \sum_{g \in G_0} \xi(g) z(g) + \xi(\sigma),$$

where $\xi(e) = 1$, $\xi(a), \xi(b), \xi(c) \in \{1, -1\}$, and $\xi(\sigma) = \overline{\xi(\sigma)} \in \mathcal{A}_\sigma$ is an orthogonal matrix. Direct computation shows

$$\begin{aligned} \Gamma(\xi)\omega(\xi^* \otimes \xi^*) &= \sum_{g,h} \xi(g+h)\overline{\xi(g)\xi(h)}\omega_0(g,h)(z(g) \otimes z(h)) \\ &\quad + \sum_g \overline{\xi(g)}(\mu_g(\xi(\sigma))v(g)\xi(\sigma)^* \otimes z(g)) \\ &\quad + \sum_g \overline{\xi(g)}(z(g) \otimes \nu_g(\xi(\sigma))\overline{v(g)}\xi(\sigma)^*) \\ &\quad + \pi(\xi) W(\xi(\sigma)^* \otimes \xi(\sigma)^*). \end{aligned}$$

This means that one could change the above triple $(\omega_0, v, W)$ to the new one $(\tilde{\omega}_0, \tilde{v}, \tilde{W})$ defined by

$$\begin{cases} \tilde{\omega}_0(g,h) = \xi(g+h)\overline{\xi(g)\xi(h)}\omega_0(g,h), \\ \tilde{v}(g) = \overline{\xi(g)}\mu_g(\xi(\sigma))v(g)\xi(\sigma)^*, \\ \tilde{W} = \pi(\xi) W(\xi(\sigma)^* \otimes \xi(\sigma)^*). \end{cases}$$

Next we have to write down the cocycle condition of $\omega$ means in terms of $\omega_0, v(g)$, and $W$. By direct computations we observe that $(\Gamma\otimes id)(\omega)(\omega\otimes 1)$ is equal to

$$\begin{aligned}
&\sum_{g,h}\omega_0(g,h)(\Gamma\otimes id)(\omega)(z(g)\otimes z(h)\otimes 1)+(\Gamma\otimes id)(\omega)(W\otimes 1)\\
&\quad+\sum_{h}(\Gamma\otimes id)(\omega)(v(h)\otimes z(h)\otimes 1)+\sum_{g}(\Gamma\otimes id)(\omega)(z(g)\otimes\overline{v(g)}\otimes 1)\\
=&\sum_{g,h}\omega_0(g,h)(z(g)\otimes(\mu_g\otimes id)(\omega)(z(h)\otimes 1))+(\pi\otimes id)(\omega)(W\otimes 1)\\
&\quad+\sum_{h}(\mu_h\otimes id)(\omega)_{13}(v(h)\otimes z(h)\otimes 1)+\sum_{g}z(g)\otimes(\nu_g\otimes id)(\omega)(\overline{v(g)}\otimes 1)\\
=&\sum_{g,h,k}\omega_0(g,h)\omega_0(g+h,k)(z(g)\otimes z(h)\otimes z(k))\\
&\quad+\sum_{g,h}\omega_0(g,h)(z(g)\otimes z(h)\otimes\overline{v(g+h)})\\
&\quad+\sum_{h,k}\mu_h(v(k))v(h)\otimes z(h)\otimes z(k)+\sum_{g,k}z(g)\otimes\nu_g(v(k))\overline{v(g)}\otimes z(k)\\
&\quad+\sum_{h}(\mu_h\otimes id)(W)_{13}(v(h)\otimes z(h)\otimes 1)+\sum_{g}z(g)\otimes(\nu_g\otimes id)(W)(\overline{v(g)}\otimes 1)\\
&\quad+\sum_{l,k}\omega_0(l,k)(\pi(z(l))W\otimes z(k))+\sum_{l}\pi(z(l))W\otimes\overline{v(l)}.
\end{aligned}$$

On the other hand, $(id\otimes\Gamma)(\omega)(1\otimes\omega)$ is equal to

$$\begin{aligned}
&\sum_{h,k}\omega_0(h,k)(id\otimes\Gamma)(\omega)(1\otimes z(h)\otimes z(k))+(id\otimes\Gamma)(\omega)(1\otimes W)\\
&\quad+\sum_{k}(id\otimes\Gamma)(\omega)(1\otimes v(k)\otimes z(k))+\sum_{h}(id\otimes\Gamma)(\omega)(1\otimes z(h)\otimes\overline{v(h)})\\
=&\sum_{h,k}\omega_0(h,k)((id\otimes\mu_k)(\omega)(1\otimes z(h))\otimes z(k))+(id\otimes\pi)(\omega)(1\otimes W)\\
&\quad+\sum_{k}((id\otimes\mu_k)(\omega)(1\otimes v(k))\otimes z(k))+\sum_{h}(id\otimes\nu_h)(\omega)_{13}(1\otimes z(h)\otimes\overline{v(h)})\\
=&\sum_{g,h,k}\omega_0(h,k)\omega_0(g,h+k)(z(g)\otimes z(h)\otimes z(k))\\
&\quad+\sum_{h,k}\omega_0(h,k)(v(h+k)\otimes z(h)\otimes z(k))\\
&\quad+\sum_{g,k}z(g)\otimes\mu_k(\overline{v(g)})v(k)\otimes z(k)+\sum_{g,h}z(g)\otimes z(h)\otimes\nu_h(\overline{v(g)})\overline{v(h)})\\
&\quad+\sum_{k}(id\otimes\mu_k)(W)(1\otimes v(k))\otimes z(k)+\sum_{h}(id\otimes\nu_h)(W)_{13}(1\otimes z(h)\otimes\overline{v(h)})\\
&\quad+\sum_{g,l}\omega_0(g,l)(z(g)\otimes\pi(z(l)))W+\sum_{l}v(l)\otimes\pi(z(l))W.
\end{aligned}$$

Thus, the cocycle relation of $\omega$ is equivalent to the following:

$$\omega_0(g,h)\omega_0(g+h,k) = \omega_0(h,k)\omega_0(g,h+k), \tag{13.3.1}$$

$$\mu_h(v(g))v(h) = \omega_0(h,g)v(g+h), \tag{13.3.2}$$

$$\nu_h(\overline{v(g)})\overline{v(h)} = \omega_0(g,h)\overline{v(g+h)}, \tag{13.3.3}$$

$$\mu_h(\overline{v(g)})v(h) = \nu_g(v(h))\overline{v(g)}, \tag{13.3.4}$$

$$(\mu_g \otimes id)(W)(v(g)\otimes 1) = (id\otimes \nu_g)(W)(1\otimes \overline{v(g)}), \tag{13.3.5}$$

$$(id\otimes \mu_g)(W)(1\otimes v(g)) = \sum_h \omega_0(h,g)\pi(z(h))W, \tag{13.3.6}$$

$$(\nu_g\otimes id)(W)(\overline{v(g)}\otimes 1) = \sum_h \omega_0(g,h)\pi(z(h))W, \tag{13.3.7}$$

$$\sum_g \pi(z(g))W\otimes \overline{v(g)} = \sum_g v(g)\otimes \pi(z(g))W. \tag{13.3.8}$$

We set $w(g) = \lambda_g v(g)$. Since $\lambda_g^* = \lambda_g$, $\mu_g|_{\mathcal{A}_\sigma} = Ad(\lambda_g)$, and $\nu_g|_{\mathcal{A}_\sigma} = Ad(\overline{\lambda_g})$, the conditions (13.3.2) and (13.3.3) are equivalent to

$$w(g)w(h) = \overline{\omega_0(g,h)}\omega_1(g,h)w(g+h), \tag{13.3.2$'$}$$

where $\omega_1$ is a 2-cocycle of $G_0$ determined by $\lambda_g\lambda_h = \omega_1(g,h)\lambda_{g+h}$. Next (13.3.4) is equivalent to

$$\lambda_h\,\overline{\lambda_g}\,\overline{w(g)}\,w(h) = \overline{\lambda_g}\,\lambda_h w(h)\,\overline{w(g)},$$

that is,

$$\overline{w(a)}w(a) = w(a)\overline{w(a)}, \tag{13.3.4.1}$$

$$\overline{w(b)}w(b) = -w(b)\overline{w(b)}, \tag{13.3.4.2}$$

$$\overline{w(c)}w(c) = -w(c)\overline{w(c)}, \tag{13.3.4.3}$$

$$\overline{w(a)}w(b) = -w(b)\overline{w(a)}, \tag{13.3.4.4}$$

$$\overline{w(a)}w(c) = -w(c)\overline{w(a)}, \tag{13.3.4.5}$$

$$\overline{w(b)}w(c) = w(c)\overline{w(b)}. \tag{13.3.4.6}$$

The trace value of $w(a)$ is 0 due to (13.3.4.4). Thus, (13.3.4.1) implies that $w(a)$ is proportional to a real orthogonal matrix, which is not a scalar. (Note that $w(a)$ is a unitary matrix with $w(a)^2 =$ scalar.) Therefore, using appropriate $\xi(\sigma)$, we may assume that $w(a)$ is proportional to either one of the following:

$$\begin{pmatrix} 0 & 1 \\ -1 & 0 \end{pmatrix} \quad or \quad \begin{pmatrix} 1 & 0 \\ 0 & -1 \end{pmatrix}.$$

However, the first case is impossible by (13.3.4.2) and (13.3.4.4). Thus, we may assume

$$w(a) = t\begin{pmatrix} 1 & 0 \\ 0 & -1 \end{pmatrix}$$

with a scalar $t$ satisfying $t^2 = \overline{\omega_0(a,a)}$. Now since $w(a)$ is proportional to $\overline{w(a)}$, the above commutation relations show that the projective representation $\{w(g)\}_g$ is irreducible. This means that $\overline{\omega_0}\omega_1$ is not a coboundary, and consequently $\omega_0$ is a coboundary. We may and do assume $\omega_0(g,h) = 1$ and $t = 1$. Then (13.3.2$'$) is equivalent to

$$w(a)^2 = w(b)^2 = w(c)^2 = 1, \tag{13.3.2.1}$$

$$w(a)w(b) = iw(c), \quad w(b)w(c) = iw(a), \quad w(c)w(a) = iw(b). \tag{13.3.2.2}$$

The only freedom of $\xi$ we can use now is as follows: $\xi(g)$ is a character with

$$\xi(a)\xi(\sigma)\begin{pmatrix}1 & 0\\ 0 & -1\end{pmatrix}\xi(\sigma)^* = \begin{pmatrix}1 & 0\\ 0 & -1\end{pmatrix}.$$

Then, (13.3.4.2), (13.3.4.4) and (13.3.2.1) imply that

$$w(b) = \begin{pmatrix}0 & s\\ \overline{s} & 0\end{pmatrix}, \quad s^4 = -1.$$

Therefore, using appropriate $\xi$, we may assume $w(b) = \lambda_b$, and so $w(g) = \lambda_g$ for $g \in G_0$. This implies $v(g) = 1$ and $W \otimes z(\sigma) = z(\sigma) \otimes W$ thanks to (13.3.8). Therefore, $W$ is a scalar. Normalization of $W$ shows $W = z(\sigma) \otimes z(\sigma)$ and we get $\omega = 1$. □

**Sketch of the proof of Proposition 13.13.** Suppose that $\mathbf{C}(G)^\omega$ is not as in the statement. Then, the classification result (Chapter 11) shows that $\mathbf{C}(G)^\omega$ would be isomorphic to $\mathcal{A} \otimes \mathbf{C}(\mathbf{Z}_2)$, where $\mathcal{A}$ is Kac-Paljutkin's 8-dimensional algebra. This means that the representation category $Rep(\mathcal{A})$ of $\mathcal{A}$ could be embedded into that of $G$, and so $Rep(\mathcal{A})$ would have a permutation symmetry. However, this would imply that there would exist a 2-cocycle $\omega$ of $\mathcal{A}$ such that $\Gamma^\omega$ is co-commutative [**33**] (or one can directly show that $\mathcal{A}$ never has a permutation symmetry using the above explicit formulas), which is a contradiction. □

Next, we determine $(\theta, \eta, \zeta)$ for $(\mathcal{A}, \mathbf{Z}_2)$.

LEMMA 13.16. *Let $Aut(\mathcal{A})$ be the Kac algebra automorphism group of $\mathcal{A}$. We define $\tau \in Aut(\mathcal{A})$ by*

$$\tau(z(e)) = z(e),\ \tau(z(a)) = z(a),\ \tau(z(b)) = z(c),\ \tau(z(c)) = z(b),$$
$$\tau|_{A_\sigma} = Ad\begin{pmatrix}0 & 1\\ 1 & 0\end{pmatrix},$$

*or equivalently, by*

$$\tau(u(e)) = u(e),\ \tau(u(a)) = u(a),\ \tau(u(b)) = u(c),\ \tau(u(c)) = u(b),$$
$$\tau(u(\sigma)_{i,j}) = u(\sigma)_{3-i,3-j}.$$

*Then, we have*

$$Aut(\mathcal{A}) = \{id,\ \tau,\ Ad(u(b)),\ Ad(u(b))\cdot\tau\} \cong \mathbf{Z}_2 \times \mathbf{Z}_2.$$

PROOF. It is easy to show that $\tau$ is a Kac algebra automorphism. Since the set of central group-likes consists of $u(e)$ and $u(a)$, every Kac algebra automorphism fixes $u(a)$, and also $z(a)$ because of the self-duality of $\mathcal{A}$. Let $\theta \in Aut(\mathcal{A})$ be a non-trivial element. Then, either $\theta$ fixes every abelian simple component or it exchanges $z(b)$ and $z(c)$. To prove the statement, it suffices to show $\theta = Ad(u(b))$ in the first case (because we can consider $\theta\cdot\tau$ in the second case). Thus, we assume $\theta(z(b)) = z(b)$. There are two possibilities:

$$\text{either } \theta(u(b)) = u(b) \text{ or } \theta(u(b)) = u(c).$$

In the first case, we have

$$\theta|_{A_\sigma} = Ad\begin{pmatrix}1 & 0\\ 0 & s\end{pmatrix} \text{ with some } s \in \mathbf{T}.$$

Since $\theta(u(\sigma)_{11}) = u(\sigma)_{11}$, by applying $\Gamma$ to the both sides we get

$$\theta(u(\sigma)_{12}) \otimes \theta(u(\sigma)_{21}) = u(\sigma)_{12} \otimes u(\sigma)_{21},$$

that is,

$$\begin{pmatrix} 0 & \bar{s}e^{\frac{\pi i}{4}} \\ se^{-\frac{\pi i}{4}} & 0 \end{pmatrix} \otimes \begin{pmatrix} 0 & \bar{s}e^{-\frac{\pi i}{4}} \\ se^{\frac{\pi i}{4}} & 0 \end{pmatrix} = \begin{pmatrix} 0 & e^{\frac{\pi i}{4}} \\ e^{-\frac{\pi i}{4}} & 0 \end{pmatrix} \otimes \begin{pmatrix} 0 & e^{-\frac{\pi i}{4}} \\ e^{\frac{\pi i}{4}} & 0 \end{pmatrix}.$$

Thus, we get $s = -1$ and $\theta = Ad(u(b))$. In the second case we have

$$\theta|_{A_\sigma} = Ad \begin{pmatrix} 0 & 1 \\ s & 0 \end{pmatrix} \text{ with some } s \in \mathbf{T}.$$

By the same reasoning as above, we get

$$\theta(u(\sigma)_{12}) \otimes \theta(u(\sigma)_{21}) = u(\sigma)_{12} \otimes u(\sigma)_{21}$$

and hence

$$\begin{pmatrix} 0 & \bar{s}e^{-\frac{\pi i}{4}} \\ se^{\frac{\pi i}{4}} & 0 \end{pmatrix} \otimes \begin{pmatrix} 0 & \bar{s}e^{\frac{\pi i}{4}} \\ se^{-\frac{\pi i}{4}} & 0 \end{pmatrix} = \begin{pmatrix} 0 & e^{\frac{\pi i}{4}} \\ e^{-\frac{\pi i}{4}} & 0 \end{pmatrix} \otimes \begin{pmatrix} 0 & e^{-\frac{\pi i}{4}} \\ e^{\frac{\pi i}{4}} & 0 \end{pmatrix}.$$

However, this is impossible, which means that this case does not occur. □

Now we determine possible $(\theta, \eta, \zeta)$'s for $(\mathcal{A}, \mathbf{Z}_2)$, and for simplicity we use the notation $\theta = \theta_1$, $\eta = \eta_1$, and $\zeta = \zeta_1$. Thanks to Theorem 13.15, we may assume $\eta = 1$, and note that $\theta$ and $\zeta$ should satisfy: $\theta$ is a Kac algebra automorphism and $\zeta$ is a group-like element such that

$$\theta(\zeta) = \zeta, \quad \theta^2 = Ad(\zeta).$$

Notice that $(\theta, 1, \zeta)$ and $(\tilde{\theta}, 1, \tilde{\zeta})$ are equivalent if and only if there exists a group-like $\xi \in \mathcal{A}$ satisfying

$$\tilde{\theta} = Ad(\xi) \cdot \theta, \quad \tilde{\zeta} = \xi\theta(\xi)\zeta.$$

From the discussions so far (Proposition 13.5, Theorem 13.15, and Lemma 13.16), we get the following:

THEOREM 13.17. *Let $\mathcal{A}$ be Kac-Paljutkin's 8-dimensional Kac algebra. Then, there are four weak equivalence classes of $(\theta, \eta, \zeta)$ as in* **1.4**. *Moreover, representatives of $(\theta, \eta, \zeta)$, algebra structure of $\mathcal{B} = \mathcal{A} \rtimes_{\theta,\zeta} \mathbf{Z}_2$, $\hat{\mathcal{B}}$, and the intrinsic groups $G(\mathcal{B})$, $G(\hat{\mathcal{B}})$ are as follows:*

*(i)* $(\theta, \eta, \zeta) = (id, 1, 1)$ *In this case, we have*

$$\mathcal{B} \cong \mathcal{A} \otimes \mathbf{C}(\mathbf{Z}_2).$$

*(ii)* $(\theta, \eta, \zeta) = (id, 1, u(a))$ *In this case we have*

$$\mathcal{B} \cong \hat{\mathcal{B}} \cong \mathbf{C}^8 \oplus M_2(\mathbf{C}) \oplus M_2(\mathbf{C}),$$
$$G(\mathcal{B}) \cong \mathbf{Z}_2 \times \mathbf{Z}_4, \quad G(\hat{\mathcal{B}}) \cong \mathbf{Z}_2 \times \mathbf{Z}_2 \times \mathbf{Z}_2.$$

*(iii)* $(\theta, \eta, \zeta) = (\tau, 1, 1)$ *In this case we have*

$$\mathcal{B} \cong \mathbf{C}^4 \oplus M_2(\mathbf{C}) \oplus M_2(\mathbf{C}) \oplus M_2(\mathbf{C}),$$
$$\hat{\mathcal{B}} \cong \mathbf{C}^8 \oplus M_2(\mathbf{C}) \oplus M_2(\mathbf{C}),$$
$$G(\mathcal{B}) \cong D_8, \quad G(\hat{\mathcal{B}}) \cong \mathbf{Z}_2 \times \mathbf{Z}_2.$$

(iv) $(\theta, \eta, \zeta) = (\tau, 1, u(a))$ *In this case we have*

$$\mathcal{B} \cong \mathbf{C}^4 \oplus M_2(\mathbf{C}) \oplus M_2(\mathbf{C}) \oplus M_2(\mathbf{C}),$$
$$\hat{\mathcal{B}} \cong \mathbf{C}^8 \oplus M_2(\mathbf{C}) \oplus M_2(\mathbf{C}),$$
$$G(\mathcal{B}) \cong D_8, \quad G(\hat{\mathcal{B}}) \cong \mathbf{Z}_2 \times \mathbf{Z}_2.$$

## 4. Examples II Group algebras

In this section, we classify $(\theta, \eta, \zeta)$ for $(\hat{G}, \mathbf{Z}_2)$ for the groups $G = D_8$, $D_{12}$, and $\mathfrak{A}_4$. Proposition 13.17 (on $\mathbf{C}(D_{12}) \rtimes \mathbf{Z}_2$) and Theorem 13.20 (on $\mathbf{C}(\mathfrak{A}_4) \rtimes \mathbf{Z}_2$) together with Corollary 13.6 provide us variable information on Kac algebras of dimension 24 whose intrinsic group is of order 12 (see the proof of Lemma 14.32 in the next chapter).

**4.1. $D_8$ case.** We begin with the dihedral group $D_8 = \mathbf{Z}_4 \rtimes \mathbf{Z}_2$ of order 8, where the action of $\mathbf{Z}_2$ on $\mathbf{Z}_4$ is

$$(0,1)(1,0)(0,1) = (3,0).$$

We define $\chi_1, \chi_2, \chi_3 = \chi_1\chi_2 \in Hom(D_8, \mathbf{T})$ by

$$\begin{cases} \chi_1((1,0)) = -1, & \chi_1((0,1)) = 1, \\ \chi_2((1,0)) = 1, & \chi_2((0,1)) = -1. \end{cases}$$

It is well-known that $D_8$ admits four 1-dimensional representations $\{1, \chi_1, \chi_2, \chi_3\}$ and one 2-dimensional irreducible representation $\sigma$ defined by

$$\sigma((1,0)) = \begin{pmatrix} 0 & -1 \\ 1 & 0 \end{pmatrix}, \quad \sigma((0,1)) = \begin{pmatrix} 1 & 0 \\ 0 & -1 \end{pmatrix}.$$

Thus, the group algebra $\mathbf{C}(D_8)$ is given as follows:

$$\begin{cases} \lambda_{(0,0)} = 1 \oplus 1 \oplus 1 \oplus 1 \oplus \begin{pmatrix} 1 & 0 \\ 0 & 1 \end{pmatrix}, \\ \lambda_{(1,0)} = 1 \oplus -1 \oplus 1 \oplus -1 \oplus \begin{pmatrix} 0 & -1 \\ 1 & 0 \end{pmatrix}, \\ \lambda_{(2,0)} = 1 \oplus 1 \oplus 1 \oplus 1 \oplus \begin{pmatrix} -1 & 0 \\ 0 & -1 \end{pmatrix}, \\ \lambda_{(3,0)} = 1 \oplus -1 \oplus 1 \oplus -1 \oplus \begin{pmatrix} 0 & 1 \\ -1 & 0 \end{pmatrix}, \\ \lambda_{(0,1)} = 1 \oplus 1 \oplus -1 \oplus -1 \oplus \begin{pmatrix} 1 & 0 \\ 0 & -1 \end{pmatrix}, \\ \lambda_{(1,1)} = 1 \oplus -1 \oplus -1 \oplus 1 \oplus \begin{pmatrix} 0 & 1 \\ 1 & 0 \end{pmatrix}, \\ \lambda_{(2,1)} = 1 \oplus 1 \oplus -1 \oplus -1 \oplus \begin{pmatrix} -1 & 0 \\ 0 & 1 \end{pmatrix}, \\ \lambda_{(3,1)} = 1 \oplus -1 \oplus -1 \oplus 1 \oplus \begin{pmatrix} 0 & -1 \\ -1 & 0 \end{pmatrix}. \end{cases}$$

We define $\tau \in Aut(D_8)$ by

$$\tau((1,0)) = \tau((1,0)), \quad \tau((0,1)) = (1,1).$$

Then, we have

$$\chi_1^\tau = \chi_3, \ \chi_2^\tau = \chi_2, \ \chi_3^\tau = \chi_1.$$

Since $Aut(D_8)$ is generated by $\tau$ and $Int(D_8)$ (Lemma 11.8), each outer automorphism of $D_8$ exchanges the second and the fourth simple components of $\mathbf{C}(D_8)$.

Let

$$N = \{(0,0),(2,0),(0,1),(2,1)\} \cong \mathbf{Z}_2 \times \mathbf{Z}_2,$$

and $\varphi \in Aut(N)$ be the restriction of $Ad((1,0))$ to $N$, which exchanges (0,1) and (2,1). We define $\tau_1, \tau_2, \tau_3 = \tau_1\tau_2 \in \hat{N}$, by

$$\begin{cases} \tau_1((0,1)) = -1, & \tau_1((2,1)) = 1, & \tau_1((2,0)) = -1, \\ \tau_2((0,1)) = 1, & \tau_2((2,1)) = -1, & \tau_2((2,0)) = -1. \end{cases}$$

We next define a normalized 2-cocycle $\omega$ of $\hat{N}$ by

$$\begin{cases} \omega(\tau,\tau) = 1 \text{ for each } \tau \in \hat{N}, \\ \omega(\tau_1,\tau_2) = \omega(\tau_2,\tau_3) = \omega(\tau_3,\tau_1) = i, \\ \omega(\tau_1,\tau_3) = \omega(\tau_3,\tau_2) = \omega(\tau_2,\tau_1) = -i. \end{cases}$$

We regard $\omega$ as an element of $Z^2(\hat{D}_8, \mathbf{T})$.

LEMMA 13.18. *Let $\theta_0$ be the algebra automorphism of $\mathbf{C}(D_8)$ exchanging the third and the fourth simple component. Then, we have*

$$\Gamma^\omega = (\theta_0 \otimes \theta_0) \cdot \Gamma \cdot \theta_0^{-1}.$$

PROOF. Note that $\omega^\varphi = \overline{\omega}$ and $\overline{\omega}\omega^\varphi = \overline{\omega}^2$ is a coboundary. We define $\mu \in \ell^\infty(\hat{N})$ by

$$\mu(1) = \mu(\tau_1) = \mu(\tau_2) = 1, \quad \mu(\tau_3) = -1.$$

Then, we observe

$$\overline{\omega}\omega^\varphi = \Gamma(\mu^*)(\mu \otimes \mu) = \Gamma^\omega(\mu^*)(\mu \otimes \mu),$$

where we regard $\mu$ as an element of $\mathbf{C}(D_8)$. Thus, we get

$$\begin{cases} \Gamma^\omega\left(\lambda_{(0,1)}\right) = \lambda_{(0,1)} \otimes \lambda_{(0,1)}, \\ \Gamma^\omega\left(\mu\lambda_{(1,0)}\right) = \Gamma(\mu)\omega^*\omega^\varphi\left(\lambda_{(1,0)} \otimes \lambda_{(1,0)}\right) = \mu\lambda_{(1,0)} \otimes \mu\lambda_{(1,0)}. \end{cases}$$

Since the element in $\mathbf{C}(N) \subset \mathbf{C}(D_8)$ corresponding to the characteristic function of the singleton $\{\tau_3\}$ is

$$\frac{1 - \lambda_{(0,1)} - \lambda_{(2,1)} + \lambda_{(2,0)}}{4},$$

we have

$$\begin{aligned} \mu &= 1 - 2 \times \frac{1 - \lambda_{(0,1)} - \lambda_{(2,1)} + \lambda_{(2,0)}}{4} \\ &= \frac{1 + \lambda_{(0,1)} + \lambda_{(2,1)} - \lambda_{(2,0)}}{2} = 1 \oplus 1 \oplus -1 \oplus -1 \oplus \begin{pmatrix} 1 & 0 \\ 0 & 1 \end{pmatrix}. \end{aligned}$$

Thus, we get

$$\theta_0\left(\lambda_{(0,1)}\right) = \lambda_{(0,1)}, \quad \theta_0\left(\lambda_{(1,0)}\right) = \mu\lambda_{(1,0)},$$

and we are done. □

THEOREM 13.19. *For the pair $(\hat{D}_8, \mathbf{Z}_2)$ there are two weak equivalence classes of $(\theta, \eta, \zeta)$ with non-trivial $\eta$. Moreover, representatives of $(\theta, \eta, \zeta)$, algebra structure of $\mathcal{B} = \mathbf{C}(D_8) \rtimes_{\theta,\zeta} \mathbf{Z}_2$, $\hat{\mathcal{B}}$, and the intrinsic groups $G(\mathcal{B})$, $G(\hat{\mathcal{B}})$ are as follows:*

*(i)* $(\theta, \eta, \zeta) = (\theta_0, \omega^*, \mu)$ *In this case we have*

$$\mathcal{B} \cong \mathbf{C}^4 \oplus M_2(\mathbf{C}) \oplus M_2(\mathbf{C}) \oplus M_2(\mathbf{C}),$$
$$\hat{\mathcal{B}} \cong \mathbf{C}^8 \oplus M_2(\mathbf{C}) \oplus M_2(\mathbf{C}),$$
$$G(\mathcal{B}) \cong D_8, \quad G(\hat{\mathcal{B}}) \cong \mathbf{Z}_2 \times \mathbf{Z}_2.$$

*(ii)* $(\theta, \eta, \zeta) = (\theta_0, \omega^*, \mu\lambda_{(2,0)})$ *In this case we have*

$$\mathcal{B} \cong \mathbf{C}^4 \oplus M_2(\mathbf{C}) \oplus M_2(\mathbf{C}) \oplus M_2(\mathbf{C}),$$
$$\hat{\mathcal{B}} \cong \mathbf{C}^8 \oplus M_2(\mathbf{C}) \oplus M_2(\mathbf{C}),$$
$$G(\mathcal{B}) \cong D_8, \quad G(\hat{\mathcal{B}}) \cong \mathbf{Z}_2 \times \mathbf{Z}_2.$$

PROOF. We assume that $\eta$ is non-trivial. Since two subgroups of $D_8$ isomorphic to $\mathbf{Z}_2 \times \mathbf{Z}_2$ are moved to the each other by a group automorphism, we may assume $\eta = \omega^*$. Thus, the above lemma shows that $\theta = \theta_0 \cdot \theta_1$, where $\theta_1$ comes from a group automorphism of $D_8$. If $\theta_1$ were outer, then $\theta$ would permute the second, the third, and the fourth simple components of $\mathcal{A}$, and so $\theta^2$ could not be inner as an algebra automorphism of $\mathbf{C}(D_8)$. Therefore, $\theta_1$ must come from an inner group automorphism, and hence we may assume $\theta = \theta_0$ (after passing to an equivalent triple if necessary). Since

$$1 = (\zeta^* \otimes \zeta^*)(\theta_0 \otimes \theta_0)(\eta)\eta\Gamma(\zeta) = ((\mu\zeta)^* \otimes (\mu\zeta)^*)\Gamma(\mu\zeta),$$

we get $\zeta = \mu\lambda_g$ for some $g \in D_8$. The relations $id = \theta_0^2 = Ad(\zeta)$ and $\theta_0(\zeta) = \zeta$ imply that $g$ is either $(0,0)$ or $(2,0)$. Thus, we get two triples. It is not so difficult to show that they are not weakly equivalent. □

### 4.2. $D_{12}$ case.

PROPOSITION 13.20. *There exits no triple $(\theta, \eta, \zeta)$ with non-trivial $\eta$ for the pair $(\hat{D}_{12}, \mathbf{Z}_2)$. Consequently, there is no non-trivial 24-dimensional Kac algebra whose intrinsic group is isomorphic to $D_{12}$.*

PROOF. Note that $D_{12} = \mathbf{Z}_6 \rtimes \mathbf{Z}_2$ where $\mathbf{Z}_2$ acts on $\mathbf{Z}_6$ as before. We set

$$N = \{(0.0), (3,0), (0,1), (3,1)\}.$$

Suppose that $(\theta, \eta, \zeta)$ is a triple with non-trivial $\eta$. Then, we may assume $\eta = \omega^*$, where $\omega$ is a non-trivial normalized 2-cocycle of $\hat{N}$, and we set $\beta = F\omega F\omega^*$. As in the proof of the previous theorem, we may assume $\beta = \Gamma(\mu^*)(\mu \otimes \mu)$ with

$$\mu = \frac{1 + \lambda_{(0,1)} + \lambda_{(3,1)} - \lambda_{(3,0)}}{2}.$$

Since $\Gamma^\omega = (\theta \otimes \theta) \cdot \Gamma \cdot \theta^{-1}$, we see that $\Gamma^\omega$ is co-commutative and hence we get

$$\begin{aligned}\omega^*(\lambda_g \otimes \lambda_g)\omega = \Gamma^\omega(\lambda_g) &= F\Gamma^\omega(\lambda_g)F \\ &= F\omega^*(\lambda_g \otimes \lambda_g)\omega F.\end{aligned}$$

This implies $\beta(\lambda_g \otimes \lambda_g) = (\lambda_g \otimes \lambda_g)\beta$ for all $g \in D_{12}$. Thus, we get

$$\Gamma(\mu\lambda_g\mu^*) = \mu\lambda_g\mu^* \otimes \mu\lambda_g\mu^* \ (g \in D_{12}),$$

which means that $\mu$ normalizes $\{\lambda_g\}_{g \in D_{12}}$. However, this contradicts the case $g = (1,0)$. The second statement follows from Corollary 13.6. □

**4.3. $\mathfrak{A}_4$ case.** We use the following presentation of $\mathfrak{A}_4 = N \rtimes \mathbf{Z}_3$:

$$\begin{cases} N = \{1, a, b, c\} \cong \mathbf{Z}_2 \times \mathbf{Z}_2, \\ \mathbf{Z}_3 = \langle d \rangle, \\ dad^{-1} = b, \ dbd^{-1} = c, \ dcd^{-1} = a. \end{cases}$$

The group algebra $\mathbf{C}(\mathfrak{A}_4)$ is isomorphic to $\mathbf{C}^3 \oplus M_3(\mathbf{C})$ with

$$\begin{cases} \lambda_a &= 1 \oplus 1 \oplus 1 \oplus \begin{pmatrix} 1 & 0 & 0 \\ 0 & -1 & 0 \\ 0 & 0 & -1 \end{pmatrix}, \\ \lambda_b &= 1 \oplus 1 \oplus 1 \oplus \begin{pmatrix} -1 & 0 & 0 \\ 0 & 1 & 0 \\ 0 & 0 & -1 \end{pmatrix}, \\ \lambda_c &= 1 \oplus 1 \oplus 1 \oplus \begin{pmatrix} -1 & 0 & 0 \\ 0 & -1 & 0 \\ 0 & 0 & 1 \end{pmatrix}, \\ \lambda_d &= 1 \oplus e^{\frac{2\pi i}{3}} \oplus e^{-\frac{2\pi i}{3}} \oplus \begin{pmatrix} 0 & 0 & 1 \\ 1 & 0 & 0 \\ 0 & 1 & 0 \end{pmatrix}. \end{cases}$$

It is routine to show

LEMMA 13.21. *We define $\theta_0 \in Aut(\mathfrak{A}_4)$ by*

$$\theta_0(a) = b, \quad \theta_0(b) = a, \quad \theta_0(c) = c, \quad \theta_0(d) = d^2.$$

*Then, we have*
*(i) $Aut(\mathfrak{A}_4) = Int(\mathfrak{A}_4) \rtimes \{id, \theta_0\}$.*
*(ii) Each non-trivial automorphism of order 2 is conjugate to either $Ad(a)$ or $\theta_0$.*

We define $\tau_1, \tau_2, \tau_3 = \tau_1\tau_2 \in \hat{N}$ by

$$\begin{cases} \tau_1(a) = -1, \quad \tau_1(b) = 1, \quad \tau_1(c) = -1, \\ \tau_2(a) = 1, \quad \tau_2(b) = -1, \quad \tau_2(c) = -1. \end{cases}$$

We also define a normalized 2-cocycle of $\hat{N}$ by

$$\begin{cases} \omega(\tau, \tau) = 1 \text{ for each } \tau \in \hat{N}, \\ \omega(\tau_1, \tau_2) = \omega(\tau_2, \tau_3) = \omega(\tau_3, \tau_1) = i, \\ \omega(\tau_1, \tau_3) = \omega(\tau_3, \tau_2) = \omega(\tau_2, \tau_1) = -i. \end{cases}$$

We regard $\omega$ as an element of $Z^2(\hat{\mathfrak{A}}_4, \mathbf{T})$. Note that $\omega$ commutes with $\lambda_d \otimes \lambda_d$ and $(\theta_0 \otimes \theta_0)(\omega) = \omega^*$. We define $\mu \in \ell^\infty(\hat{N})$ by

$$\mu(1) = \mu(\tau_1) = \mu(\tau_2) = 1, \quad \mu(\tau_3) = -1.$$

THEOREM 13.22. *For $(\hat{\mathfrak{A}}_4, \mathbf{Z}_2)$ there are exactly two weak equivalence classes of $(\theta, \eta, \zeta)$ with non-trivial $\eta$ for $(\hat{\mathfrak{A}}_4, \mathbf{Z}_2)$. Moreover, representatives of $(\theta, \eta, \zeta)$, algebra structure of $\mathcal{B} = \mathbf{C}(\mathfrak{A}_4) \rtimes_{\theta,\zeta} \mathbf{Z}_2$, $\hat{\mathcal{B}}$, and the intrinsic groups $G(\mathcal{B})$, $G(\hat{\mathcal{B}})$ are as follows:*
*(i) $(\theta, \eta, \zeta) = (id, \omega, \mu\lambda_c)$ In this case we have*

$$\mathcal{B} \cong \mathbf{C}^6 \oplus M_3(\mathbf{C}) \oplus M_3(\mathbf{C}),$$
$$\hat{\mathcal{B}} \cong \mathbf{C}^{12} \oplus M_2(\mathbf{C}) \oplus M_2(\mathbf{C}) \oplus M_2(\mathbf{C}),$$
$$G(\mathcal{B}) \cong \mathfrak{A}_4, \quad G(\hat{\mathcal{B}}) \cong \mathbf{Z}_6.$$

(*ii*) $(\theta, \eta, \zeta) = (\theta_0, \omega, 1)$ *In this case we have*

$$\mathcal{B} \cong \mathbf{C}^2 \oplus M_2(\mathbf{C}) \oplus M_3(\mathbf{C}) \oplus M_3(\mathbf{C}),$$
$$\hat{\mathcal{B}} \cong \mathbf{C}^{12} \oplus M_2(\mathbf{C}) \oplus M_2(\mathbf{C}) \oplus M_2(\mathbf{C}),$$
$$G(\mathcal{B}) \cong \mathfrak{A}_4, \quad G(\hat{\mathcal{B}}) \cong \mathbf{Z}_2.$$

PROOF. Let $(\theta, \eta, \mu)$ be a triple with non-trivial $\eta$. Then, we may assume $\eta = \omega$ and $\omega^2 = \Gamma(\mu^*)(\mu \otimes \mu)$ with

$$\mu = \frac{1 + \lambda_a + \lambda_b - \lambda_c}{2} = 1 \oplus 1 \oplus 1 \oplus \begin{pmatrix} 1 & 0 & 0 \\ 0 & 1 & 0 \\ 0 & 0 & -1 \end{pmatrix}.$$

Since $\omega$ commutes with $\lambda_d \otimes \lambda_d$, we have $\Gamma^{\omega^*} = \Gamma$, and so $\theta$ comes from a group automorphism. Thus, thanks to Lemma 13.21, we may assume that $\theta$ is either $id$ or $\theta_0$.

(*i*) Assume $\theta = id$. Then, we have

$$1 = (\zeta^* \otimes \zeta^*)\eta^2\Gamma(\zeta) = ((\mu\zeta)^* \otimes (\mu\zeta)^*)\Gamma(\mu\zeta).$$

Thus, $\zeta = \mu\lambda_g$ for some $g \in \mathfrak{A}_4$. Note $id = \theta^2 = Ad(\zeta)$. Therefore, $\zeta$ is central, and so $g = c$ and

$$\zeta = 1 \oplus 1 \oplus 1 \oplus \begin{pmatrix} -1 & 0 & 0 \\ 0 & -1 & 0 \\ 0 & 0 & -1 \end{pmatrix}.$$

The rest of the statements follows from Corollary 13.6.

(*ii*) Assume $\theta = \theta_0$. Then, we have

$$1 = (\zeta^* \otimes \zeta^*)(\theta_0 \otimes \theta_0)(\eta)\eta\Gamma(\zeta) = (\zeta^* \otimes \zeta^*)\Gamma(\zeta),$$

showing that $\zeta$ is a group-like. Since $id = \theta_0^2 = Ad(\zeta)$, $\zeta$ is central and we get $\zeta = 1$. The rest of the statements follows from Corollary 13.6. □

CHAPTER 14

# Classification of Kac algebras of dimension 24

In this chapter, classification of Kac algebras of dimension 24 is presented. Our strategy for classification will be explained at the beginning of §3.

## 1. Preliminaries

By semi-direct product groups $N \rtimes H$, we shall always mean those where the action of $H$ on $N$ is non-trivial. Also, throughout the chapter $\mathfrak{S}_n$ and $\mathfrak{A}_n$ denote the symmetric and alternating groups of degree $n$ respectively.

We begin with a list of groups of order 24 (see [**7**] for example).

LEMMA 14.1. *Groups of order* 24 *are classified as follows:*
*Abelian case:*

$$\mathbf{Z}_8 \times \mathbf{Z}_3 = \mathbf{Z}_{24}, \quad \mathbf{Z}_2 \times \mathbf{Z}_4 \times \mathbf{Z}_3, \quad \mathbf{Z}_2 \times \mathbf{Z}_2 \times \mathbf{Z}_2 \times \mathbf{Z}_3.$$

*Non-abelian case:*

(i) *When both of a Sylow* 2*-subgroup and a Sylow* 3*-subgroup are normal:*
 (1) $D_8 \times \mathbf{Z}_3$.
 (2) $Q_8 \times \mathbf{Z}_3$.

(ii) *When a Sylow* 2*-subgroup is normal but a Sylow* 3*-subgroup is not:*
 (1) $(\mathbf{Z}_2 \times \mathbf{Z}_2 \times \mathbf{Z}_2) \rtimes \mathbf{Z}_3 = \mathbf{Z}_2 \times \mathfrak{A}_4$ *with the* $\mathbf{Z}_3$ *action given by*

$$\begin{pmatrix} 1 & 0 & 0 \\ 0 & 0 & 1 \\ 0 & 1 & 1 \end{pmatrix}.$$

 (2) $Q_8 \rtimes \mathbf{Z}_3$, *where* $Q_8$ *is the quaternion group* $\{\pm 1, \pm i, \pm j, \pm k\}$ *and the* $\mathbf{Z}_3$ *action is given by the cyclic permutation of* $\{i, j, k\}$.

(iii) *When a Sylow* 3*-subgroup is normal but a Sylow* 2*-subgroup is not:*
 (1) $\mathbf{Z}_3 \rtimes \mathbf{Z}_8$.
 (2) $\mathbf{Z}_3 \rtimes (\mathbf{Z}_2 \times \mathbf{Z}_4) = \mathfrak{S}_3 \times \mathbf{Z}_4$.
 (3) $\mathbf{Z}_3 \rtimes (\mathbf{Z}_2 \times \mathbf{Z}_4) = (\mathbf{Z}_3 \rtimes \mathbf{Z}_4) \times \mathbf{Z}_2$.
 (4) $\mathbf{Z}_3 \rtimes (\mathbf{Z}_2 \times \mathbf{Z}_2 \times \mathbf{Z}_2) = \mathfrak{S}_3 \times \mathbf{Z}_2 \times \mathbf{Z}_2$.
 (5) $\mathbf{Z}_3 \rtimes Q_8$.
 (6) $\mathbf{Z}_3 \rtimes D_8 = (\mathbf{Z}_3 \times \mathbf{Z}_4) \rtimes \mathbf{Z}_2 = D_{24}$.
 (7) $\mathbf{Z}_3 \rtimes D_8 = (\mathbf{Z}_3 \rtimes \mathbf{Z}_4) \rtimes \mathbf{Z}_2$, *where* $\mathbf{Z}_2$ *acts on both of* $\mathbf{Z}_3$ *and* $\mathbf{Z}_4$ *non-trivially.*

(iv) *When Sylow subgroups are not normal:*

$$\mathfrak{S}_4.$$

It is easy to show the following:

LEMMA 14.2. *Let $G$ be a group $G$ of order 24 with subgroups $N$, $H$ satisfying*

$$\#N = 4, \quad \#H = 6, \quad N \cap H = \{e\}.$$

*(i) If $G = \mathbf{Z}_3 \times D_8$, then $N$ is normal in $G$.*
*(ii) There is no such pair of subgroups for the groups*

$$\mathbf{Z}_3 \times Q_8, \ Q_8 \rtimes \mathbf{Z}_3, \ \mathbf{Z}_3 \rtimes Q_8.$$

*(iii) If $G = \mathbf{Z}_3 \rtimes L$ with $L$ abelian, then $H$ is normal in $G$.*

LEMMA 14.3.
*(i) Let $\mathfrak{S}_3 = \langle a, b |\ a^3 = b^2 = (ba)^2 = 1\rangle$, $\mathbf{Z}_4 = \langle c |\ c^4 = 1\rangle$. Then, a product group $G = \mathfrak{S}_3 \cdot \mathbf{Z}_4$ is classified as follows:*
(1) $G = \mathfrak{S}_3 \times \mathbf{Z}_4$.
(2) $G = \mathfrak{S}_3 \rtimes \mathbf{Z}_4$ *with* $cac^{-1} = a^2$, $cb = bc$.
(3) $G = \mathbf{Z}_4 \rtimes \mathfrak{S}_3 \cong D_{24}$ *with* $ac = ca$, $bcb = c^3$.
(4) $G = (\mathbf{Z}_3 \rtimes \mathbf{Z}_4) \rtimes \mathbf{Z}_2$, *where $\mathbf{Z}_2$ acts on both of $\mathbf{Z}_3$ and $\mathbf{Z}_4$ non-trivially, and $a, b, c$ are expressed as*

$$a = (1, 0, 0), \ b = (0, 0, 1), \ c = (0, 1, 0).$$

(5) $G = \mathfrak{S}_4$, *where $\mathfrak{S}_3$ is the set of the permutations of the first three letters and $\mathbf{Z}_4 = \langle(1234)\rangle$.*

*(ii) Let $\mathfrak{S}_3 = \langle a, b |\ a^3 = b^2 = (ba)^2 = 1\rangle$ and $c$, $d$ be two generators of $\mathbf{Z}_2 \times \mathbf{Z}_2$. A product group $G = \mathfrak{S}_3 \cdot (\mathbf{Z}_2 \times \mathbf{Z}_2)$ is classified as follows:*
(1) $G = \mathfrak{S}_3 \times (\mathbf{Z}_2 \times \mathbf{Z}_2)$.
(2) $G = \mathfrak{S}_3 \rtimes (\mathbf{Z}_2 \times \mathbf{Z}_2)$ *with* $cac^{-1} = a^2$, $cb = bc$, $da = ad$, $db = bd$.
(3) $G = (\mathbf{Z}_2 \times \mathbf{Z}_2) \rtimes \mathfrak{S}_3$, *where $\mathfrak{S}_3$ acts as permutations of $\mathbf{Z}_2 \times \mathbf{Z}_2 \setminus \{e\}$.*
(4) $G = (\mathbf{Z}_2 \times \mathbf{Z}_2) \rtimes \mathfrak{S}_3$ *with* $ac = ca$, $ad = da$, $bcb = d$.
(5) $G = D_{24} = (\mathbf{Z}_3 \times \mathbf{Z}_4) \rtimes \mathbf{Z}_2$, *where $\mathbf{Z}_2$ acts on both of $\mathbf{Z}_3$ and $\mathbf{Z}_4$ non-trivially, and $a, b, c, d$ are expressed as*

$$a = (1, 0, 0), \ b = (0, 0, 1), \ c = (0, 2, 0), \ d = (0, 1, 1).$$

(6) $G = (\mathbf{Z}_3 \rtimes \mathbf{Z}_4) \rtimes \mathbf{Z}_2$, *where $\mathbf{Z}_2$ acts on both $\mathbf{Z}_3$ and $\mathbf{Z}_4$ non-trivially, and $a, b, c, d$ are expressed as*

$$a = (1, 0, 0), \ b = (0, 0, 1), \ c = (0, 1, 1), \ d = (0, 3, 1).$$

*(iii) Let $a$ and $b$ be generators of $\mathbf{Z}_6$ and $\mathbf{Z}_4$ respectively. Then, a product group $G = \mathbf{Z}_6 \cdot \mathbf{Z}_4$ is classified as follows:*
(1) $G = \mathbf{Z}_6 \times \mathbf{Z}_4$.
(2) $G = \mathbf{Z}_6 \rtimes \mathbf{Z}_4$ *with* $bab^{-1} = a^{-1}$.
(3) $G = \mathbf{Z}_4 \rtimes \mathbf{Z}_6$ *with* $aba^{-1} = b^{-1}$.
(4) $G = (\mathbf{Z}_3 \rtimes \mathbf{Z}_4) \rtimes \mathbf{Z}_2$, *where $\mathbf{Z}_2$ acts on both of $\mathbf{Z}_3$ and $\mathbf{Z}_4$ non-trivially, and $a$, $b$ can be expressed as*

$$a = (1, 1, 1), \ b = (0, 1, 0).$$

*(iv) Let $a$ be a generator of $\mathbf{Z}_6$ and $b$, $c$ generators of $\mathbf{Z}_2 \times \mathbf{Z}_2$. Then, a product group $G = \mathbf{Z}_6 \cdot (\mathbf{Z}_2 \times \mathbf{Z}_2)$ is classified as follows:*
(1) $G = \mathbf{Z}_6 \times (\mathbf{Z}_2 \times \mathbf{Z}_2)$.
(2) $G = \mathbf{Z}_6 \rtimes (\mathbf{Z}_2 \times \mathbf{Z}_2)$ *with* $bab^{-1} = a^{-1}$, $ca = ac$.
(3) $G = (\mathbf{Z}_2 \times \mathbf{Z}_2) \rtimes \mathbf{Z}_6$ *with* $aba^{-1} = c$, $aca^{-1} = b$.
(3) $G = (\mathbf{Z}_2 \times \mathbf{Z}_2) \rtimes \mathbf{Z}_6$ *with* $aba^{-1} = c$, $aca^{-1} = bc$.

(4) $G = (\mathbf{Z}_3 \rtimes \mathbf{Z}_4) \rtimes \mathbf{Z}_2$, *where* $\mathbf{Z}_2$ *acts on both* $\mathbf{Z}_3$ *and* $\mathbf{Z}_4$ *non-trivially, and* $a, b, c$ *are expressed as*

$$a = (1,1,1), \ b = (0,0,1), \ c = (0,2,1).$$

(5) $G = \mathbf{Z}_2 \times \mathfrak{A}_4 = \mathbf{Z}_2 \times ((\mathbf{Z}_2 \times \mathbf{Z}_2) \rtimes \mathbf{Z}_3)$, *where* $\mathbf{Z}_3$ *action is given by*

$$\begin{pmatrix} 0 & 1 \\ 1 & 1 \end{pmatrix},$$

*and* $a, b, c$ *are expressed as*

$$a = (1,0,0,1), \ b = (0,1,0,0), \ c = (1,1,1,0).$$

PROOF. For non-semi-direct product cases, it suffices to examine the cases (1) of (ii), (6),(7) of (iii), and (iv) in Lemma 14.1 thanks to Lemma 14.2. Using the fact that every subgroup of $\mathfrak{S}_4$ isomorphic to $\mathfrak{S}_3$ is conjugate to the set of permutations of the first 3 letters, it is easy to show the only possible product decompositions o $\mathfrak{S}_4$ of this type are (5) of (i) and (3) of (ii).

The group $\mathbf{Z}_2 \times \mathfrak{A}_4$ gives rise to only $\mathbf{Z}_6 \cdot (\mathbf{Z}_2 \times \mathbf{Z}_2)$. It is routine to show that there are three non-semi-direct product decompositions of this type and they are conjugate to each other by automorphisms of $\mathbf{Z}_2 \times \mathfrak{A}_4$.

Let $H$ be a subgroup of $\mathbf{Z}_3 \rtimes D_8$ of order 4. Since the second component $D_8$ is a Sylow 2-subgroup, we may assume that $H$ is a subgroup of the second component of the semi-direct product (up to conjugacy). If $k_0$ is an order two element of $K$, there exists $g_0 \in G$ such that $g_0 k_0 g_0^{-1}$ belongs to the second component thanks to the Sylow theorem. This implies that there exists $g_1$ in the first component such that $k_1 = g_1 k_0 g_1^{-1}$ belongs to the intersection of the second component and $K$. Since $D_8 = \langle k_1 \rangle \cdot H$ gives a product decomposition of $D_8$, we conclude that every product decomposition $\mathbf{Z}_3 \rtimes D_8 = K \cdot H$ with $\#H = 4$, $\#K = 6$, $H \cap K = \{e\}$ comes from an appropriate decomposition $D_8 = \mathbf{Z}_2 \cdot H$. We use the following presentation of $D_8$:

$$D_8 = \langle x, y |\ x^4 = y^2 = (xy)^2 = 1 \rangle.$$

Then, possible product decompositions of $D_8$ are

$$\begin{cases} \langle x \rangle \cdot \langle y \rangle, \quad \langle x \rangle \cdot \langle xy \rangle, \quad \langle x \rangle \cdot \langle x^2 y \rangle, \quad \langle x \rangle \cdot \langle x^3 y \rangle, \\ \{1, x^2, y, x^2 y\} \cdot \langle xy \rangle, \quad \{1, x^2, y, x^2 y\} \cdot \langle x^3 y \rangle, \\ \{1, x^2, xy, x^3 y\} \cdot \langle y \rangle, \quad \{1, x^2, xy, x^3 y\} \cdot \langle x^2 y \rangle. \end{cases}$$

The group $D_{24}$ gives (5) of (ii) while $(\mathbf{Z}_3 \rtimes \mathbf{Z}_4) \rtimes \mathbf{Z}_2$ gives (4) of (i), (6) of (ii), (4) of (iii), and (4) of (iv). □

LEMMA 14.4. *Let* $G = \mathbf{Z}_4 \cdot \mathfrak{S}_3$ *be a product group as in (4) of (i) in Lemma* 14.3, *and* $(\alpha, \beta)$ *be its near action on a factor* $\mathcal{R}$. *We set*

$$\mathcal{P} = \mathcal{R} \rtimes \mathbf{Z}_4 \supseteq \mathcal{L} = \mathcal{R} \rtimes_{\alpha_{c^2}} \mathbf{Z}_2 \supseteq \mathcal{Q} = \mathcal{R}^{\beta}.$$

*Then, we have*

$$\mathcal{P} = \mathcal{L} \rtimes \mathbf{Z}_2, \quad \mathcal{Q} = \mathcal{L}^{G_1},$$

*where* $G_1$ *is either* $\mathbf{Z}_2 \times \mathfrak{S}_3 \cong D_{12}$ *or* $\mathbf{Z}_3 \rtimes \mathbf{Z}_4$. *Consequently, the Kac algebra associated to the above near action arises from a semi-direct product group* $G_1 \rtimes \mathbf{Z}_2$ *possibly with a cocycle* $(\eta, \zeta)$. *Furthermore,* $\mathcal{L}$ *is a unique intermediate subfactor between* $\mathcal{P}$ *and* $\mathcal{R}$ *such that* $\mathcal{L} \supseteq \mathcal{Q}$ *is of depth* 2.

PROOF. The result follows from the fact that $[\alpha_{c^2}]$ commutes with $[\beta_{\mathfrak{S}_3}]$. We note that $G_1$ is determined by an element in $H^2((\mathfrak{S}_3, \mathbf{Z}_2), \mathbf{T})$ corresponding to $(\beta, \alpha|_{\{1,c^2\}})$ describing the exact sequence

$$1 \longrightarrow \hat{\mathbf{Z}}_2 \longrightarrow G_1 \longrightarrow \mathfrak{S}_3 \longrightarrow 1.$$

An intermediate subfactors comes from a subgroup (see [**57**] for example). Hence, the uniqueness of $\mathcal{L}$ follows from the fact that $[\alpha_{c^2}]$ is the only non-trivial element of $[\alpha_{\mathbf{Z}_4}]$ normalizing $[\beta_{\mathfrak{S}_3}]$. □

LEMMA 14.5. *Let $G = (\mathbf{Z}_2 \times \mathbf{Z}_2) \cdot \mathfrak{S}_3$ be a product group as in* (5) *of (ii) in Lemma* 14.3, *and $(\alpha, \beta)$ be its near action on a factor $\mathcal{R}$. We set*

$$\mathcal{P} = \mathcal{R} \rtimes (\mathbf{Z}_2 \times \mathbf{Z}_2) \supseteq \mathcal{L} = \mathcal{R} \rtimes_{\alpha_c} \mathbf{Z}_2 \supseteq \mathcal{Q} = \mathcal{R}^{\mathfrak{S}_3}.$$

*Then, we have*

$$\mathcal{P} = \mathcal{L} \rtimes \mathbf{Z}_2, \quad \mathcal{Q} = \mathcal{L}^{G_1},$$

*where $G_1$ is either $\mathbf{Z}_2 \times \mathfrak{S}_3 \cong D_{12}$ or $\mathbf{Z}_3 \rtimes \mathbf{Z}_4$. Consequently, the Kac algebra associated to the above near action arises from a semi-direct product group $G_1 \rtimes \mathbf{Z}_2$ possibly with a cocycle $(\eta, \zeta)$. Furthermore, $\mathcal{L}$ is a unique intermediate subfactor between $\mathcal{P}$ and $\mathcal{R}$ such that $\mathcal{L} \supseteq \mathcal{Q}$ is of depth* 2.

PROOF. This follows from the fact that $[\alpha_c]$ commutes with $[\beta_{\mathfrak{S}_3}]$, and $G_1$ is determined by the element in $H^2((\mathfrak{S}_3, \mathbf{Z}_2), \mathbf{T})$ corresponding to $(\beta, \alpha|_{\{1,c\}})$. □

LEMMA 14.6. *Let $G = (\mathbf{Z}_2 \times \mathbf{Z}_2) \cdot \mathfrak{S}_3$ be a product group as in* (6) *of (ii) in Lemma* 14.3, *and $(\alpha, \beta)$ be its near action on a factor $\mathcal{R}$. We set*

$$\mathcal{P} = \mathcal{R} \rtimes (\mathbf{Z}_2 \times \mathbf{Z}_2) \supseteq \mathcal{L} = \mathcal{R} \rtimes_{\alpha_{cd}} \mathbf{Z}_2 \supseteq \mathcal{Q} = \mathcal{R}^{\mathfrak{S}_3}.$$

*Then, we have*

$$\mathcal{P} = \mathcal{L} \rtimes \mathbf{Z}_2, \quad \mathcal{Q} = \mathcal{L}^{G_1},$$

*where $G_1$ is either $\mathbf{Z}_2 \times \mathfrak{S}_3 \cong D_{12}$ or $\mathbf{Z}_3 \rtimes \mathbf{Z}_4$. Consequently, the Kac algebra associated to the above near action arises from a semi-direct product group $G_1 \rtimes \mathbf{Z}_2$ possibly with a cocycle $(\eta, \zeta)$. Furthermore, $\mathcal{L}$ is a unique intermediate subfactor between $\mathcal{P}$ and $\mathcal{R}$ such that $\mathcal{L} \supseteq \mathcal{Q}$ is of depth* 2.

PROOF. This follows from the fact that $[\alpha_{cd}]$ commutes with $[\beta_{\mathfrak{S}_3}]$, and $G_1$ is determined by the element in $H^2((\mathfrak{S}_3, \mathbf{Z}_2), \mathbf{T})$ corresponding to $(\beta, \alpha|_{\{1,cd\}})$. □

LEMMA 14.7. *Let $G = \mathbf{Z}_4 \cdot \mathbf{Z}_6$ be a product group as in* (4) *of (iii) in Lemma* 14.3, *and $(\alpha, \beta)$ be its near action on a factor $\mathcal{R}$. We set*

$$\mathcal{P} = \mathcal{R} \rtimes \mathbf{Z}_4 \supseteq \mathcal{L} = \mathcal{R} \rtimes_{\alpha_{b^2}} \mathbf{Z}_2 \supseteq \mathcal{Q} = \mathcal{R}^{\mathbf{Z}_6}.$$

*Then, we have*

$$\mathcal{P} = \mathcal{L} \rtimes \mathbf{Z}_2, \quad \mathcal{Q} = \mathcal{L}^{G_1},$$

*where $G_1$ is either of the following:*

$$\mathbf{Z}_2 \times \mathbf{Z}_2 \times \mathbf{Z}_3, \quad \mathbf{Z}_4 \times \mathbf{Z}_3.$$

*Consequently, the Kac algebra associated to the above near action arises from a semi-direct product group $G_1 \rtimes \mathbf{Z}_2$ possibly with a cocycle $(\eta, \zeta)$.*

PROOF. This follows from the fact that $[\alpha_{b^2}]$ commutes with $[\beta_{\mathbf{Z}_6}]$, and $G_1$ is determined by the element in $H^2((\mathbf{Z}_6, \mathbf{Z}_2), \mathbf{T})$ corresponding to $(\beta, \alpha|_{\{1,b^2\}})$. □

LEMMA 14.8. *Let $G = (\mathbf{Z}_2 \times \mathbf{Z}_2) \cdot \mathbf{Z}_6$ be a product group as in (4) of (iv) in Lemma 14.3, and $(\alpha, \beta)$ be its near action on a factor $\mathcal{R}$. We set*

$$\mathcal{P} = \mathcal{R} \rtimes_\alpha (\mathbf{Z}_2 \times \mathbf{Z}_2) \supseteq \mathcal{L} = \mathcal{R} \rtimes_{\alpha_{bc}} \mathbf{Z}_2 \supseteq \mathcal{Q} = \mathcal{R}^\beta$$

*Then, we have*

$$\mathcal{P} = \mathcal{L} \rtimes \mathbf{Z}_2, \quad \mathcal{Q} = \mathcal{L}^{G_1},$$

*where $G_1$ is either of the following:*

$$\mathbf{Z}_2 \times \mathbf{Z}_2 \times \mathbf{Z}_3, \quad \mathbf{Z}_4 \times \mathbf{Z}_3.$$

*Consequently, the Kac algebra associated to the above near action arises from a semi-direct product group $G_1 \rtimes \mathbf{Z}_2$ possibly with a cocycle $(\eta, \zeta)$.*

PROOF. This follows from the fact that $[\alpha_{bc}]$ commutes with $[\beta_{\mathbf{Z}_6}]$, and $G_1$ is determined by the element in $H^2((\mathbf{Z}_6, \mathbf{Z}_2), \mathbf{T})$ corresponding to $(\beta, \alpha|_{\{1,bc\}})$. □

LEMMA 14.9. *Let $G = (\mathbf{Z}_2 \times \mathbf{Z}_2) \cdot \mathbf{Z}_6$ be a product group as in (5) of (iv) in Lemma 14.3, and $(\alpha, \beta)$ be its near action on a factor $\mathcal{R}$. We set*

$$\mathcal{P} = \mathcal{R} \rtimes_\alpha (\mathbf{Z}_2 \times \mathbf{Z}_2) \supseteq \mathcal{L} = \mathcal{R}^{(\beta_{a^3}, \mathbf{Z}_2)} \supseteq \mathcal{Q} = \mathcal{R}^\beta$$

*Then, we have*

$$\mathcal{P} = \mathcal{L} \rtimes G_1, \quad \mathcal{Q} = \mathcal{L}^{\mathbf{Z}_3},$$

*where $G_1$ is either of the following:*

$$\mathbf{Z}_2 \times \mathbf{Z}_2 \times \mathbf{Z}_2, \quad \mathbf{Z}_4 \times \mathbf{Z}_2.$$

*Consequently, the Kac algebra associated to the above near action arises from a semi-direct product group $G_1 \rtimes \mathbf{Z}_3$ possibly with a cocycle $(\eta, \zeta)$.*

PROOF. This follows from the fact that $[\beta_{a^3}]$ commutes with $[\alpha_{\mathbf{Z}_2 \times \mathbf{Z}_2}]$, and $G_1$ is determined by the element $H^2((\mathbf{Z}_2 \times \mathbf{Z}_2, \mathbf{Z}_2), \mathbf{T})$ corresponding to $(\alpha, \beta|_{\{1,a^3\}})$. □

LEMMA 14.10. *A non-trivial Kac algebra $\mathcal{A}$ of dimension 24 having a simple component of rank 4 does not arise from a semi-direct product group $N \rtimes H$ (with possibly a cocycle $(\eta, \zeta) \in Z^2((N, H), \mathbf{T})$). In this case the dual Kac algebra $\hat{\mathcal{A}}$ does not arise from a semi-direct product group either.*

PROOF. Suppose $\ell^\infty(N) \rtimes_\zeta H \cong \mathcal{A}$, and let $N = \cup_{i \in I} \mathcal{O}_i$ be the decomposition of $N$ into the $H$-orbits. Since we have the decomposition

$$\ell^\infty(N) \rtimes_\zeta H = \oplus_{i \in I} \ell^\infty(\mathcal{O}_i) \rtimes H$$

and $M_4(\mathbf{C})$ appears in one of the above summands, there exists an orbit $\mathcal{O}_i$ such that

$$(\#H)^2 \geq \#\mathcal{O}_i \times \#H \geq 16,$$

showing $\#H \geq 4$.

If $\#H = 4$, then we would have $\#N = 6$ and $\#\mathcal{O}_i = 4$. Therefore, $N$ is either $\mathbf{Z}_6$ or $\mathfrak{S}_3$, and $N$ is decomposed into three $H$-orbits (one of which contains 4 elements). However, this is impossible (note $Aut(\mathbf{Z}_6) \cong \mathbf{Z}_2$ and $Aut(\mathfrak{S}_3) = Int(\mathfrak{S}_3) \cong \mathfrak{S}_3$). If $\#H = 6$, then $\#N = 4$ and $\#\mathcal{O}_i = 3$, i.e., $N$ consists of two $H$-orbits (the trivial one and the other containing 3 elements). In this situation, $H$ (of order 6) acts transitively on the non-trivial orbit $\mathcal{O}_i$ and we get

$$\ell^\infty(N) \rtimes_\zeta H \cong \mathbf{C}(H) \oplus M_3(\mathbf{C}) \oplus M_3(\mathbf{C}),$$

which is a contradiction again. Finally, if $\#H = 8, 12$, then $N$ is the cyclic group $\mathbf{Z}_2$ or $\mathbf{Z}_3$. Therefore, in either case we have

$$\hat{\mathcal{A}} = \oplus_{h \in H} \mathbf{C}_{\eta_h}(N) = \oplus_{h \in H} \mathbf{C}(N) = \ell^\infty(H) \otimes \mathbf{C}(N).$$

This means that $\mathcal{A}$ is co-commutative, contradicting the assumption.

We now assume $\oplus_{h \in H} \mathbf{C}_{\eta_h}(N) \cong \mathcal{A}$ with a simple component of rank 4. This means that $N$ would have an irreducible projective representation of dimension 4. This implies that $M_4(\mathbf{C})$ would be the image of a homomorphism from the twisted group ring of $N$. Comparing the dimensions of the both algebras, we would get $\#N \geq 16$, which is impossible. □

It is straight-forward to show the following:

LEMMA 14.11. *Assume that $\mathcal{A}$ is a non-commutative Kac algebra of dimension 24. Then, possible algebra structures are*:

$$\begin{aligned}
&\mathbf{C}^{12} \oplus M_2(\mathbf{C}) \oplus M_2(\mathbf{C}) \oplus M_2(\mathbf{C}),\\
&\mathbf{C}^{8} \oplus M_4(\mathbf{C}),\\
&\mathbf{C}^{8} \oplus M_2(\mathbf{C}) \oplus M_2(\mathbf{C}) \oplus M_2(\mathbf{C}) \oplus M_2(\mathbf{C}),\\
&\mathbf{C}^{6} \oplus M_3(\mathbf{C}) \oplus M_3(\mathbf{C}),\\
&\mathbf{C}^{4} \oplus M_2(\mathbf{C}) \oplus M_4(\mathbf{C}),\\
&\mathbf{C}^{4} \oplus M_2(\mathbf{C}) \oplus M_2(\mathbf{C}) \oplus M_2(\mathbf{C}) \oplus M_2(\mathbf{C}) \oplus M_2(\mathbf{C}),\\
&\mathbf{C}^{3} \oplus M_2(\mathbf{C}) \oplus M_2(\mathbf{C}) \oplus M_2(\mathbf{C}) \oplus M_3(\mathbf{C}),\\
&\mathbf{C}^{2} \oplus M_2(\mathbf{C}) \oplus M_3(\mathbf{C}) \oplus M_3(\mathbf{C}).
\end{aligned}$$

## 2. Computation of $H^2((N, H), \mathbf{T})/\sim$

Throughout this section, by $\omega_0$ and $\omega_1$ we denote the 2-cocycles of $\mathbf{Z}_2 \times \mathbf{Z}_2$ defined by

$$\begin{aligned}
&\omega_0((x_1, x_2), (x_1, x_2)) = 1,\\
&\omega_0((1,0),(0,1)) = \omega_0((0,1),(1,1)) = \omega_0((1,1),(1,0)) = i,\\
&\omega_0((1,0),(1,1)) = \omega_0((1,1),(0,1)) = \omega_0((0,1),(1,0)) = -i,\\
&\omega_1((x_1, x_2), (y_1, y_2)) = (-1)^{x_1 y_2}.
\end{aligned}$$

**2.1. $(\mathbf{Z}_2 \times \mathbf{Z}_2 \times \mathbf{Z}_3) \rtimes \mathbf{Z}_2$ case.** Let $N = \mathbf{Z}_2 \times \mathbf{Z}_2 \times \mathbf{Z}_3$ with generators

$$a = (1, 0, 0), \;\; b = (0, 1, 0), c = (0, 0, 1).$$

We define $\tau_1, \tau_2, \tau_3 \in Aut(N)$ by

$$\begin{aligned}
\tau_1(a) &= b, & \tau_1(b) &= a, & \tau_1(c) &= c,\\
\tau_2(a) &= b, & \tau_2(b) &= a, & \tau_2(c) &= 2c,\\
\tau_3(a) &= a, & \tau_3(b) &= b, & \tau_3(c) &= 2c.
\end{aligned}$$

We use the same notations $\omega_0$ and $\omega_1$ for the 2-cocycles of $N$ coming from the quotient of $N$ by $\mathbf{Z}_3$. Then, it is easy to see $H^2(N, \mathbf{T}) \cong \mathbf{Z}_2$ with a generator $[\omega_0] = [\omega_1]$. Let $\chi_1, \chi_2, \chi_3 = \chi_1\chi_2 \in \hat{N}$ be given by

$$\begin{cases} \chi_1(a) = -1, & \chi_1(b) = 1, \quad \chi_1(c) = 1,\\ \chi_2(a) = 1, & \chi_2(b) = -1, \quad \chi_2(c) = 1, \end{cases}$$

and we set $\chi_3 = \chi_1\chi_2$. For simplicity, we use the notations

$$\eta(n_1, n_2) = \eta_1(n_1, n_2), \;\; \zeta_n = \zeta_n(1, 1), \;\; \xi(n) = \xi(1, n).$$

LEMMA 14.12.
(*i*) *For* $N = \mathbf{Z}_2 \times \mathbf{Z}_2 \times \mathbf{Z}_3$ *with the* $\mathbf{Z}_2$ *action* $\tau_1$ *we have*

$$H^2((N,\mathbf{Z}_2),\mathbf{T}) = \{[(1,1)],[(\omega_0,1)]\}.$$

*Only* $(\omega_0,1)$ *gives rise to a non-trivial Kac algebra, which is isomorphic to the tensor product of the Kac-Paljutkin 8-dimensional Kac algebra and* $\mathbf{C}(\mathbf{Z}_3)$.
(*ii*) *For* $N = \mathbf{Z}_2 \times \mathbf{Z}_2 \times \mathbf{Z}_3$ *with the* $\mathbf{Z}_2$ *action* $\tau_2$ *we have*

$$H^2((N,\mathbf{Z}_2),\mathbf{T}) = \{[(1,1)],[(\omega_0,1)]\}.$$

*Only* $(\omega_0,1)$ *gives rise to a non-trivial Kac algebra:*

$$\begin{aligned}
&\mathcal{A} = \ell^\infty(N) \rtimes_{\tau_2} \mathbf{Z}_2 \cong \mathbf{C}^4 \oplus M_2(\mathbf{C}) \oplus M_2(\mathbf{C}) \oplus M_2(\mathbf{C}) \oplus M_2(\mathbf{C}) \oplus M_2(\mathbf{C}),\\
&\hat{\mathcal{A}} = \mathbf{C}(N) \oplus \mathbf{C}_{\omega_0}(N) \cong \mathbf{C}^{12} \oplus M_2(\mathbf{C}) \oplus M_2(\mathbf{C}) \oplus M_2(\mathbf{C}),\\
&G(\mathcal{A}) \cong \mathbf{Z}_2 \times \mathbf{Z}_2 \times \mathbf{Z}_3, \quad G(\hat{\mathcal{A}}) \cong \mathbf{Z}_2 \times \mathbf{Z}_2.
\end{aligned}$$

(*iii*) *For* $N = \mathbf{Z}_2 \times \mathbf{Z}_2 \times \mathbf{Z}_3$ *with the* $\mathbf{Z}_2$ *action* $\tau_3$ *we have*

$$\begin{aligned}
H^2((N,\mathbf{Z}_2),\mathbf{T}) &= \{[(1,1)],[(1,\chi_1)],[(1,\chi_2)],[(1,\chi_3)],\\
&\qquad [(\omega_1,1)],[(\omega_1,\chi_1)],[(\omega_1,\chi_2)],[(\omega_1,\chi_3)]\}\\
&= \{[(1,1)],[(1,\chi_1)],[(1,\chi_2)],[(1,\chi_3)],\\
&\qquad [(\omega_0,\zeta^1)],[(\omega_0,\zeta^2)],[(\omega_0,\zeta^3)],[(\omega_0,\zeta^4)]\}\\
&\cong \mathbf{Z}_2 \times \mathbf{Z}_2 \times \mathbf{Z}_2,
\end{aligned}$$

*where* $\zeta^i$ $(i = 1,2,3,4)$ *are given by*

$$\begin{cases}
\zeta^1((1,0,x)) = -1, & \zeta^1((0,1,x)) = 1, & \zeta^1((1,1,x)) = 1,\\
\zeta^2((1,0,x)) = 1, & \zeta^2((0,1,x)) = -1, & \zeta^2((1,1,x)) = 1,\\
\zeta^3((1,0,x)) = 1, & \zeta^3((0,1,x)) = 1, & \zeta^3((1,1,x)) = -1,\\
\zeta^4((1,0,x)) = -1, & \zeta^4((0,1,x)) = -1, & \zeta^4((1,1,x)) = -1.
\end{cases}$$

*Only cases with non-trivial* $\eta$ *give rise to a non-trivial Kac algebras and* $[(\omega_0,\zeta^i)]$ $(i = 1,2,3)$ *are equivalent. Let* $\mathcal{A} = \ell^\infty(N) \rtimes_{\tau_3,\zeta} \mathbf{Z}_2$. *Then, we get*

(1) $(\eta,\zeta) = (\omega_1,1)$ *case* $([(\omega_1,1)] = [(\omega_0,\zeta^3)])$.

$$\begin{aligned}
&\mathcal{A} \cong \mathbf{C}^8 \oplus M_2(\mathbf{C}) \oplus M_2(\mathbf{C}) \oplus M_2(\mathbf{C}) \oplus M_2(\mathbf{C}),\\
&\hat{\mathcal{A}} \cong \mathbf{C}^{12} \oplus M_2(\mathbf{C}) \oplus M_2(\mathbf{C}) \oplus M_2(\mathbf{C}),\\
&G(\mathcal{A}) \cong \mathbf{Z}_2 \times \mathbf{Z}_2 \times \mathbf{Z}_3, \quad G(\hat{\mathcal{A}}) \cong D_8.
\end{aligned}$$

(2) $(\eta,\zeta) = (\omega_0,\zeta^4)$ *case.*

$$\begin{aligned}
&\mathcal{A} \cong \mathbf{C}^8 \oplus M_2(\mathbf{C}) \oplus M_2(\mathbf{C}) \oplus M_2(\mathbf{C}) \oplus M_2(\mathbf{C}),\\
&\hat{\mathcal{A}} \cong \mathbf{C}^{12} \oplus M_2(\mathbf{C}) \oplus M_2(\mathbf{C}) \oplus M_2(\mathbf{C}),\\
&G(\mathcal{A}) \cong \mathbf{Z}_2 \times \mathbf{Z}_2 \times \mathbf{Z}_3, \quad G(\hat{\mathcal{A}}) \cong Q_8.
\end{aligned}$$

PROOF. (i) It is easy to show $(\omega_0,1) \in Z^2((N,\mathbf{Z}_2),\mathbf{T})$. Thus we may assume that $\eta$ is either 1 or $\omega_0$ and $\zeta \in Z^2(\mathbf{Z}_2,\hat{N})$. Considering the corresponding extensions, we see that $H^2(\mathbf{Z}_2,\hat{N})$ is trivial, and so

$$H^2((N,\mathbf{Z}_2),\mathbf{T}) = \{[(1,1)],[(\omega_0,1)]\}.$$

If $(\eta,\zeta) = (\omega_0,1)$, the third component of $N$ is not involved either in $\tau_1$ or $(\eta,\zeta)$. Therefore, the resulting Kac algebra is as in the statement.
(ii) In the same way as above, we can get $H^2((N,\mathbf{Z}_2),\mathbf{T})$. If $(\eta,\zeta) = (\omega_0,1)$, we

have $G(\mathcal{A}) = \hat{N}$. Let $N_0 = \langle(1,1,0)\rangle \cong \mathbf{Z}_2$. Then, $\omega_0$ restricted to $N_0$ describes the exact sequence

$$1 \longrightarrow \hat{\mathbf{Z}}_2 \longrightarrow G(\hat{\mathcal{A}}) \longrightarrow N_0 \longrightarrow 1,$$

and so $G(\hat{\mathcal{A}}) \cong \mathbf{Z}_2 \times \mathbf{Z}_2$.
(iii) It is easy to show $(\omega_1, 1) \in Z^2((N, \mathbf{Z}_2), \mathbf{T})$. Since

$$H^2(\mathbf{Z}_2, \hat{N}) = \{[1], [\chi_1], [\chi_2], [\chi_3]\},$$

we obtain

$$\begin{aligned} H^2((N,\mathbf{Z}_2),\mathbf{T}) &= \{[(1,1)], [(1,\chi_1)], [(1,\chi_2)], [(1,\chi_3)], \\ &\qquad [(\omega_1,1)], [(\omega_1,\chi_1)], [(\omega_1,\chi_2)], [(\omega_1,\chi_3)]\}. \end{aligned}$$

Direct computation shows $[(\omega_1, 1)] = [(\omega_0, \zeta^3)]$, which gives the second expression of $H^2((N, \mathbf{Z}_2), \mathbf{T})$. Since the group automorphism of $N$ permuting $a, b, ab$ leaves $\omega_0$ invariant, $[(\omega_0, \zeta^i)]$ $(i = 1, 2, 3)$ are equivalent. If $(\eta, \zeta) = (\omega_1, 1)$, $\omega$ restricted to $N_0 := \langle a, b\rangle$ describes the exact sequence

$$1 \longrightarrow \hat{\mathbf{Z}}_2 \longrightarrow G(\hat{\mathcal{A}}) \longrightarrow N_0 \longrightarrow 1.$$

Thus, we get $G(\hat{\mathcal{A}}) \cong D_8$. If $(\eta, \zeta) = (\omega_0, \zeta^4)$, we have $\xi(n)\xi(n^{\tau_3}) = \zeta^4(n)$, where $\xi(n) = i$ for $n \neq e$ and $\xi(e) = 1$. Let

$$\eta'(n_1, n_2) = \omega_0(n_1, n_2)\xi(n_1)\xi(n_2)\overline{\xi(n_1 n_2)}.$$

Then, $\eta'$ describes the exact sequence

$$1 \longrightarrow \hat{\mathbf{Z}}_2 \longrightarrow G(\hat{\mathcal{A}}) \longrightarrow N_0 \longrightarrow 1,$$

and the values of $\eta'$ on $N_0$ are given as follows:

| | $e$ | $a$ | $b$ | $ab$ |
|---|---|---|---|---|
| $e$ | 1 | 1 | 1 | 1 |
| $a$ | 1 | $-1$ | $-1$ | 1 |
| $b$ | 1 | 1 | $-1$ | $-1$ |
| $ab$ | 1 | $-1$ | 1 | $-1$ |

Thus $G(\hat{\mathcal{A}}) \cong Q_8$. □

**2.2. $\mathfrak{A}_4 \rtimes \mathbf{Z}_2$ case.** We use the following presentation of $\mathfrak{A}_4 = (\mathbf{Z}_2 \times \mathbf{Z}_2) \rtimes \mathbf{Z}_3$:

$$\begin{cases} \mathbf{Z}_2 \times \mathbf{Z}_2 = \{1,\ a,\ b,\ c\}, \\ \mathbf{Z}_3 = \langle d\rangle, \\ a^d = b,\ b^d = c,\ c^d = a. \end{cases}$$

We define $\theta_0, \theta_1 \in Aut(\mathfrak{A}_4)$ by $\theta_1 = Ad(a)$ and

$$\theta_0(a) = b, \quad \theta_0(b) = a, \quad \theta_0(c) = c, \quad \theta_0(d) = d^2.$$

LEMMA 14.13.
*(i) Every order 2 element in $Aut(\mathfrak{A}_4)$ is conjugate to either $\theta_0$ or $\theta_1$.*
*(ii) We have*

$$\begin{cases} \ell^\infty(\mathfrak{A}_4) \rtimes_{\theta_0} \mathbf{Z}_2 \cong \mathbf{C}^4 \oplus M_2(\mathbf{C}) \oplus M_2(\mathbf{C}) \oplus M_2(\mathbf{C}) \oplus M_2(\mathbf{C}) \oplus M_2(\mathbf{C}), \\ \ell^\infty(\mathfrak{A}_4) \rtimes_{\theta_1} \mathbf{Z}_2 \cong \mathbf{C}^8 \oplus M_2(\mathbf{C}) \oplus M_2(\mathbf{C}) \oplus M_2(\mathbf{C}) \oplus M_2(\mathbf{C}). \end{cases}$$

*(iii) We define $\omega \in Z^2(\mathfrak{A}_4, \mathbf{T})$ by*

$$\omega((x_1, y_1), (x_2, y_2)) = \omega_0\left(x_1, x_2^{y_1^{-1}}\right) \quad \textit{for } x_1, x_2 \in \mathbf{Z}_2 \times \mathbf{Z}_2 \textit{ and } y_1, y_2 \in \mathbf{Z}_3.$$

*Then,* $H^2(\mathfrak{A}_4,\mathbf{T})\cong\mathbf{Z}_2$ *with a generator* $[\omega]$.

PROOF. It is routine to show (i) and (ii). It is know that $H^2(\mathfrak{A}_4,\mathbf{T})\cong\mathbf{Z}_2$ holds (see [**69**]), and $\omega$ can be obtained from the following projective representation of $\mathfrak{A}_4$:

$$\pi(a)=\begin{pmatrix}1 & 0\\ 0 & -1\end{pmatrix},\qquad \pi(b)=\begin{pmatrix}0 & e^{\frac{\pi i}{4}}\\ e^{-\frac{\pi i}{4}} & 0\end{pmatrix},$$
$$\pi(c)=\begin{pmatrix}0 & e^{-\frac{\pi i}{4}}\\ e^{\frac{\pi i}{4}} & 0\end{pmatrix},\quad \pi(d)=\frac{1}{\sqrt{2}}\begin{pmatrix}-e^{\frac{\pi i}{4}} & -i\\ -i & -e^{-\frac{\pi i}{4}}\end{pmatrix}.$$

□

LEMMA 14.14. *Let* $\mathfrak{A}_4\rtimes\mathbf{Z}_2$ *be a semi-direct product group with the* $\mathbf{Z}_2$ *action given by* $\theta_0$. *Then,*

$$H^2((\mathfrak{A}_4,\mathbf{Z}_2),\mathbf{T})=\{[(1,1)],[(\omega,1)]\}\cong\mathbf{Z}_2.$$

*Let* $\mathcal{A}=\ell^\infty(\mathfrak{A}_4)\rtimes_{\theta_0}\mathbf{Z}_2$. *Then,*

$$\mathcal{A}\cong\mathbf{C}^4\oplus M_2(\mathbf{C})\oplus M_2(\mathbf{C})\oplus M_2(\mathbf{C})\oplus M_2(\mathbf{C})\oplus M_2(\mathbf{C}),$$

*and* $\hat{\mathcal{A}}$, $G(\mathcal{A})$, *and* $G(\hat{\mathcal{A}})$ *are given as follows:*
*(i)* $(\eta,\zeta)=(1,1)$ *case.*

$$\hat{\mathcal{A}}\cong\mathbf{C}^6\oplus M_3(\mathbf{C})\oplus M_3(\mathbf{C}),$$
$$G(\mathcal{A})\cong\mathfrak{S}_3,\quad G(\hat{\mathcal{A}})\cong\mathbf{Z}_2\times\mathbf{Z}_2.$$

*(ii)* $(\eta,\zeta)=(\omega,1)$ *case.*

$$\hat{\mathcal{A}}\cong\mathbf{C}^3\oplus M_2(\mathbf{C})\oplus M_2(\mathbf{C})\oplus M_3(\mathbf{C}),$$
$$G(\mathcal{A})\cong\mathbf{Z}_3,\quad G(\hat{\mathcal{A}})\cong\mathbf{Z}_2\times\mathbf{Z}_2.$$

PROOF. Note that $H^2(\mathbf{Z}_2,Hom(\mathfrak{A}_4,\mathbf{T}))\cong H^2(\mathbf{Z}_2,\mathbf{Z}_3)$ is trivial no matter how $\mathbf{Z}_2$ acts. Thus, we do not have freedom coming from $\zeta$. To obtain the statement, it suffices to show $(\omega,1)\in Z^2((\mathfrak{A}_4,\mathbf{Z}_2),\mathbf{T})$. Indeed, we have

$$\begin{aligned}
&\omega((x_1,y_1),(x_2,y_2))\omega\left((x_1,y_1)^{\theta_0},(x_2,y_2)^{\theta_0}\right)\\
&=\omega((x_1,y_1),(x_2,y_2))\omega\left(\left(x_1^{\theta_0},y_1^2\right),\left(x_2^{\theta_0},y_2^2\right)\right)\\
&=\omega_0\left(x_1,x_2^{y_1^{-1}}\right)\omega_0\left(x_1^{\theta_0},x_2^{\theta_0 y_1}\right)=\omega_0\left(x_1,x_2^{y_1^{-1}}\right)\overline{\omega_0\left(x_1,x_2^{\theta_0 y_1\theta_0}\right)}\\
&=\omega_0\left(x_1,x_2^{y_1^{-1}}\right)\overline{\omega_0\left(x_1,x_2^{y_1^{-1}}\right)}=1.
\end{aligned}$$

It is easy to obtain $\mathcal{A}$, $\hat{\mathcal{A}}$, $G(\mathcal{A})$, and $G(\hat{\mathcal{A}})$. □

**2.3.** $D_{12}\rtimes\mathbf{Z}_2$ **case.** It is routine to show the following:

LEMMA 14.15.
*(i) Let* $D_{12}=\mathbf{Z}_6\rtimes\mathbf{Z}_2$ *be the dihedral group of order* 12 *with the* $\mathbf{Z}_2$ *action on* $\mathbf{Z}_6$ *defined by*

$$(0,1)(x,0)(0,1)=(6-x,0).$$

*Then,* $Aut(D_{12})$ *is isomorphic to* $D_{12}$ *itself with generators*

$$(x_1,x_2)^\varphi=(x_1+x_2,x_2)\ \textit{and}\ (x_1,x_2)^\psi=(6-x_1,x_2)$$

*satisfying*

$$\varphi^6=\psi^2=id,\ \psi\varphi\psi=\varphi^{-1}.$$

*(ii) The conjugacy classes of period two automorphisms of* $D_{12}$ *and centralizers of their representatives are as follows:*

$$\{\varphi^3\}, \quad \{\psi, \varphi^2\psi, \varphi^4\psi\}, \quad \{\varphi\psi, \varphi^3\psi, \varphi^5\psi\},$$

$$\begin{aligned} C_{Aut(D_{12})}(\varphi^3) &= Aut(D_{12}), \\ C_{Aut(D_{12})}(\psi) &= \{id, \varphi^3, \psi, \varphi^3\psi\}, \\ C_{Aut(D_{12})}(\varphi\psi) &= \{id, \varphi^3, \varphi\psi, \varphi^4\psi\}. \end{aligned}$$

*(iii) Let* $\tau_1 = \varphi\psi$, $\tau_2 = \psi$, $\tau_3 = \varphi^3$. *Then, we have*

$$\begin{aligned} (x_1, x_2)^{\tau_1} &= (6 - x_1 - x_2, x_2), \\ (x_1, x_2)^{\tau_2} &= (6 - x_1, x_2), \\ (x_1, x_2)^{\tau_3} &= (x_1 + 3x_2, x_2), \end{aligned}$$

*and, the fixed-point subgroups* $D_{12}^{\tau_i}$ $(i = 1, 2, 3)$ *are*

$$\begin{aligned} D_{12}^{\tau_1} &= \{(0,0), (3,0)\}, \\ D_{12}^{\tau_2} &= \{(0,0), (3,0), (0,1), (3,1)\}, \\ D_{12}^{\tau_3} &= \langle (1,0) \rangle. \end{aligned}$$

We define $\omega \in Z^2(D_{4n}, \mathbf{T})$ by

$$\begin{cases} \omega((x,0),(y,0)) &= \omega((x,0),(y,1)) &= 1, \\ \omega((x,1),(y,0)) &= \omega((x,1),(y,1)) &= e^{\frac{\pi i}{n} y} \end{cases}$$

for $0 \le x, y \le 2n - 1$.

The next two results were already shown in Chapter 11 (see Lemma 11.9) and in Chapter 7,§6.

LEMMA 14.16.
*(i) We have* $H^2(D_{4n}, \mathbf{T}) \cong \mathbf{Z}_2$ *with a generator* $[\omega]$.
*(ii) The twisted group algebra* $\mathbf{C}_\omega(D_{4n})$ *is isomorphic to*

$$M_2(\mathbf{C}) \oplus M_2(\mathbf{C}) \oplus \cdots \oplus M_2(\mathbf{C}).$$

LEMMA 14.17. *For* $N = D_{12}$ *with the* $\mathbf{Z}_2$ *action* $\tau_1$ *we have* $H^2((N, \mathbf{Z}_2), \mathbf{T}) \cong \mathbf{Z}_2$. *The corresponding Kac algebras are non-trivial in both cases. Without a cocycle, the algebra structure is given by*

$$\begin{aligned} \mathcal{A} &\cong \mathbf{C}^4 \oplus M_2(\mathbf{C}) \oplus M_2(\mathbf{C}) \oplus M_2(\mathbf{C}) \oplus M_2(\mathbf{C}) \oplus M_2(\mathbf{C}), \\ \hat{\mathcal{A}} &\cong \mathbf{C}^8 \oplus M_2(\mathbf{C}) \oplus M_2(\mathbf{C}) \oplus M_2(\mathbf{C}) \oplus M_2(\mathbf{C}), \end{aligned}$$

*and the intrinsic groups are*

$$G(\mathcal{A}) \cong D_8, \ G(\hat{\mathcal{A}}) \cong \mathbf{Z}_2 \times \mathbf{Z}_2.$$

*On the other hand, with a non-trivial cocycle, we have*

$$\mathcal{A} \cong \hat{\mathcal{A}} \cong \mathbf{C}^4 \oplus M_2(\mathbf{C}) \oplus M_2(\mathbf{C}) \oplus M_2(\mathbf{C}) \oplus M_2(\mathbf{C}) \oplus M_2(\mathbf{C}),$$

*and*

$$G(\mathcal{A}) \cong G(\hat{\mathcal{A}}) \cong \mathbf{Z}_2 \times \mathbf{Z}_2.$$

We define $\chi_1, \chi_2, \chi_3 = \chi_1\chi_2 \in Hom(D_{12}, \mathbf{T})$ by

$$\chi_1((x_1, x_2)) = (-1)^{x_1}, \quad \chi_2((x_1, x_2)) = (-1)^{x_2}.$$

Note that $\tau_2$ acts on $Hom(D_{12}, \mathbf{T})$ trivially.

LEMMA 14.18. *For $N = D_{12}$ with the $\mathbf{Z}_2$ action $\tau_2$ we have*

$$\begin{aligned} H^2((N,\mathbf{Z}_2),\mathbf{T}) &= \{[(1,1)],[(1,\chi_1)],[(1,\chi_2)],[(1,\chi_3)],\\ &\qquad [(\omega,1)],[(\omega,\chi_1)],[(\omega,\chi_2)],[(\omega,\chi_3)]\}\\ &\cong \mathbf{Z}_2\times\mathbf{Z}_2\times\mathbf{Z}_2,\\ \#(H^2((N,\mathbf{Z}_2),\mathbf{T})/\sim) &= 6. \end{aligned}$$

*The algebra structure of the corresponding Kac algebra $\mathcal{A}$ is*

$$\mathcal{A}\cong \ell^\infty(N)\rtimes_\zeta \mathbf{Z}_2 \cong \mathbf{C}^8\oplus M_2(\mathbf{C})\oplus M_2(\mathbf{C})\oplus M_2(\mathbf{C})\oplus M_2(\mathbf{C}).$$

*If $[\eta_1]$ is trivial, the algebra structure of $\hat{\mathcal{A}}$ is*

$$\hat{\mathcal{A}}\cong \mathbf{C}(D_8)\oplus\mathbf{C}(D_8)\cong \mathbf{C}^8\oplus M_2(\mathbf{C})\oplus M_2(\mathbf{C})\oplus M_2(\mathbf{C})\oplus M_2(\mathbf{C}).$$

*If $[\eta_1]$ is not trivial, then we have*

$$\hat{\mathcal{A}}\cong \mathbf{C}(D_8)\oplus\mathbf{C}_{\eta_1}(D_8)\cong \mathbf{C}^4\oplus M_2(\mathbf{C})\oplus M_2(\mathbf{C})\oplus M_2(\mathbf{C})\oplus M_2(\mathbf{C})\oplus M_2(\mathbf{C}).$$

*The representatives of $H^2((N,\mathbf{Z}_2),\mathbf{T})/\sim$ and the intrinsic groups are as follows:*

*(i)* $(\eta,\zeta)=(1,1)$,

$$G(\mathcal{A})\cong\mathbf{Z}_2\times\mathbf{Z}_2\times\mathbf{Z}_2,\quad G(\hat{\mathcal{A}})\cong\mathbf{Z}_2\times\mathbf{Z}_2\times\mathbf{Z}_2.$$

*(ii)* $(\eta,\zeta)=(1,\chi_1)$,

$$G(\mathcal{A})\cong\mathbf{Z}_4\times\mathbf{Z}_2,\quad G(\hat{\mathcal{A}})\cong\mathbf{Z}_4\times\mathbf{Z}_2.$$

*(iii)* $(\eta,\zeta)=(1,\chi_2)$,

$$G(\mathcal{A})\cong\mathbf{Z}_4\times\mathbf{Z}_2,\quad G(\hat{\mathcal{A}})\cong\mathbf{Z}_4\times\mathbf{Z}_2.$$

*(iv)* $(\eta,\zeta)=(\omega,1)$,

$$G(\mathcal{A})\cong\mathbf{Z}_2\times\mathbf{Z}_2,\quad G(\hat{\mathcal{A}})\cong D_8.$$

*(v)* $(\eta,\zeta)=(\omega,\chi_1)$,

$$G(\mathcal{A})\cong\mathbf{Z}_2\times\mathbf{Z}_2,\quad G(\hat{\mathcal{A}})\cong D_8.$$

*(vi)* $(\eta,\zeta)=(\omega,\chi_3)$,

$$G(\mathcal{A})\cong\mathbf{Z}_2\times\mathbf{Z}_2,\quad G(\hat{\mathcal{A}})\cong Q_8.$$

PROOF. We may assume either $\eta=1$ or $\eta=\omega$. Since $\omega\omega^{\tau_2}=1$, $\zeta$ belongs to $Z^2(\mathbf{Z}_2, Hom(N,\mathbf{T}))$, where the action of $\mathbf{Z}_2$ on $Hom(N,\mathbf{T}))$ is trivial. Thus,

$$H^2(\mathbf{Z}_2, Hom(N,\mathbf{T}))\cong\mathbf{Z}_2\times\mathbf{Z}_2,$$

and we get $H^2((N,\mathbf{Z}_2),\mathbf{T})$. To prove $H^2((N,\mathbf{Z}_2),\mathbf{T})\cong\mathbf{Z}_2\times\mathbf{Z}_2\times\mathbf{Z}_2$, it suffices to show that $[(\omega^2,1)]$ is trivial. Indeed, with $\xi(x_1,x_2)=e^{\frac{\pi i}{3}x_1}$ we have

$$\begin{cases} \omega(n_1,n_2)^2=\xi(n_1)\xi(n_2)\overline{\xi(n_1n_2)},\\ \xi(n)\xi(n^{\tau_2})=1,\end{cases}$$

showing that $(\omega^2,1)$ is a coboundary. Note that $C_{Aut(D_8)}(\tau_2)=\langle\tau_2,\tau_3\rangle$ thanks to Lemma 14.15. To obtain the equivalence relation in $H^2((N,\mathbf{Z}_2),\mathbf{T})$, it suffices to compute the action of $\tau_3$ on $H^2((N,\mathbf{Z}_2),\mathbf{T})$. We set

$$\xi(x_1,x_2)=\begin{cases}1 & \text{if } x_2=0,\\ i & \text{if } x_2=1.\end{cases}$$

Then, we get the following for $n,n_1,n_2\in N$:

$$\omega((n_1,n_2)^{\tau_3})=\omega((n_1,n_2))\xi(n_1)\xi(n_2)\overline{\xi(n_1n_2)},$$

$$\xi(n)\xi(n^{\tau_3}) = \chi_2(n).$$

Direct computation shows

$$\chi_1^{\tau_3} = \chi_3, \quad \chi_2^{\tau_3} = \chi_2, \quad \chi_3^{\tau_3} = \chi_1.$$

Thus, $[(1,\chi_1)] \sim [(1,\chi_3)]$, $[(\omega,1)] \sim [(\omega,\chi_2)]$. When $\eta = 1$, $\zeta$ describes the exact sequence

$$1 \longrightarrow Hom(N,\mathbf{T}) \longrightarrow G(\mathcal{A}) \longrightarrow \mathbf{Z}_2 \longrightarrow 1.$$

When $\eta = \omega$, $G(\mathcal{A}) = Hom(N,\mathbf{T})$ and so we get the results for $G(\mathcal{A})$. When $\zeta = 1$, $\eta$ restricted to $N^{\tau_2}$ describes the exact sequence

$$1 \longrightarrow \hat{\mathbf{Z}}_2 \longrightarrow G(\hat{\mathcal{A}}) \longrightarrow N^{\tau_2} \longrightarrow 1.$$

Thus, we get the result for (i). Since the values of $\omega$ on $N^{\tau_2}$ are given by

| | $(0,0)$ | $(3,0)$ | $(0,1)$ | $(3,1)$ |
|---|---|---|---|---|
| $(0,0)$ | 1 | 1 | 1 | 1 |
| $(3,0)$ | 1 | 1 | 1 | 1 |
| $(0,1)$ | 1 | $-1$ | 1 | $-1$ |
| $(3,1)$ | 1 | $-1$ | 1 | $-1$ |

the corresponding extension is $D_8$ in case (iv). We take $\xi_i : N \longrightarrow \mathbf{T}$, $i = 1,2,3$ satisfying

$$\xi_1((3,0)) = i, \quad \xi_1((0,1)) = 1, \quad \xi_1((3,1)) = i,$$
$$\xi_2((3,0)) = 1, \quad \xi_2((0,1)) = i, \quad \xi_2((3,1)) = i,$$
$$\xi_3((3,0)) = i, \quad \xi_3((0,1)) = i, \quad \xi_1((3,1)) = 1.$$

Then, we get $\xi_i(n)\xi_i(n^{\tau_2}) = \chi_i(n)$ for $n \in N^{\tau_2}$. Thus, when $\zeta = \chi_i$, we have

$$\eta'(n_1,n_2) := \xi_i(n_1)\xi_i(n_2)\overline{\xi_i(n_1n_2)}\eta(n_1,n_2)$$

describes the exact sequence

$$1 \longrightarrow \hat{\mathbf{Z}}_2 \longrightarrow G(\hat{\mathcal{A}}) \longrightarrow N^{\tau_2} \longrightarrow 1.$$

In cases (ii) and (iii), the corresponding extensions are non-trivial and abelian, and so $G(\hat{\mathcal{A}}) \cong \mathbf{Z}_4 \times \mathbf{Z}_2$. In case (v), the values of $\eta'$ are given as follows:

| | $(0,0)$ | $(3,0)$ | $(0,1)$ | $(3,1)$ |
|---|---|---|---|---|
| $(0,0)$ | 1 | 1 | 1 | 1 |
| $(3,0)$ | 1 | $-1$ | 1 | $-1$ |
| $(0,1)$ | 1 | $-1$ | 1 | $-1$ |
| $(3,1)$ | 1 | 1 | 1 | 1 |

This shows $G(\hat{\mathcal{A}}) \cong D_8$. In case (vi), the values of $\eta'$ are given as follows:

| | $(0,0)$ | $(3,0)$ | $(0,1)$ | $(3,1)$ |
|---|---|---|---|---|
| $(0,0)$ | 1 | 1 | 1 | 1 |
| $(3,0)$ | 1 | $-1$ | $-1$ | 1 |
| $(0,1)$ | 1 | 1 | $-1$ | $-1$ |
| $(3,1)$ | 1 | $-1$ | 1 | $-1$ |

This shows $G(\hat{\mathcal{A}}) \cong Q_8$. □

LEMMA 14.19. *For $N = D_{12}$ with the $\mathbf{Z}_2$ action $\tau_3$ we have*

$$H^2((N,\mathbf{Z}_2),\mathbf{T}) \cong \mathbf{Z}_2.$$

*The algebra structure of the corresponding Kac algebra $\mathcal{A}$ is*

$$\mathcal{A} \cong \ell^\infty(N) \rtimes_\zeta \mathbf{Z}_2 \cong \mathbf{C}^{12} \oplus M_2(\mathbf{C}) \oplus M_2(\mathbf{C}) \oplus M_2(\mathbf{C}).$$

*Without a cocycle, the dual Kac algebra and the intrinsic groups are as follows:*

$$\hat{\mathcal{A}} \cong \mathbf{C}(D_{12}) \oplus \mathbf{C}(D_{12}) \cong \mathbf{C}^8 \oplus M_2(\mathbf{C}) \oplus M_2(\mathbf{C}) \oplus M_2(\mathbf{C}) \oplus M_2(\mathbf{C}),$$
$$G(\mathcal{A}) \cong D_8, \quad G(\hat{\mathcal{A}}) \cong \mathbf{Z}_2 \times \mathbf{Z}_2 \times \mathbf{Z}_3.$$

*With a cocycle, we have*

$$\begin{aligned}\hat{\mathcal{A}} &\cong \mathbf{C}(D_{12}) \oplus \mathbf{C}_\omega(D_{12}) \\ &\cong \mathbf{C}^4 \oplus M_2(\mathbf{C}) \oplus M_2(\mathbf{C}) \oplus M_2(\mathbf{C}) \oplus M_2(\mathbf{C}) \oplus M_2(\mathbf{C}),\end{aligned}$$
$$G(\mathcal{A}) \cong \mathbf{Z}_2 \times \mathbf{Z}_2, \quad G(\hat{\mathcal{A}}) \cong \mathbf{Z}_2 \times \mathbf{Z}_2 \times \mathbf{Z}_3.$$

PROOF. Note that since $\tau_3$ exchanges $\chi_1$ and $\chi_3$ $H^2(\mathbf{Z}_2, Hom(N,\mathbf{T}))$ is trivial. Thus, there is no freedom coming from $\zeta$ in $H^2((N,\mathbf{Z}_2),\mathbf{T})$. If $\eta = \omega$, then $\zeta$ should satisfy

$$\zeta_n = \zeta_{n^{\tau_3}}, \quad \omega(n_1,n_2)\omega(n_1^{\tau_3}, n_2^{\tau_3}) = \zeta_{n_1n_2}\overline{\zeta_{n_1}\zeta_{n_2}}.$$

If we set $\zeta^0_{(x,y)} = e^{(-\frac{x}{3}+\frac{y}{4})2\pi i}$, then $\zeta = \zeta^0$ actually satisfies the above and we get

$$H^2((N,\mathbf{Z}_2),\mathbf{T}) = \{[(1,1)], [(\omega,\zeta^0)]\}.$$

It is easy to determine the intrinsic groups in the trivial cocycle case. On the other hand, if $(\eta,\zeta) = (\omega,\zeta^0)$, then we have

$$G(\mathcal{A}) = Hom(N,\mathbf{T}) \cong \mathbf{Z}_2 \times \mathbf{Z}_2.$$

Note that $N^{\tau_3} = \langle(1,0)\rangle$. We take $\xi : N \longrightarrow \mathbf{T}$ satisfying $\xi((k,0)) = e^{\frac{2\pi i}{3}k}$ for $0 \le k \le 5$. Then, $\zeta_n = \xi(n)\xi(n^{\tau_3})$ holds for $n \in N^{\tau_3}$, and the restriction of

$$\eta'(n_1,n_2) := \xi(n_1)\xi(n_2)\overline{\xi(n_1n_2)}\eta(n_1,n_2)$$

to $N^{\tau_3}$ describes the exact sequence

$$1 \longrightarrow \hat{\mathbf{Z}}_2 \longrightarrow G(\hat{\mathcal{A}}) \longrightarrow N^{\tau_3} \longrightarrow 1.$$

Thus we get $G(\hat{\mathcal{A}}) \cong \mathbf{Z}_2 \times \mathbf{Z}_2 \times \mathbf{Z}_3$. □

COROLLARY 14.20. *The Kac algebra $\mathcal{A}$ obtained as the crossed product by $\tau_1$ with the non-trivial cocycle as in Lemma* 14.17 *is self-dual.*

PROOF. Let $R$ be a factor equipped with a near action $(\alpha,\beta)$ of $(D_{12},\mathbf{Z}_2)$ carrying the invariant $(\eta,\zeta)$ as above. We set $P = R \rtimes_\alpha D_{12}$, $Q = R^\beta$, and $L = R \rtimes_\alpha \mathbf{Z}_6$. Then, $Q$ is the fixed-point subalgebra of $L$ under an action of a group $G$, where $G$ is determined by the exact sequence

$$1 \longrightarrow \hat{\mathbf{Z}}_6 \longrightarrow G \longrightarrow \mathbf{Z}_2 \longrightarrow 1.$$

This extension is described by the restriction of $(\zeta,\eta)$ to $(\mathbf{Z}_6,\mathbf{Z}_2)$, and direct computation shows $G \cong D_{12}$. Thus, the dual Kac algebra $\hat{\mathcal{A}}$ is also obtained by $\ell^\infty(D_{12}) \rtimes \mathbf{Z}_2$ with some action and some invariant. Since

$$\mathcal{A} \cong \hat{\mathcal{A}} \cong \mathbf{C}^4 \oplus M_2(\mathbf{C}_2) \oplus M_2(\mathbf{C}_2) \oplus M_2(\mathbf{C}_2) \oplus M_2(\mathbf{C}_2) \oplus M_2(\mathbf{C}_2)$$

occurs in only one case in the above three lemmas, we conclude that $\mathcal{A}$ is self-dual. □

**2.4. $(\mathbf{Z}_3 \rtimes \mathbf{Z}_4) \rtimes \mathbf{Z}_2$ case.** Let

$$N = \mathbf{Z}_3 \rtimes \mathbf{Z}_4 = \langle a, b |\ a^3 = b^4 = bab^{-1}a = 1\rangle.$$

We define $\chi \in Hom(N, \mathbf{T})$ by

$$\chi(a) = 1, \quad \chi(b) = i.$$

It is easy to show the following:

LEMMA 14.21. *We define $\varphi, \psi \in Aut(N)$ by $\psi = Ad(b)$ and*

$$\varphi(a) = a, \quad \varphi(b) = ab^{-1}.$$

*(i) We have*

$$Aut(N) = \langle \varphi, \psi |\ \varphi^6 = \psi^2 = (\varphi\psi)^2 = id\rangle \cong D_{12}.$$

*In particular, every order two automorphism of $N$ is conjugate to either of the following:*

$$\theta_1 = \psi,\ \theta_2 = \varphi^3 \cdot \psi,\ \theta_3 = \varphi^3.$$

*Explicit formulas for $\theta_i$ $(i = 1, 2, 3)$ and their fixed-point subgroups $N^{\theta_i}$ $(i = 1, 2, 3)$ are given as follows:*

$$\begin{aligned} a^{\theta_1} &= a^2, & b^{\theta_1} &= b, & N^{\theta_1} &= \langle b\rangle \cong \mathbf{Z}_4, \\ a^{\theta_2} &= a^2, & b^{\theta_2} &= b^3, & N^{\theta_2} &= \langle b^2\rangle \cong \mathbf{Z}_2, \\ a^{\theta_3} &= a, & b^{\theta_3} &= b^3, & N^{\theta_3} &= \langle a, b^2\rangle \cong \mathbf{Z}_3 \times \mathbf{Z}_2. \end{aligned}$$

*(ii) We have $Hom(N, \mathbf{T}) = \langle \chi\rangle \cong \mathbf{Z}_4$.*
*(iii) We have $\mathbf{C}(N) \cong \mathbf{C}^4 \oplus M_2(\mathbf{C}) \oplus M_2(\mathbf{C})$.*
*(iv) The cohomology group $H^2(N, \mathbf{T})$ is trivial.*

LEMMA 14.22. *For $N$ with the $\mathbf{Z}_2$ action $\theta_1$ we have*

$$H^2((N, \mathbf{Z}_2), \mathbf{T}) = \{[(1, 1)], [(1, \chi)]\}.$$

*Let $\mathcal{A} = \ell^\infty(N) \rtimes_\zeta \mathbf{Z}_2$. Then, the algebra structure of $\mathcal{A}$, $\hat{\mathcal{A}}$ and the intrinsic groups $G(\mathcal{A})$, $G(\hat{\mathcal{A}})$ are as follows:*
*(i) $(\eta, \zeta) = (1, 1)$ case.*

$$\mathcal{A} \cong \hat{\mathcal{A}} \cong \mathbf{C}^8 \oplus M_2(\mathbf{C}) \oplus M_2(\mathbf{C}) \oplus M_2(\mathbf{C}) \oplus M_2(\mathbf{C}),$$

$$G(\mathcal{A}) \cong G(\hat{\mathcal{A}}) \cong \mathbf{Z}_2 \times \mathbf{Z}_4.$$

*(ii) $(\eta, \zeta) = (1, \chi)$ case.*

$$\mathcal{A} \cong \hat{\mathcal{A}} \cong \mathbf{C}^8 \oplus M_2(\mathbf{C}) \oplus M_2(\mathbf{C}) \oplus M_2(\mathbf{C}) \oplus M_2(\mathbf{C}),$$

$$G(\mathcal{A}) \cong G(\hat{\mathcal{A}}) \cong \mathbf{Z}_8.$$

PROOF. Since $H^2(N, \mathbf{T})$ is trivial, we have

$$H^2((N, \mathbf{Z}_2), \mathbf{T}) \cong H^2(\mathbf{Z}_2, Hom(N, \mathbf{T})) \cong H^2(\mathbf{Z}_2, \mathbf{Z}_4)$$

with the trivial $\mathbf{Z}_2$ action on $\mathbf{Z}_4$. Thus, we get the first statement. It is easy to compute the algebra structure of $\mathcal{A}$ and $\hat{\mathcal{A}}$. Since $\zeta$ determines the extension

$$\{1\} \longrightarrow Hom(N, \mathbf{T}) \longrightarrow G(\mathcal{A}) \longrightarrow \mathbf{Z}_2 \longrightarrow \{1\},$$

we get $G(\mathcal{A})$. Let $N_0 = \langle b\rangle$ be the fixed-point subgroup of $N$ under $\theta_1$. It is easy to see that in the first case $G(\hat{\mathcal{A}})$ is a trivial extension of $N_0$ by $\hat{\mathbf{Z}}_2$. In the second case we take $\xi :\ N \longrightarrow \mathbf{T}$ satisfying $\xi(b^k) = e^{k\frac{\pi i}{4}}$ for $0 \le k \le 3$. Then, we have

$\zeta_n = \xi(n)^2$ for $n \in N_0$. Let $\eta'(n_1,n_2) = \xi(n_1 n_2)\overline{\xi(n_1)\xi(n_2)}$. Since the restriction of $\eta'$ to $N_0$ describes the exact sequence

$$\{1\} \longrightarrow \hat{\mathbf{Z}}_2 \longrightarrow G(\hat{\mathcal{A}}) \longrightarrow N_0 \longrightarrow \{1\},$$

we get the result. □

LEMMA 14.23. *For $N$ with the $\mathbf{Z}_2$ action $\theta_2$ we have*

$$H^2((N,\mathbf{Z}_2),\mathbf{T}) = \{[(1,1)],[(1,\chi^2)]\}.$$

*Let $\mathcal{A} = \ell^\infty(N) \rtimes_\zeta \mathbf{Z}_2$. Then, the algebra structure of $\mathcal{A}$ is*

$$\mathcal{A} \cong \mathbf{C}^4 \oplus M_2(\mathbf{C}) \oplus M_2(\mathbf{C}) \oplus M_2(\mathbf{C}) \oplus M_2(\mathbf{C}) \oplus M_2(\mathbf{C}).$$

*The algebra structure of $\hat{\mathcal{A}}$ and the intrinsic groups $G(\mathcal{A})$, $G(\hat{\mathcal{A}})$ are as follows:*
*(i) $(\eta,\zeta) = (1,1)$ case.*

$$\hat{\mathcal{A}} \cong \mathbf{C}^8 \oplus M_2(\mathbf{C}) \oplus M_2(\mathbf{C}) \oplus M_2(\mathbf{C}) \oplus M_2(\mathbf{C}),$$
$$G(\mathcal{A}) \cong D_8, \quad G(\hat{\mathcal{A}}) \cong \mathbf{Z}_2 \times \mathbf{Z}_2.$$

*(ii) $(\eta,\zeta) = (1,\chi^2)$ case.*

$$\hat{\mathcal{A}} \cong \mathbf{C}^8 \oplus M_2(\mathbf{C}) \oplus M_2(\mathbf{C}) \oplus M_2(\mathbf{C}) \oplus M_2(\mathbf{C}),$$
$$G(\mathcal{A}) \cong Q_8, \quad G(\hat{\mathcal{A}}) \cong \mathbf{Z}_2 \times \mathbf{Z}_2.$$

PROOF. Since $H^2(N,\mathbf{T})$ is trivial, we have

$$H^2((N,\mathbf{Z}_2),\mathbf{T}) \cong H^2(\mathbf{Z}_2, Hom(N,\mathbf{T})) \cong H^2(\mathbf{Z}_2,\mathbf{Z}_4)$$

with a non-trivial $\mathbf{Z}_2$ action on $\mathbf{Z}_4$. Thus, we get the first statement. It is easy to compute the algebra structure of $\mathcal{A}$, $\hat{\mathcal{A}}$. Since $\zeta$ determines the extension

$$\{1\} \longrightarrow Hom(N,\mathbf{T}) \longrightarrow G(\mathcal{A}) \longrightarrow \mathbf{Z}_2 \longrightarrow \{1\},$$

we get the description of $G(\mathcal{A})$. Let $N_0 = \langle b^2 \rangle$ be the fixed-point subgroup of $N$ under $\theta_2$. Since $\zeta$ restricted to $N_0$ is trivial, $G(\hat{\mathcal{A}})$ is a trivial extension of $N_0$ by $\hat{\mathbf{Z}}_2$ in the both cases and we get the result. □

LEMMA 14.24. *For $N$ with the $\mathbf{Z}_2$ action $\theta_3$ we have*

$$H^2((N,\mathbf{Z}_2),\mathbf{T}) = \{[(1,1)],[(1,\chi^2)]\}.$$

*Let $\mathcal{A} = \ell^\infty(N) \rtimes_\zeta \mathbf{Z}_2$. Then, the algebra structure of $\mathcal{A}$ is*

$$\mathcal{A} \cong \mathbf{C}^{12} \oplus M_2(\mathbf{C}) \oplus M_2(\mathbf{C}) \oplus M_2(\mathbf{C}).$$

*The algebra structure of $\hat{\mathcal{A}}$ and the intrinsic groups $G(\mathcal{A})$, $G(\hat{\mathcal{A}})$ are as follows:*
*(i) $(\eta,\zeta) = (1,1)$ case.*

$$\hat{\mathcal{A}} \cong \mathbf{C}^8 \oplus M_2(\mathbf{C}) \oplus M_2(\mathbf{C}) \oplus M_2(\mathbf{C}) \oplus M_2(\mathbf{C}),$$
$$G(\mathcal{A}) \cong D_8, \quad G(\hat{\mathcal{A}}) \cong \mathbf{Z}_2 \times \mathbf{Z}_2 \times \mathbf{Z}_3.$$

*(ii) $(\eta,\zeta) = (1,\chi^2)$ case.*

$$\hat{\mathcal{A}} \cong \mathbf{C}^8 \oplus M_2(\mathbf{C}) \oplus M_2(\mathbf{C}) \oplus M_2(\mathbf{C}) \oplus M_2(\mathbf{C}),$$
$$G(\mathcal{A}) \cong Q_8, \quad G(\hat{\mathcal{A}}) \cong \mathbf{Z}_2 \times \mathbf{Z}_2 \times \mathbf{Z}_3.$$

PROOF. Since $H^2(N,\mathbf{T})$ is trivial, we have

$$H^2((N,\mathbf{Z}_2),\mathbf{T}) \cong H^2(\mathbf{Z}_2, Hom(N,\mathbf{T})) \cong H^2(\mathbf{Z}_2,\mathbf{Z}_4)$$

with a non-trivial $\mathbf{Z}_2$ action on $\mathbf{Z}_4$. Thus, we get the first statement. It is easy to show the algebra structure of $\mathcal{A}$ and $\hat{\mathcal{A}}$. Since $\zeta$ determines the extension

$$\{1\} \longrightarrow Hom(N,\mathbf{T}) \longrightarrow G(\mathcal{A}) \longrightarrow \mathbf{Z}_2 \longrightarrow \{1\},$$

we get the description of $G(\mathcal{A})$. Let $N_0 = \langle a, b^2\rangle$ be the fixed-point subgroup of $N$ under $\theta_1$. Since $\zeta$ restricted to $N_0$ is trivial, $G(\hat{\mathcal{A}})$ is a trivial extension of $N_0$ by $\hat{\mathbf{Z}}_2$ in the both cases and we get the result. □

**2.5. $(\mathbf{Z}_2 \times \mathbf{Z}_2 \times \mathbf{Z}_2) \rtimes \mathbf{Z}_3$ case.**

LEMMA 14.25. *Let $N = \mathbf{Z}_2 \times \mathbf{Z}_2 \times \mathbf{Z}_2$. Then, a non-trivial automorphism $\tau \in Aut(N)$ of period 3 is conjugate to*

$$\begin{pmatrix} 1 & 0 & 0 \\ 0 & 0 & 1 \\ 0 & 1 & 1 \end{pmatrix}.$$

PROOF. Comparing the number of the orbits of $\tau$ and the order of $N$, we can see that the order of the fixed-point subgroup $N^\tau$ under $\tau$ is 2. Let $a$ be a generator of $N^\tau$ and take $b \notin N^\tau$. Then, $a, b, \tau(b)$ generate $N$. Since $b + \tau(b) + \tau^2(b) \in N^\tau$, it is either 0 or $a$. By replacing $b$ with $a + b$ in the second case, we may assume $b + \tau(b) + \tau^2(b) = 0$. Thus, the matrix expression of $\tau$ in terms of $(a, b, \tau(b))$ has the desired form. □

Let $N$ and $\tau$ be as above. We set $\omega_{12}, \omega_{23}, \omega_{13} \in Z^2(N,\mathbf{T})$ by

$$\begin{cases} \omega_{12}((x_1,x_2,x_3),(y_1,y_2,y_3)) &= \omega_1((x_1,x_2),(y_1,y_2)), \\ \omega_{23}((x_1,x_2,x_3),(y_1,y_2,y_3)) &= \omega_0((x_2,x_3),(y_2,y_3)), \\ \omega_{13}((x_1,x_2,x_3),(y_1,y_2,y_3)) &= \omega_1((x_1,x_3),(y_1,y_3)). \end{cases}$$

Note that these are generators of $H^2(N,\mathbf{T})$. The action of $\tau$ on these are as follows:

$$\omega_{12}^\tau = \omega_{13}, \quad \omega_{13}^\tau = \omega_{12}\omega_{13}, \quad \omega_{23}^\tau = \omega_{23}.$$

It is easy to show

LEMMA 14.26. *Let $N = \mathbf{Z}_2 \times \mathbf{Z}_2 \times \mathbf{Z}_2$ with the above $\mathbf{Z}_3$ action $\tau$. Then, we have*

$$\begin{aligned} H^2((N,\mathbf{Z}_3),\mathbf{T}) &= \{[(1,1)],[(\omega_{12},1)],[(\omega_{13},1)],[(\omega_{12}\omega_{13},1)]\} \\ &\cong \mathbf{Z}_2 \times \mathbf{Z}_2, \end{aligned}$$

*and the three non-trivial elements are equivalent. Let $\mathcal{A} = \ell^\infty(N) \rtimes_\tau \mathbf{Z}_3$ in the non-trivial case. Then, we have*

$$\begin{aligned} &\mathcal{A} = \mathbf{C}^6 \oplus M_3(\mathbf{C}) \oplus M_3(\mathbf{C}), \\ &\hat{\mathcal{A}} = \mathbf{C}^8 \oplus M_2(\mathbf{C}) \oplus M_2(\mathbf{C}) \oplus M_2(\mathbf{C}) \oplus M_2(\mathbf{C}), \\ &G(\mathcal{A}) \cong \mathbf{Z}_2 \times \mathbf{Z}_2 \times \mathbf{Z}_2, \quad G(\hat{\mathcal{A}}) \cong \mathbf{Z}_6. \end{aligned}$$

**2.6. $Q_3 \rtimes \mathbf{Z}_3$ case.**

LEMMA 14.27. *For $Q_3 \rtimes \mathbf{Z}_3$ as in Lemma 14.1 $H^2((Q_8,\mathbf{Z}_3),\mathbf{T})$ is trivial. Let $\mathcal{A} = \ell^\infty(Q_3) \rtimes \mathbf{Z}_3$. Then, we have*

$$\begin{aligned}&\mathcal{A} \cong \mathbf{C}^6 \oplus M_3(\mathbf{C}) \oplus M_3(\mathbf{C}),\\ &\hat{\mathcal{A}} \cong \mathbf{C}^{12} \oplus M_2(\mathbf{C}) \oplus M_2(\mathbf{C}) \oplus M_2(\mathbf{C}),\\ &G(\mathcal{A}) \cong \mathfrak{A}_4, \quad G(\hat{\mathcal{A}}) \cong \mathbf{Z}_6.\end{aligned}$$

PROOF. Since $H^2(Q_8,\mathbf{T})$ is trivial, we have

$$H^2((Q_8,\mathbf{Z}_3),\mathbf{T}) \cong H^2(\mathbf{Z}_3, Hom(Q_8,\mathbf{T})) \cong H^2(\mathbf{Z}_3, \mathbf{Z}_2 \times \mathbf{Z}_2)$$

with the non-trivial $\mathbf{Z}_3$ action on $\mathbf{Z}_2\times\mathbf{Z}_2$. Considering the corresponding extensions, we conclude that the above is actually trivial. The rest of the statement follows from direct computation. □

**2.7. $\mathfrak{S}_3 \rtimes \mathbf{Z}_4$ case.** Let

$$\mathfrak{S}_3 = \langle a,b|\ a^3 = b^2 = (ab)^2 = 1\rangle, \quad \mathbf{Z}_4 = \langle c|\ c^4 = 1\rangle.$$

We define $\chi \in Hom(\mathfrak{S}_3,\mathbf{T})$ by $\chi(a) = 1$, $\chi(b) = -1$. We consider $\mathfrak{S}_3 \rtimes \mathbf{Z}_4$ with the action of $\mathbf{Z}_4$ given by

$$a^c = a^2, \quad b^c = b.$$

LEMMA 14.28. *For $\mathfrak{S}_3 \rtimes \mathbf{Z}_4$ just explained we have*

$$H^2((\mathfrak{S}_3,\mathbf{Z}_4),\mathbf{T}) = \{[(1,1)],[(1,\zeta^0)]\},$$

*where $\zeta^0 \in Z^2(\mathbf{Z}_4, Hom(\mathfrak{S}_3),\mathbf{T})$ is defined by*

$$\zeta^0(c^i,c^j) = \begin{cases}1 & \text{if } i+j<4,\\ \chi & \text{if } i+j\geq 4.\end{cases}$$

*Let $\mathcal{A} = \ell^\infty(\mathfrak{S}_3) \rtimes_\zeta \mathbf{Z}_4$. Then, we have*

$$\mathcal{A} \cong \hat{\mathcal{A}} \cong \mathbf{C}^8 \oplus M_2(\mathbf{C}) \oplus M_2(\mathbf{C}) \oplus M_2(\mathbf{C}) \oplus M_2(\mathbf{C}),$$

*and $G(\mathcal{A})$, $G(\hat{\mathcal{A}})$ are given as follows:*
*(i) $(\eta,\zeta) = (1,1)$ case.*

$$G(\mathcal{A}) \cong G(\hat{\mathcal{A}}) \cong \mathbf{Z}_2 \times \mathbf{Z}_4.$$

*(ii) $(\eta,\zeta) = (1,\zeta^0)$ case.*

$$G(\mathcal{A}) \cong G(\hat{\mathcal{A}}) \cong \mathbf{Z}_8.$$

PROOF. Since $H^2(\mathfrak{S}_3,\mathbf{T})$ is trivial, we get

$$H^2((\mathfrak{S}_3,\mathbf{Z}_4),\mathbf{T}) \cong H^2(\mathbf{Z}_4, Hom(\mathfrak{S}_3,\mathbf{T})) \cong H^2(\mathbf{Z}_4,\mathbf{Z}_2).$$

Thus, we get the first part of the statement. It is easy to get the algebra structure of $\mathcal{A}$ and $\hat{\mathcal{A}}$. Since $\zeta$ describes the exact sequence

$$1 \longrightarrow Hom(\mathfrak{S}_3,\mathbf{T}) \longrightarrow G(\mathcal{A}) \longrightarrow \mathbf{Z}_4 \longrightarrow 1,$$

we get the description of $G(\mathcal{A})$. If $(\eta,\zeta) = (1,1)$, then $G(\hat{\mathcal{A}})$ is a trivial extension of $\hat{\mathbf{Z}}_4$ by $N_0 = \langle b\rangle$. Thus, we get $G(\hat{\mathcal{A}}) \cong \mathbf{Z}_2 \times \mathbf{Z}_4$. If $(\eta,\zeta) = (1,\zeta^0)$, we take $\xi : \mathbf{Z}_4 \times \mathfrak{S}_3 \longrightarrow \mathbf{T}$ satisfying

$$\xi(c^j, a^k b) = e^{\frac{\pi i}{4} j}\ (j = 0,1,2,3).$$

Then, we have $\zeta_n(h_1, h_2) = \xi(h_1, n)\xi(h_2, n^{h_1})\overline{\xi(h_1h_2, n)}$ for $n \in N_0$ and $h_1, h_2 \in \mathbf{Z}_4$. Thus, $\eta'_h(n_1, n_2) = \xi(h, n_1)\xi(h, n_2)\overline{\xi(h, n_1n_2)}$ restricted to $N_0$ describes the exact sequence

$$1 \longrightarrow \hat{\mathbf{Z}}_4 \longrightarrow G(\hat{\mathcal{A}}) \longrightarrow N_0 \longrightarrow 1,$$

showing $G(\hat{\mathcal{A}}) \cong \mathbf{Z}_8$. □

**2.8. $\mathfrak{S}_3 \rtimes (\mathbf{Z}_2 \times \mathbf{Z}_2)$ case.** Let $\mathfrak{S}_3$ and $\chi$ be as above and $c$, $d$ be generators of $\mathbf{Z}_2 \times \mathbf{Z}_2$. We consider $\mathfrak{S}_3 \rtimes (\mathbf{Z}_2 \times \mathbf{Z}_2)$ with the action of $\mathbf{Z}_2 \times \mathbf{Z}_2$ given by

$$a^c = a^2, \quad b^c = b, \quad a^d = a, \quad b^d = b.$$

LEMMA 14.29. *For this $\mathfrak{S}_3 \rtimes (\mathbf{Z}_2 \times \mathbf{Z}_2)$ we have*

$$H^2((\mathfrak{S}_3, \mathbf{Z}_2 \times \mathbf{Z}_2), \mathbf{T}) \cong H^2(\mathbf{Z}_2 \times \mathbf{Z}_2, Hom(\mathfrak{S}_3, \mathbf{T})) \cong \mathbf{Z}_2 \times \mathbf{Z}_2 \times \mathbf{Z}_2.$$

*Here, instead of giving $(\eta, \zeta)$ explicitly, we describe the corresponding extensions*

$$1 \longrightarrow Hom(\mathfrak{S}_3, \mathbf{T}) \longrightarrow G \longrightarrow \mathbf{Z}_2 \times \mathbf{Z}_2 \longrightarrow 1.$$

*Assume that $\tilde{c}$ and $\tilde{d}$ are the corresponding lifts of $c$ and $d$ to $G$ respectively. Then, $(x_1, x_2, x_3) \in \mathbf{Z}_2 \times \mathbf{Z}_2 \times \mathbf{Z}_2$ corresponds to*

$$\tilde{c}^2 = \chi^{x_1}, \quad \tilde{d}^2 = \chi^{x_2}, \quad \tilde{c}\tilde{d} = \chi^{x_3}\tilde{d}\tilde{c}.$$

*The two cases $(0, 1, 0)$, $(1, 1, 0)$ give equivalent cocycles and so do the two cases $(0, 0, 1)$, $(1, 0, 1)$ (in the sense of $H^2((\mathfrak{S}_3, \mathbf{Z}_2 \times \mathbf{Z}_2), \mathbf{T})/\sim$).*

*Let $\mathcal{A} = \ell^\infty(\mathfrak{S}_3) \rtimes_\zeta (\mathbf{Z}_2 \times \mathbf{Z}_2)$. Then, we have*

$$\hat{\mathcal{A}} \cong \mathbf{C}^8 \oplus M_2(\mathbf{C}) \oplus M_2(\mathbf{C}) \oplus M_2(\mathbf{C}) \oplus M_2(\mathbf{C}),$$

*and*

$$\mathcal{A} \cong \begin{cases} \mathbf{C}^8 \oplus M_2(\mathbf{C}) \oplus M_2(\mathbf{C}) \oplus M_2(\mathbf{C}) \oplus M_2(\mathbf{C}) & \text{when } x_3 = 0, \\ \mathbf{C}^4 \oplus M_2(\mathbf{C}) \oplus M_2(\mathbf{C}) \oplus M_2(\mathbf{C}) \oplus M_2(\mathbf{C}) \oplus M_2(\mathbf{C}) & \text{when } x_3 = 1. \end{cases}$$

*Furthermore, the intrinsic groups $G(\mathcal{A})$, $G(\hat{\mathcal{A}})$ are given as follows:*
*(i) $(x_1, x_2, x_3) = (0, 0, 0)$ case.*

$$G(\mathcal{A}) \cong G(\hat{\mathcal{A}}) \cong \mathbf{Z}_2 \times \mathbf{Z}_2 \times \mathbf{Z}_2.$$

*This $\mathcal{A}$ is isomorphic to $\mathcal{A}_0 \otimes \mathbf{C}(\mathbf{Z}_2)$ with the unique Kac algebra $\mathcal{A}_0$ of dimension 12 (see §5 in Chapter 7) satisfying*

$$\begin{cases} \mathcal{A}_0 \cong \hat{\mathcal{A}}_0 \cong \mathbf{C}^4 \oplus M_2(\mathbf{C}) \oplus M_2(\mathbf{C}), \\ G(\mathcal{A}_0) \cong G(\hat{\mathcal{A}}_0) \cong \mathbf{Z}_2 \times \mathbf{Z}_2. \end{cases}$$

*(ii) $(x_1, x_2, x_3) = (1, 0, 0)$ case.*

$$G(\mathcal{A}) \cong G(\hat{\mathcal{A}}) \cong \mathbf{Z}_2 \times \mathbf{Z}_4.$$

*This $\mathcal{A}$ is isomorphic to $\mathcal{A}_1 \otimes \mathbf{C}(\mathbf{Z}_2)$ with the unique Kac algebra $\mathcal{A}_1$ of dimension 12 (see §5 in Chapter 7) satisfying*

$$\begin{cases} \mathcal{A}_1 \cong \hat{\mathcal{A}}_1 \cong \mathbf{C}^4 \oplus M_2(\mathbf{C}) \oplus M_2(\mathbf{C}), \\ G(\mathcal{A}_1) \cong G(\hat{\mathcal{A}}_1) \cong \mathbf{Z}_4. \end{cases}$$

*(iii) $(x_1, x_2, x_3) = (0, 1, 0)$ case.*

$$G(\mathcal{A}) \cong G(\hat{\mathcal{A}}) \cong \mathbf{Z}_2 \times \mathbf{Z}_4.$$

*(iv) $(x_1, x_2, x_3) = (0, 0, 1)$ case.*

$$G(\mathcal{A}) \cong D_8, \quad G(\hat{\mathcal{A}}) \cong \mathbf{Z}_2 \times \mathbf{Z}_2.$$

*(v)* $(x_1,x_2,x_3)=(0,1,1)$ *case.*

$$G(\mathcal{A})\cong D_8,\quad G(\hat{\mathcal{A}})\cong \mathbf{Z}_2\times\mathbf{Z}_2.$$

*(vi)* $(x_1,x_2,x_3)=(1,1,1)$ *case.*

$$G(\mathcal{A})\cong Q_8,\quad G(\hat{\mathcal{A}})\cong \mathbf{Z}_2\times\mathbf{Z}_2.$$

PROOF. Since $H^2(\mathfrak{S}_3,\mathbf{T})$ is trivial, we get

$$H^2((\mathfrak{S}_3,\mathbf{Z}_2\times\mathbf{Z}_2),\mathbf{T})\cong H^2(\mathbf{Z}_2\times\mathbf{Z}_2,Hom(\mathfrak{S}_3,\mathbf{T}))\cong H^2(\mathbf{Z}_2\times\mathbf{Z}_2,\mathbf{Z}_2),$$

and hence $H^2((\mathfrak{S}_3,\mathbf{Z}_2\times\mathbf{Z}_2),\mathbf{T})=\mathbf{Z}_2\times\mathbf{Z}_2\times\mathbf{Z}_2$. The equivalence relation of $H^2((\mathfrak{S}_3,\mathbf{Z}_2\times\mathbf{Z}_2),\mathbf{T})$ comes from those automorphisms exchanging $c$ and $cd$. This amounts to change $(x_1,x_2,x_3)$ to $(x_1+x_2+x_3,x_2,x_3)$, and we get the desired equivalence. It is easy to get $\mathcal{A}$, $\hat{\mathcal{A}}$, and $G(\mathcal{A})$. If $x_3=0$, using a similar argument as in Lemma 14.4, 14.5, or 14.6, we can show that $\mathcal{A}$ arises from either $D_{12}\rtimes\mathbf{Z}_2$ or $(\mathbf{Z}_3\rtimes\mathbf{Z}_4)\rtimes\mathbf{Z}_2$. Thus, Lemma 14.19 and Lemma 14.22 show $G(\mathcal{A})\cong G(\hat{\mathcal{A}})$. □

**2.9.** $(\mathbf{Z}_2\times\mathbf{Z}_2)\rtimes\mathfrak{S}_3$ **case.**

LEMMA 14.30. *Let* $(\mathbf{Z}_2\times\mathbf{Z}_2)\rtimes\mathfrak{S}_3$ *be the semi-direct product group, where* $\mathfrak{S}_3$ *acts as the permutations of the three non-trivial elements* $a,\ b,\ c\in\mathbf{Z}_2\times\mathbf{Z}_2$. *Then, we have*

$$H^2((\mathbf{Z}_2\times\mathbf{Z}_2,\mathfrak{S}_3),\mathbf{T})=\{[(1,1)],[(\eta^0,1)]\}$$

*with*

$$\eta^0_g=\begin{cases}1 & \text{if } g\in\mathfrak{S}_3 \text{ is an even permutation,}\\ \omega_0 & \text{if } g\in\mathfrak{S}_3 \text{ is an odd permutation.}\end{cases}$$

*Let* $\mathcal{A}=\ell^\infty(\mathbf{Z}_2\times\mathbf{Z}_2)\rtimes\mathfrak{S}_3$. *Just the case with the non-trivial* $\eta$ *gives rise to a non-trivial Kac algebra, and in this case we have*

$$\begin{aligned}&\mathcal{A}\cong\mathbf{C}^2\oplus M_2(\mathbf{C})\oplus M_3(\mathbf{C})\oplus M_3(\mathbf{C}),\\ &\hat{\mathcal{A}}\cong\mathbf{C}^{12}\oplus M_2(\mathbf{C})\oplus M_2(\mathbf{C})\oplus M_2(\mathbf{C}),\\ &G(\mathcal{A})\cong\mathfrak{A}_4,\quad G(\hat{\mathcal{A}})\cong\mathbf{Z}_2.\end{aligned}$$

PROOF. We note $\omega_0^g=\overline{\omega_0}$ for $g\in\mathfrak{S}_3$ odd while even permutations leave $\omega_0$ invariant. This implies $(\eta^0,1)\in Z^2((\mathbf{Z}_2\rtimes\mathbf{Z}_2,\mathfrak{S}_3),\mathbf{T})$. Direct computation based on Lemma 14.1 shows that $H^2(\mathfrak{S}_3,\widehat{\mathbf{Z}_2\times\mathbf{Z}_2})$ is trivial, and $H^2((\mathbf{Z}_2\rtimes\mathbf{Z}_2,\mathfrak{S}_3),\mathbf{T})$, $\mathcal{A}$, $\hat{\mathcal{A}}$, $G(\mathcal{A})$, and $G(\hat{\mathcal{A}})$ can be easily determined. □

**2.10. Trivial cases.** Finally, we remark the next easy lemma, which enables us to eliminate many cases from our consideration in the next two sections.

LEMMA 14.31.
*(i) Assume that* $N\rtimes H$ *is a (possibly direct) semi-direct product group with* $N$ *abelian. The Kac algebra* $\mathcal{A}=\ell^\infty(N)\rtimes_\zeta H$ *(obtained as a twisted crossed product) associated with* $(1,\zeta)\in Z^2((N,H),\mathbf{T})$ *is co-commutative. In particular, if* $N$ *is an abelian group with trivial* $H^2(N,\mathbf{T})$ *(for example cyclic groups), then the above* $\mathcal{A}$ *is always co-commutative.*
*(ii) When* $N$ *is either* $\mathbf{Z}_4\times\mathbf{Z}_2$ *or* $D_8$, *the Kac algebra* $\mathcal{A}=\ell^\infty(N)\rtimes_\zeta\mathbf{Z}_3$ *is always abelian.*

PROOF. (i) $\mathcal{A}$ is co-commutative because of

$$\hat{\mathcal{A}} = \oplus_{h \in H} \mathbf{C}(N) = \ell^\infty(H) \otimes \mathbf{C}(N).$$

(ii) Since $N$ admits no non-trivial action of $\mathbf{Z}_3$, the result follows from the description of $\mathcal{A}$ in terms of isotropy groups (see Theorem 7 in [**20**] for example). □

## 3. Reduction to the semi-direct product case

Among the eight possibilities listed in Lemma 14.11, we will see that the fifth case

$$\mathcal{A} \cong \mathbf{C}^4 \oplus M_2(\mathbf{C}) \oplus M_4(\mathbf{C})$$

is impossible as long as $\mathcal{A}$ is non-trivial (Proposition 14.37), and there are seven remaining cases. For "almost all" of them we will show that $\mathcal{A}$ can be captured via our construction with a certain semi-direct product. More precisely, our arguments go as follows:

(a) In the second case

$$\mathcal{A} \cong \mathbf{C}^8 \oplus M_4(\mathbf{C})$$

the Kac algebra $\mathcal{A}$ must be the bicrossed product arising from the product group $\mathfrak{S}_4 = \mathfrak{S}_3 \cdot \mathbf{Z}_3$ (Theorem 14.33).

(b) The third case

$$\mathcal{A} \cong \mathbf{C}^8 \oplus M_2(\mathbf{C}) \oplus M_2(\mathbf{C}) \oplus M_2(\mathbf{C}) \oplus M_2(\mathbf{C})$$

arises from certain semi-direct products **unless $\#G(\mathcal{A})$ is a multiple of** 3 (Proposition 14.38).

(c) The sixth case

$$\mathcal{A} \cong \mathbf{C}^4 \oplus M_2(\mathbf{C}) \oplus M_2(\mathbf{C}) \oplus M_2(\mathbf{C}) \oplus M_2(\mathbf{C}) \oplus M_2(\mathbf{C})$$

arises from certain semi-direct products **if the underlying algebra structure of the dual Kac algebra $\hat{\mathcal{A}}$ is of the same type** (Proposition 14.39).

(d) The fourth, seventh and eighth cases

$$\begin{aligned} &\mathcal{A} \cong \mathbf{C}^6 \oplus M_3(\mathbf{C}) \oplus M_3(\mathbf{C}) \\ &\mathcal{A} \cong \mathbf{C}^3 \oplus M_2(\mathbf{C}) \oplus M_2(\mathbf{C}) \oplus M_2(\mathbf{C}) \oplus M_3(\mathbf{C}) \\ &\mathcal{A} \cong \mathbf{C}^2 \oplus M_2(\mathbf{C}) \oplus M_3(\mathbf{C}) \oplus M_3(\mathbf{C}) \end{aligned}$$

always come from certain semi-product products (thanks to Proposition 14.34, Lemma 14.35, and Proposition 14.36 respectively).

We also point out

(e) The first case

$$\mathcal{A} \cong \mathbf{C}^{12} \oplus M_2(\mathbf{C}) \oplus M_2(\mathbf{C}) \oplus M_2(\mathbf{C}).$$

does arise from semi-direct products as well. However, this case is more involved (see Lemma 14.32 and the paragraph after the lemma), and Theorem 13.22 (on the twisted crossed product $\mathbf{C}(\mathfrak{A}_4) \rtimes \mathbf{Z}_2$) is actually used to get the conclusion.

(f) In Cases (b), (c) (corresponding to Cases III, V in Theorem 14.40) some additional assumption was made. But, this difficulty can be overcome by considering respective dual Kac algebras (see the proof of Theorem 14.40).

LEMMA 14.32. *If $\mathcal{A}$ is a non-trivial Kac algebra with the underlying algebra structure*

$$\mathbf{C}^{12} \oplus M_2(\mathbf{C}) \oplus M_2(\mathbf{C}) \oplus M_2(\mathbf{C}),$$

*then the intrinsic group $G(\hat{\mathcal{A}})$ is either $\mathfrak{A}_4$ or $\mathbf{Z}_2 \times \mathbf{Z}_2 \times \mathbf{Z}_3$.*
*(i) When $G(\hat{\mathcal{A}}) = \mathfrak{A}_4$, we have*

$$\hat{\mathcal{A}} = \mathbf{C}(\mathfrak{A}_4) \rtimes \mathbf{Z}_2.$$

*Hence, $\mathcal{A}$ is classified by weak equivalence classes of $(\theta, \eta, \zeta)$ for $(\mathfrak{A}_4, \mathbf{Z}_2)$ with non-trivial $[\eta_1] \in H^2(\hat{\mathfrak{A}}_4, \mathbf{T})$, and consequently by Theorem 13.22.*
*(ii) When $G(\hat{\mathcal{A}}) = \mathbf{Z}_2 \times \mathbf{Z}_2 \times \mathbf{Z}_3$, we have*

$$\hat{\mathcal{A}} = \ell^\infty(\mathbf{Z}_2 \times \mathbf{Z}_2 \times \mathbf{Z}_3) \rtimes_{\theta,\zeta} \mathbf{Z}_2.$$

*Hence, $\mathcal{A}$ is classified by $(\mathbf{Z}_2 \times \mathbf{Z}_2 \times \mathbf{Z}_3) \rtimes \mathbf{Z}_2$ with non-trivial action and the equivalence class of $[(\eta, \zeta)] \in H^2((\mathbf{Z}_2 \times \mathbf{Z}_2 \times \mathbf{Z}_3, \mathbf{Z}_2), \mathbf{T})$ with non-trivial $[\eta_1] \in H^2(\mathbf{Z}_2 \times \mathbf{Z}_2 \times \mathbf{Z}_3, \mathbf{T})$.*

PROOF. Let $G = G(\mathcal{A})$. Thanks to Corollary 13.6, such $\mathcal{A}$ is classified by $(\theta, \eta, \zeta)$ for $(\mathbf{C}(G), \mathbf{Z}_2)$ with non-trivial $\eta$. Note that a group of order 12 is one of the following:

$$\mathbf{Z}_4 \times \mathbf{Z}_3, \quad \mathbf{Z}_2 \times \mathbf{Z}_2 \times \mathbf{Z}_3, \quad \mathbf{Z}_3 \rtimes \mathbf{Z}_4, \quad D_{12}, \quad \mathfrak{A}_4.$$

Since $\eta_1$ is not trivial, $G$ should have a non-trivial ergodic action with full multiplicity, and so it has a subgroup isomorphic to $\mathbf{Z}_2 \times \mathbf{Z}_2$. Also recall that $G = D_{12}$ is impossible thanks to Proposition 13.20. Consequently, $G$ must be either $\mathbf{Z}_2 \times \mathbf{Z}_2 \times \mathbf{Z}_3$ or $\mathfrak{A}_4$. □

The abelian case ($G = \mathbf{Z}_2 \times \mathbf{Z}_2 \times \mathbf{Z}_3$) is easy while the non-abelian case ($G = \mathfrak{A}_4$) requires Theorem 13.22. In this theorem two Kac algebras are obtained, and note that their duals are in the fourth and eighth cases. Therefore, Kac algebras in the case (i) of the above lemma also come from semi-direct products.

THEOREM 14.33. *Assume that $\mathcal{A}$ is a non-trivial Kac algebra with the underlying algebra structure*

$$\mathcal{A} \cong \mathbf{C}^8 \oplus M_4(\mathbf{C}).$$

*Then, it must come from the bicrossed product associated with the product group $\mathfrak{S}_4 = \mathfrak{S}_3 \cdot \mathbf{Z}_4$. The dual Kac algebras and the intrinsic groups are*

$$\hat{\mathcal{A}} \cong \mathbf{C}^2 \oplus M_2(\mathbf{C}) \oplus M_3(\mathbf{C}) \oplus M_3(\mathbf{C}),$$
$$G(\mathcal{A}) \cong \mathbf{Z}_2, \quad G(\hat{\mathcal{A}}) \cong D_8.$$

PROOF. Let $\mathcal{P} \supseteq \mathcal{Q}$ be a depth 2 inclusion of factors corresponding to the dual Kac algebra and $\rho$ be the inclusion map. Then, the irreducible decomposition of $\rho\bar{\rho}$ is as follows:

$$[\rho\bar{\rho}] = \oplus_{g \in G(\hat{\mathcal{A}})} [\tau_g] \oplus 4[\pi]$$

with $d(\pi) = 4$. We take representatives $\{\tau_g\}_{g \in G(\hat{\mathcal{A}})}$ satisfying $\tau_g \cdot \rho = \rho$. Note $H^2(G(\hat{\mathcal{A}}), \mathbf{T})$ is not trivial thanks to Lemma 11.14,(i), and consequently $G(\hat{\mathcal{A}})$ is isomorphic to either of the following:

$$\mathbf{Z}_2 \times \mathbf{Z}_2 \times \mathbf{Z}_2, \quad \mathbf{Z}_2 \times \mathbf{Z}_4, \quad D_8.$$

*(i)* $G(\hat{\mathcal{A}}) \cong D_8$ or $\mathbf{Z}_2 \times \mathbf{Z}_4$ case.
We fix a subgroup $G_0 \subseteq G(\hat{\mathcal{A}})$ isomorphic to $\mathbf{Z}_4$. Let $\mathcal{R}$ be the fixed-point subalgebra under $G_0$ and $\nu : \mathcal{R} \hookrightarrow \mathcal{P}$ the inclusion map. Since $H^2(G_0, \mathbf{T})$ is trivial, we may assume $\tau_g \cdot \pi = \pi$ for $g \in G_0$. We claim that there exists an automorphism $\varphi \in Aut(\mathcal{R})$ satisfying $[\pi] = [\nu\varphi\overline{\nu}]$. Take a $\mathcal{P}$-$\mathcal{R}$ sector $\mu$ with $d(\mu) = 2$ satisfying $\pi = \nu\overline{\mu}$. Since $\mu\overline{\mu}$ is contained in $\overline{\pi}\pi$, it is decomposed into $\{\tau_g\}_{g\in G(\hat{\mathcal{A}})} \cup \{\pi\}$, and actually is decomposed into automorphisms because it cannot contain $\pi$. By duality, $\overline{\mu}\mu$ is decomposed into automorphisms as well. Since $\pi$ is contained in $\pi\overline{\pi} = \nu\overline{\mu}\mu\overline{\nu}$, we get the claim. Let $H$ be the dual group of $G_0$ and $\alpha$ an action of $H$ on $\mathcal{R}$ satisfying $\mathcal{P} = \mathcal{R} \rtimes_\alpha H$. Let $\rho_1 : \mathcal{Q} \hookrightarrow \mathcal{R}$ be the inclusion map with the irreducible decomposition $[\rho_1\overline{\rho}_1] = \oplus_{i\in I}[\sigma_i]$. Then, since $\rho\overline{\rho} = \nu\rho_1\overline{\rho}_1\overline{\nu}$, there exists $\sigma_i$ such that $\nu\sigma_i\overline{\nu}$ contains $\pi$. Therefore, we get

$$\begin{aligned} 0 \neq \dim(\nu\varphi\overline{\nu}, \nu\sigma_i\overline{\nu}) &= \dim(\varphi, \overline{\nu}\nu\sigma_i\overline{\nu}\nu) \\ &= \dim(\varphi, \oplus_{h,k\in H}\alpha_h\sigma_i\alpha_k). \end{aligned}$$

This shows that $\sigma_i$ is an automorphism and $\pi = \nu\sigma_i\overline{\nu}$. Since the multiplicity of $\pi$ in $\rho\overline{\rho}$ is 4, $\rho_1\overline{\rho}_1$ contains at least 5 automorphisms, and so $\rho_1\overline{\rho}_1$is decomposed into automorphisms. Thus, there exist a group $K$ of order 6 and an action $\beta$ of $K$ on $\mathcal{R}$ such that $\mathcal{Q}$ is the fixed-point algebra under $\beta$. Since $\mathcal{P} \supseteq \mathcal{Q}$ is of depth 2, $[\beta_K][\alpha_H]$ forms a product group of order 24, which is not semi-direct product thanks to Lemma 14.10. Lemma 14.3, Lemma 14.4, and Lemma 14.7 imply that there is only one possibility:

$$K \cong \mathfrak{S}_3, \quad [\alpha_H] \cdot [\beta_K] \cong \mathfrak{S}_4.$$

We regard $H$ and $K$ as subgroups of $\mathfrak{S}_4$. We would like to show that our near action $(\alpha, \beta)$ actually comes from a genuine action of $\mathfrak{S}_4$. If our product group were a semi-direct product group, it would suffice to show that $H^2((\mathbf{Z}_4, \mathfrak{S}_3), \mathbf{T})$ is trivial. The corresponding result for the general product group case is actually obtained Masuoka in [**55**]. Since we have not formulated the cohomology group for the general product group case, we borrow only a key observation from [**55**]. Namely, in that paper Masuoka shows that the following map is injective:

$$H^3(\mathfrak{S}_4, \mathbf{T}) \longrightarrow H^3(\mathbf{Z}_4, \mathbf{T}) \oplus H^3(\mathfrak{S}_3, \mathbf{T}),$$

where the map comes from the restriction of the cocycle to the both subgroups $\mathbf{Z}_4$ and $\mathfrak{S}_3$. Let $\omega$ be the 3-cocycle in $Z^3(\mathfrak{S}_4, \mathbf{T})$ corresponding to the $\mathfrak{S}_4$-kernel arising from our near action. Since $\alpha$ and $\beta$ are genuine actions, restriction of $\omega$ to the two subgroups are coboundaries, and so is $\omega$ thanks to Masuoka's observation above. Thus, there exists an action $\theta$ of $\mathfrak{S}_4$ on $\mathcal{R}$ and unitaries $\{u_h\}_{h\in H}$, $\{v_k\}_{k\in K}$ satisfying

$$\theta_h = Ad(u_h) \cdot \alpha_h, \quad \theta_k = Ad(v_k) \cdot \beta_k.$$

Since $H^2(\mathbf{Z}_4, \mathbf{T})$ and $H^2(\mathfrak{S}_3, \mathbf{T})$ are trivial, we may assume that $\{u_h\}_{h\in H}$ is an $\alpha$-cocycle and $\{v_k\}_{k\in K}$ is a $\beta$-cocycle. Therefore, Connes theorem ([**4**]) shows that there exist unitaries $u, v \in \mathcal{R}$ satisfying $u_h = u\alpha_h(u^*)$, $v_k = v\beta_k(v^*)$. By replacing $\theta_g$ by $Ad(v^*) \cdot \theta_g \cdot Ad(v)$ if necessary, we may assume $v = 1$ from the beginning and $\theta_k = \beta_k$. Since the crossed product does not depends on cocycle perturbation, we get $\mathcal{P} = \mathcal{R} \rtimes_\theta H$, which shows that $\mathcal{A}$ is the desired one.
*(ii)* $G(\hat{\mathcal{A}}) \cong \mathbf{Z}_2 \times \mathbf{Z}_2 \times \mathbf{Z}_2$ case.
Let $\mathcal{R}$ be the fixed-point algebra of $\mathcal{P}$ under $\tau$ and $\nu : \mathcal{R} \hookrightarrow \mathcal{P}$ and $\rho_1 : \mathcal{Q} \hookrightarrow \mathcal{R}$ the inclusion maps. Note that $[\mathcal{R} : \mathcal{Q}] = 3$ and the principal graph of $\mathcal{R} \supseteq \mathcal{Q}$ is

either $A_5$ or $D_4$. Let $H$ be the dual group of $G(\hat{\mathcal{A}})$ and $\alpha$ its action on $\mathcal{R}$ satisfying $\mathcal{P} = \mathcal{R} \rtimes_\alpha H$. Suppose first that the principal graph of $\mathcal{R} \supseteq \mathcal{Q}$ is $D_4$. Then, there exists an action $\beta$ of $\mathbf{Z}_3$ on $\mathcal{R}$ satisfying $\mathcal{Q} = \mathcal{R}^\beta$. Thus, $[\alpha_H]\cdot[\beta_{\mathbf{Z}_3}]$ forms a product group of order 24. Since $H \cong \mathbf{Z}_2 \times \mathbf{Z}_2 \times \mathbf{Z}_2$, $[\alpha_H]\cdot[\beta_{\mathbf{Z}_3}]$ is a semi-direct product thanks to Lemma 14.1, which contradicts Lemma 14.10. Now, we assume that the principal graph of $\mathcal{R} \supseteq \mathcal{Q}$ is $A_5$. Then, the irreducible decomposition of $\rho_1\bar{\rho}_1$ and fusion rules of its descendant sectors are given by

$$[\rho_1\bar{\rho}_1] = [id] \oplus [\sigma],$$
$$[\sigma][\sigma] = [id] \oplus [\theta] \oplus [\sigma],$$
$$[\theta][\theta] = [id], \quad [\theta][\sigma] = [\sigma][\theta] = [\sigma]$$

with $d(\theta) = 1$ and $d(\sigma) = 2$. We claim that $\theta$ is contained in $\bar{\nu}\nu$. Since $[\rho\bar{\rho}] = [\nu\bar{\nu}] \oplus [\nu\sigma\bar{\nu}]$, we have $[\nu\sigma\bar{\nu}] = 4[\pi]$ and hence

$$16 = \dim(\nu\sigma\bar{\nu}, \nu\sigma\bar{\nu}) = \dim(\sigma, \oplus_{h_1,h_2}\alpha_{h_1}\sigma\alpha_{h_2}).$$

This implies that $[\alpha_h\sigma] = [\sigma]$ for some $h \in H \setminus \{e\}$, which shows the claim. We set $\mathcal{L} = \mathcal{R} \rtimes_\theta \mathbf{Z}_2$. Then, thanks to [**26**], we get

$$\mathcal{P} = \mathcal{L} \rtimes (\mathbf{Z}_2 \times \mathbf{Z}_2), \quad \mathcal{Q} = \mathcal{L}^{\mathfrak{S}_3}.$$

However, Lemma 14.4, Lemma 14.5, and Lemma 14.10 show that this is impossible. □

PROPOSITION 14.34. *Let $\mathcal{A}$ be a non-trivial Kac algebra with the underlying algebraic structure*

$$\mathbf{C}^6 \oplus M_3(\mathbf{C}) \oplus M_3(\mathbf{C}).$$

*Then, one of the following holds*:
*(i) There exists $[(\eta,\zeta)] \in H^2((\mathbf{Z}_2 \times \mathbf{Z}_2 \times \mathbf{Z}_2, \mathbf{Z}_3), \mathbf{T})$ such that*

$$\mathcal{A} = \ell^\infty(\mathbf{Z}_2 \times \mathbf{Z}_2 \times \mathbf{Z}_2) \rtimes_\zeta \mathbf{Z}_3$$

*with the $\mathbf{Z}_3$ action in* (1) *of (ii) in Lemma* 14.1.
*(ii) There exists $[(\eta,\zeta)] \in H^2((Q_8, \mathbf{Z}_3), \mathbf{T})$ such that*

$$\mathcal{A} = \ell^\infty(Q_8) \rtimes_\zeta \mathbf{Z}_3$$

*with the $\mathbf{Z}_3$ action in* (2) *of (ii) in Lemma* 14.1.
*(iii) There exists $[(\eta,\zeta)] \in H^2((\mathfrak{A}_4, \mathbf{Z}_2), \mathbf{T})$ such that*

$$\mathcal{A} = \mathbf{C}(\mathfrak{S}_4) \oplus \mathbf{C}_\eta(\mathfrak{S}_4),$$

*where $\mathbf{Z}_2$ acts on the Sylow* 2*-subgroup of $\mathfrak{A}_4$ non-trivially.*

PROOF. Let $\mathcal{P} \supseteq \mathcal{Q}$ be a depth 2 inclusion of factors corresponding to the dual Kac algebra and $\rho$ be the inclusion map. Then, the irreducible decomposition of $\rho\bar{\rho}$ is as follows:

$$[\rho\bar{\rho}] = \oplus_{g\in G(\hat{\mathcal{A}})}[\tau_g] \oplus 3[\pi_1] \oplus 3[\pi_2]$$

with $d(\pi_1) = d(\pi_2) = 3$. We take representatives $\{\tau_g\}_{g\in G(\hat{\mathcal{A}})}$ satisfying $\tau_g \cdot \rho = \rho$. Since $G(\hat{\mathcal{A}})$ is isomorphic to either $\mathbf{Z}_6$ or $\mathfrak{S}_3$, there exists a unique order 3 subgroup $G_0$. Let $\mathcal{R}$ be the fixed-point subalgebra of $\mathcal{P}$ under $G_0$ action and $\nu : \mathcal{R} \hookrightarrow \mathcal{P}$ the inclusion map. Then, as before we can show that there exists automorphisms $\varphi_1, \varphi_2 \in Aut(\mathcal{R})$ satisfying $[\pi_i] = [\nu\varphi_i\bar{\nu}]$, $i = 1, 2$. Using the same argument as in the proof of the previous lemma, we can show that there exist a group $K$ of order 8 and its action $\beta$ on $\mathcal{R}$ such that $\mathcal{Q}$ is the fixed-point subalgebra of $\mathcal{R}$ under $\beta$. Let

$H$ be the dual group of $G_0$ and $\alpha$ its action on $\mathcal{R}$ satisfying $\mathcal{P} = \mathcal{R} \rtimes_\beta H$. Then, $G := [\alpha_H][\beta_K]$ forms a product group of order 24. If $[\alpha_H]$ were normal in $G$, $\mathcal{A}$ would be abelian thanks to Lemma 14.31. Thus, one of the following holds:

$$G = (\mathbf{Z}_2 \times \mathbf{Z}_2 \times \mathbf{Z}_2) \rtimes \mathbf{Z}_3, \quad G = Q_8 \rtimes \mathbf{Z}_3, \quad G = \mathfrak{S}_4.$$

The first and the second cases correspond to (i) and (ii) respectively. We assume $G = \mathfrak{S}_4$ and regards $H$ and $K$ as its subgroups. Since $H$ and $K$ are Sylow subgroups, we may assume $H = \langle(123)\rangle$ and

$$K = \{e, (1234), (13)(24), (1432), (13), (14)(23), (24), (12)(34)\}.$$

Let $K_0 = \{e, (13)(24), (14)(23), (12)(34)\} \subset K$ and $\mathcal{L}$ be the fixed-point subalgebra of $\mathcal{R}$ under $K_0$ action. We note that $H$ normalizes $K_0$. Since $H$ acts transitively on $K_0 \setminus \{e\}$, $K_0 \rtimes H$ admits no cocycle $(\eta, \zeta)$. Therefore, we conclude $\mathcal{P} = \mathcal{L} \rtimes \mathfrak{S}_4 \supseteq \mathcal{Q} = \mathcal{L}^{\mathbf{Z}_2}$. Let $\alpha'$ and $\beta'$ the corresponding actions of $\mathfrak{A}_4$ and $\mathbf{Z}_2$ on $\mathcal{L}$. If $[\beta'_{\mathbf{Z}_2}]$ commuted with $[\alpha'|_{\mathbf{Z}_2 \times \mathbf{Z}_2}]$, $\mathcal{Q}$ would be a fixed-point subalgebra of $\mathcal{R}$ under an abelian group action, which is a contradiction. Thus, we get (iii). □

LEMMA 14.35. *Assume that $\mathcal{A}$ is a non-trivial Kac algebra with the underlying algebra structure*

$$\mathcal{A} \cong \mathbf{C}^3 \oplus M_2(\mathbf{C}) \oplus M_2(\mathbf{C}) \oplus M_2(\mathbf{C}) \oplus M_3(\mathbf{C}).$$

*In this case $\mathcal{A}$ is classified by $\mathfrak{A}_4 \rtimes \mathbf{Z}_2$ and an equivalence class of $[(\eta, \zeta)] \in H^2((\mathfrak{A}_4, \mathbf{Z}_2), \mathbf{T})$ giving rise to*

$$\mathbf{C}(\mathfrak{A}_4) \oplus \mathbf{C}_\eta(\mathfrak{A}_4) \cong \mathcal{A}.$$

*Here, the action of $\mathbf{Z}_2$ on $\mathfrak{A}_4$ is $\theta_0$ in* **2.2**.

PROOF. Let $\mathcal{P} \supseteq \mathcal{Q}$ be a depth 2 inclusion of factors corresponding to the dual Kac algebra and $\rho$ be the inclusion map. Then, the irreducible decomposition of $\rho\bar{\rho}$ is as follows:

$$[\rho\bar{\rho}] = \oplus_{g \in G(\hat{\mathcal{A}})}[\tau_g] \oplus 2[\pi_1] \oplus 2[\pi_2] \oplus 3[\sigma]$$

with $d(\pi_1) = d(\pi_2) = 2$ and $d(\sigma) = 3$. Note that $\sigma$ is a unique sector of dimension 3 and it is self-conjugate. Since $G(\hat{\mathcal{A}}) \cong \mathbf{Z}_3$, the principal graph of $\mathcal{P} \supseteq \pi_1(\mathcal{P})$ is $E_6^{(1)}$ and hence we get

$$[\tau_g][\sigma] = [\sigma][\tau_g] = [\sigma], \quad [\sigma][\sigma] = \oplus_{g \in G(\hat{\mathcal{A}})}[\tau_g] \oplus 2[\sigma].$$

Thus, $\{\tau_g\}_{g \in G(\hat{\mathcal{A}})} \cup \{\sigma\}$ is closed under conjugation and the irreducible decomposition of products. Therefore, there exists a unique intermediate subfactor $\mathcal{R}$ such that the canonical endomorphism into $\mathcal{R}$ is decomposed as

$$\oplus_{g \in G(\hat{\mathcal{A}})}[\tau_g] \oplus 3[\sigma].$$

The classification of 12-dimensional Kac algebras shows that there exists an action $\alpha$ of $\mathfrak{A}_4$ on $\mathcal{R}$ satisfying $\mathcal{P} = \mathcal{R} \rtimes_\alpha \mathfrak{A}_4$. Since $[\mathcal{R} : \mathcal{Q}] = 2$, there exists an action $\beta$ of $\mathbf{Z}_2$ on $\mathcal{R}$ satisfying $\mathcal{Q} = \mathcal{R}^\beta$ so that the first part of the lemma is shown.

Note that the action of $[\beta_{\mathbf{Z}_2}]$ on $[\alpha_{\mathfrak{A}_4}]$ cannot be trivial and it is conjugate to either $\theta_0$ or $\theta_1$ in §2-2. It remains to show that it is actually conjugate to $\theta_0$. To show this by contradiction, let us assume that it is conjugate to $\theta_1$ and we set $\mathcal{L} = \mathcal{R} \rtimes_\alpha (\mathbf{Z}_2 \times \mathbf{Z}_2)$, where $\mathbf{Z}_2 \times \mathbf{Z}_2$ is the unique Sylow 2-subgroup of $\mathfrak{A}_4$. Then, there exists an action $\tilde{\alpha}$ of $\mathbf{Z}_3$ on $\mathcal{L}$ and an action $\tilde{\beta}$ of some group $G_0$ on $\mathcal{L}$ satisfying $\mathcal{P} = \mathcal{L} \rtimes_{\tilde{\alpha}} \mathbf{Z}_3$ and $\mathcal{Q} = \mathcal{L}^{\tilde{\beta}}$. Since $\theta_1$ is trivial on $\mathbf{Z}_2 \times \mathbf{Z}_2$, $G_0$ is isomorphic to either $\mathbf{Z}_2 \times \mathbf{Z}_2 \times \mathbf{Z}_2$ or $\mathbf{Z}_4 \times \mathbf{Z}_2$. Let $G = [\tilde{\alpha}_{\mathbf{Z}_3}] \cdot [\tilde{\beta}_{G_0}]$. If we had $G \cong \mathfrak{S}_4$, then $G_0$ would be isomorphic to the Sylow 2-subgroup of $\mathfrak{S}_4$, and it is impossible. Thus, thanks to Lemma 14.1, the above product decomposition of $G$ is actually a semi-direct product. Note that $[\tilde{\alpha}_{\mathbf{Z}_3}]$ cannot be normal because it would imply that $\mathcal{A}$ is abelian. If $[\tilde{\beta}_{G_0}]$ were normal, then we would get $\mathcal{A} \cong \ell^\infty(G_0) \rtimes \mathbf{Z}_3$, which is impossible thanks to Lemma 14.26. Thus, we have shown that the action of $[\beta_{\mathbf{Z}_2}]$ on $[\alpha_{\mathfrak{A}_4}]$ is conjugate to $\theta_0$ of **2.2**. $\square$

PROPOSITION 14.36. *Assume that $\mathcal{A}$ is a non-trivial Kac algebra with the underlying algebra structure*

$$\mathcal{A} \cong \mathbf{C}^2 \oplus M_2(\mathbf{C}) \oplus M_3(\mathbf{C}) \oplus M_3(\mathbf{C}).$$

*Then, either of the following holds:*
*(i) $\mathcal{A}$ comes from the bicrossed product associated with $\mathfrak{S}_4 = \mathfrak{S}_3 \cdot \mathbf{Z}_4$.*
*(ii) There exists $[(\eta, \zeta)] \in H^2((\mathbf{Z}_2 \times \mathbf{Z}_2, \mathfrak{S}_3), \mathbf{T})$ with non-trivial $\eta$ such that*

$$\mathcal{A} = \ell^\infty(\mathbf{Z}_2 \times \mathbf{Z}_2) \rtimes_\zeta \mathfrak{S}_3.$$

PROOF. Let $\mathcal{P} \supseteq \mathcal{Q}$ be a depth 2 inclusion of factors corresponding to the dual Kac algebra and $\rho$ be the inclusion map. Then, the irreducible decomposition of $\rho\bar{\rho}$ is as follows:

$$[\rho\bar{\rho}] = \oplus_{g \in G(\hat{\mathcal{A}})}[\tau_g] \oplus 2[\pi] \oplus 3[\sigma_1] \oplus 3[\sigma_2]$$

with $d(\pi) = 2$ and $d(\sigma_1) = d(\sigma_2) = 3$. Since $\{\tau_g\}_{g \in G(\hat{\mathcal{A}})} \cup \{\pi\}$ is closed under conjugation and the irreducible decomposition of products, there exists a unique intermediate subfactor $\mathcal{R}$ and an action $\alpha$ of $\mathfrak{S}_3$ on $\mathcal{R}$ such that $\mathcal{P} = \mathcal{R} \rtimes_\alpha \mathfrak{S}_3$. Let $\rho_1 : \mathcal{R} \hookrightarrow \mathcal{P}$ and $\rho_2 : \mathcal{Q} \hookrightarrow \mathcal{R}$ be the inclusion maps.

Note that $[\mathcal{R} : \mathcal{Q}] = 4$ and we would like to show that $\rho_2\bar{\rho}_2$ is decomposed into automorphisms. To this end, we need to eliminate the possibility of principal graphs $E_8^{(1)}, E_7^{(1)}, E_6^{(1)}, D_k^{(1)}$. (see the figures in Chapter 10, §5)
(a) The principal graph $E_8^{(1)}$ is impossible by the argument in the proof of Proposition 10.1,(iii).
(b) Suppose that the principal graph is $E_7^{(1)}$, and let $[\rho_2\bar{\rho}_2] = [id] \oplus [\mu]$ be the irreducible decomposition. Then, $\mu$ is a self-conjugate sector of dimension 3, which never absorbs outer automorphisms through product. Thus, we have

$$[\rho\bar{\rho}] = [\rho_1][\bar{\rho}_1] \oplus [\rho_1\mu\bar{\rho}_1] = \oplus_{g \in G(\hat{\mathcal{A}})}[\tau_g] \oplus 2[\pi] \oplus [\rho_1\mu\bar{\rho}_1],$$

and so $[\rho_1\mu\bar{\rho}_1] = 3[\sigma_1] \oplus 3[\sigma_2]$. This implies

$$\begin{aligned} 18 = \dim(\rho_1\mu\bar{\rho}_1, \rho_1\mu\bar{\rho}_1) &= \dim(\bar{\rho}_1\rho_1\mu, \mu\bar{\rho}_1\rho_1) \\ &= \dim(\oplus_{h \in \mathfrak{S}_3}\alpha_k\mu, \oplus_{k \in \mathfrak{S}_3}\mu\alpha_k), \end{aligned}$$

which is impossible because $\{[\alpha_h\mu]\}_{h \in \mathfrak{S}_3}$ are mutually distinct.
(c) Suppose that the principal graph is $E_6^{(1)}$, and let $[\rho_2\bar{\rho}_2] = [id] \oplus [\mu]$ be the irreducible decomposition. We get the same formula for the dimension of intertwiners as in (b), and this time $\mu$ absorbs some $\alpha_{g \neq e}$. Therefore, an intermediate subfactor $\mathcal{L} = \mathcal{R} \rtimes \mathbf{Z}_3$ can be taken between $\mathcal{P}$ and $\mathcal{R}$ so that we get

$$\mathcal{P} \supseteq \mathcal{L} = \mathcal{R} \rtimes \mathbf{Z}_3 \supseteq \mathcal{R} \supseteq \mathcal{Q}.$$

Thus, the Goldman-type theorem for $E_6^{(1)}$ ([**19**] or [**30**]) guarantees the existence of an $\mathfrak{A}_4$ action $\beta$ on $\mathcal{L}$ satisfying $\mathcal{Q} = \mathcal{L}^\beta$. This fact and $[\mathcal{P} : \mathcal{L}] = 2$ mean that our Kac algebra arises from $\mathfrak{A}_4 \rtimes \mathbf{Z}_2$ considered in **2.2** of §2. But, the underlying algebra

structure computed there is different from the present one, which is a contradiction.
(d) Suppose that the principal graph is $D_k^{(1)}$ $(k \geq 5)$, and let $[\rho_2\bar{\rho}_2] = [id] \oplus [\theta] \oplus [\mu]$ be the irreducible decomposition with $d(\theta) = 1$ and $d(\mu) = 2$. Considering the irreducible decomposition of $\rho\bar{\rho}$ we get $\dim(\rho_1\theta\bar{\rho}_1, \rho_1\mu\bar{\rho}_1) \neq 1$. However, the Frobenius reciprocity implies

$$\begin{aligned}\dim(\rho_1\theta\bar{\rho}_1, \rho_1\mu\bar{\rho}_1) &= \dim(\bar{\rho}_1\rho_1\theta, \mu\bar{\rho}_1\rho_1)\\ &= \dim(\oplus_{h\in\mathfrak{S}_3}\alpha_k\theta, \oplus_{k\in\mathfrak{S}_3}\mu\alpha_k) = 0,\end{aligned}$$

which is a contradiction.

In the preceding paragraph we have seen that $\rho_2\bar{\rho}_2$ is indeed decomposed into automorphisms, which means $\mathcal{Q} = \mathcal{R}^{(\beta,K)}$ with a group $K$ of order 4 and its action $\beta$. If the product $[\alpha_{\mathfrak{S}_3}]\cdot[\beta_K]$ is not a semi-direct product group, then Lemma 14.4, Lemma 14.5, Lemma 14.16, Lemma 14.17, Lemma 14.18, and Lemma 14.19 show that $K = \mathbf{Z}_4$ and our Kac algebra is the dual of the one in Theorem 14.33. On the other hand, if $[\alpha_{\mathfrak{S}_3}]\cdot[\beta_K]$ is a semi-direct product, then it is easy to see that the only allowed case is $(\mathbf{Z}_2 \times \mathbf{Z}_2) \rtimes \mathfrak{S}_3$, where $\mathfrak{S}_3$ acts as permutations of the non-neutral three elements. □

PROPOSITION 14.37. *No non-trivial Kac algebra admits the underlying algebra*

$$\mathbf{C}^4 \oplus M_2(\mathbf{C}) \oplus M_4(\mathbf{C}).$$

PROOF. Suppose that $\mathcal{A}$ is a non-trivial Kac algebra whose algebra structure is as above. Let $\mathcal{P} \supseteq \mathcal{Q}$ be a depth 2 inclusion of factors corresponding to the dual Kac algebra, and $\rho$ be the inclusion map. Then, the irreducible decomposition of $\rho\bar{\rho}$ is

$$[\rho\bar{\rho}] = \oplus_{g\in G(\hat{\mathcal{A}})}[\tau_g] \oplus 2[\pi] \oplus 4[\sigma],$$

where $\pi$ and $\sigma$ are self-conjugate sectors with $d(\pi) = 2$ and $d(\sigma) = 4$. We claim that there is no intermediate subfactor $\mathcal{P} \supseteq \mathcal{R} \supseteq \mathcal{Q}$ with $[\mathcal{P} : \mathcal{R}] = 3$. If there were such a subfactor $\mathcal{R}$ (with the canonical endomorphism $\gamma_{\mathcal{R}} : \mathcal{P} \longrightarrow \mathcal{R}$), then we would have $[\gamma_{\mathcal{R}}] = [id] \oplus [\pi]$ and the principal graph of $\mathcal{P} \supseteq \mathcal{R}$ must be $A_5$. However, since $[\tau_g][\pi] = [\pi]$ for all $g \in G(\hat{\mathcal{A}})$, we would have $[\pi^2] = \oplus_{g\in G(\hat{\mathcal{A}})}[\tau_g]$, which is not the fusion rule for $A_5$. Thus, the claim has been established. Since $\{\tau_g\}_{G(\hat{\mathcal{A}})} \cup \{\pi\}$ is closed under conjugation and the irreducible decomposition of products, there exists an intermediate subfactor $\mathcal{P} \supseteq \mathcal{L} \supseteq \mathcal{Q}$ satisfying $[\mathcal{P} : \mathcal{L}] = 8$ and $[\mathcal{L} : \mathcal{Q}] = 3$. Thus, the above claim applied to the dual Kac algebra implies that $\hat{\mathcal{A}}$ is not isomorphic to $\mathcal{A}$ as an algebra. Thanks to Lemma 14.10 and classification results we have obtained so far, we may assume that the dual Kac algebra $\hat{\mathcal{A}}$ is one of the following:

$$\mathbf{C}^8 \oplus M_2(\mathbf{C}) \oplus M_2(\mathbf{C}) \oplus M_2(\mathbf{C}) \oplus M_2(\mathbf{C}),$$

$$\mathbf{C}^4 \oplus M_2(\mathbf{C}) \oplus M_2(\mathbf{C}) \oplus M_2(\mathbf{C}) \oplus M_2(\mathbf{C}) \oplus M_2(\mathbf{C}).$$

The first case is impossible because of $[\mathcal{P} : \mathcal{Q} \rtimes G(\mathcal{A})] = 3$ (and the above claim). We assume that the second holds. Then, we have the following the irreducible decomposition:

$$[\bar{\rho}\rho] = \oplus_{h\in G(\mathcal{A})}[\theta_h] \oplus (\oplus_{i=1}^5 2[\mu_i])$$

with $d\mu_i = 2$. Note $G(\mathcal{A})$ acts on $\{[\mu_i]\}_{i=1}^5$ by left multiplication. Since $\#G(\mathcal{A}) = 4$, there exists $i$ such that $[\theta_h][\pi_i] = [\pi_i]$ for each $h \in G(\mathcal{A})$. Lemma 11.14 shows that such $i$ is unique, and so $\{\theta_h\}_{h\in G(\mathcal{A})} \cup \{\pi_i\}$ is closed under conjugation and the irreducible decomposition of products. This means that there exists an intermediate subfactor $\mathcal{P} \supseteq \mathcal{R} \supseteq \mathcal{Q}$ with $[\mathcal{P} : \mathcal{R}] = 3$, which is a contradiction. $\square$

PROPOSITION 14.38. *Let $\mathcal{A}$ be a Kac algebra with the underlying algebra*

$$\mathbf{C}^8 \oplus M_2(\mathbf{C}) \oplus M_2(\mathbf{C}) \oplus M_2(\mathbf{C}) \oplus M_2(\mathbf{C}),$$

*and we assume that 3 does not divide $\#G(\mathcal{A})$.*
*(i) If $G(\hat{\mathcal{A}})$ is abelian, $\mathcal{A}$ is classified by $\mathfrak{S}_3 \rtimes H$ with $H$ abelian of order 4 and the equivalence class of $[(\eta, \zeta)] \in H^2((\mathfrak{S}_3, H), \mathbf{T})$ giving*

$$\ell^\infty(\mathfrak{S}_3) \rtimes_\zeta H \cong \mathbf{C}^8 \oplus M_2(\mathbf{C}) \oplus M_2(\mathbf{C}) \oplus M_2(\mathbf{C}) \oplus M_2(\mathbf{C}).$$

*(ii) If $G(\hat{\mathcal{A}})$ is non-abelian, $\mathcal{A}$ is classified by $N \rtimes \mathbf{Z}_2$ with $N = D_{12}$ or $N = \mathbf{Z}_3 \rtimes \mathbf{Z}_4$ and the equivalence class of $[(\eta, \zeta)] \in H^2((N, \mathbf{Z}_2), \mathbf{T})$ giving*

$$\ell^\infty(N) \rtimes_\zeta \mathbf{Z}_2 \cong \mathbf{C}^8 \oplus M_2(\mathbf{C}) \oplus M_2(\mathbf{C}) \oplus M_2(\mathbf{C}) \oplus M_2(\mathbf{C}).$$

PROOF. Let $\mathcal{P} \supseteq \mathcal{Q}$ be a depth 2 inclusion of factors corresponding to the dual Kac algebra $\hat{\mathcal{A}}$. Then, we have $[\mathcal{P}^{G(\hat{\mathcal{A}})} : \mathcal{Q}] = 3$ and the principal graph of $\mathcal{P}^{G(\hat{\mathcal{A}})} \supseteq \mathcal{Q}$ is either $A_5$ or $D_4$ thanks to the classification of subfactor of index 3. It is actually $A_5$. In fact, if it were $D_4$, then $\mathcal{P}^{G(\hat{\mathcal{A}})}$ would be the crossed product of $\mathcal{Q}$ by a $\mathbf{Z}_3$ action and $G(\mathcal{A})$ (of order 8) would contain a subgroup isomorphic to $\mathbf{Z}_3$, which is a contradiction. Since the principal graph of $\mathcal{P}^{G(\hat{\mathcal{A}})} \supseteq \mathcal{Q}$ is $A_5$, the arguments in the proof of Proposition 14.36 together with the Goldman-type theorem give us a system of sectors contained in $\bar{\rho}\rho$ isomorphic to the dual of $\mathfrak{S}_3$. Thus, there exists a unique intermediate subfactor $\mathcal{P} \supseteq \mathcal{R} \supseteq \mathcal{P}^{G(\hat{\mathcal{A}})}$ such that $\mathcal{Q}$ is the fixed-point subalgebra of $\mathcal{R}$ under a $\mathfrak{S}_3$ action. We denote this action by $\beta$. Being an intermediate subfactor of $\mathcal{P} \supseteq \mathcal{P}^{G(\hat{\mathcal{A}})}$, $\mathcal{R}$ is the fixed-point subalgebra of $\mathcal{P}$ under a subgroup of $G(\hat{\mathcal{A}})$ of order 4. By duality, we have $\mathcal{P} = \mathcal{R} \rtimes_\alpha H$ with a group $H$ of order 4 and its action $\alpha$. We remark that $\mathcal{R}$ is characterized by the conditions that $\mathcal{Q}$ is the fixed-point subalgebra of $\mathcal{R}$ under an $\mathfrak{S}_3$ action and $\mathcal{P}$ is the crossed product of $\mathcal{R}$ by an action of a group of order 4. Indeed, if $\mathcal{R}_1$ also satisfies the two conditions, then we have $\mathcal{R}_1 \supseteq \mathcal{P}^{G(\hat{\mathcal{A}})} \supseteq \mathcal{Q}$. Since the action of $\mathfrak{S}_3$ above is determined by the inclusion $\mathcal{P}^{G(\hat{\mathcal{A}})} \supseteq \mathcal{Q}$ (see [**26**]), we conclude $\mathcal{R} = \mathcal{R}_1$.

We set $G = [\alpha_H] \cdot [\beta_{\mathfrak{S}_3}]$, and claim that $[\alpha_H]$ is not normal in $G$. Indeed, if it were the case, we would get

$$\mathbf{C}^8 \oplus M_2(\mathbf{C}) \oplus M_2(\mathbf{C}) \oplus M_2(\mathbf{C}) \oplus M_2(\mathbf{C}) \cong \mathcal{A} \cong \oplus_{g \in \mathfrak{S}_3} \mathbf{C}_{\eta_g}(H).$$

However, this would imply that $H \cong \mathbf{Z}_2 \times \mathbf{Z}_2$ and there would exist exactly 4 elements $g \in \mathfrak{S}_3$ satisfying $[\eta_g] \neq 0$ in $H^2(H, \mathbf{T})$, which is impossible. Therefore either $\mathfrak{S}_3$ is normal or $G = [\alpha_H] \cdot [\beta_{\mathfrak{S}_3}]$ is not a semi-direct product. Assume $g_0 \in \mathfrak{S}_3$ satisfy $\mathcal{P}^{G(\hat{\mathcal{A}})} = \mathcal{R}^{\beta_{g_0}}$. Then, it is easy to show that $G(\hat{\mathcal{A}})$ is abelian if and only if $[\beta_{g_0}]$ commutes with $[\alpha_H]$. On the other hand, thanks to the list of the product groups of order 24 in Lemma 14.3, we can see that there exists an non-trivial element in $[\beta_{\mathfrak{S}_3}]$ commuting with $[\alpha_H]$ if and only if $[\beta_{\mathfrak{S}_3}]$ is normal in $G$. Therefore, we get the statement for abelian $G(\hat{\mathcal{A}})$. Now we consider the case where $G(\hat{\mathcal{A}})$ is not abelian. If $[\beta_{\mathfrak{S}_3}]$ were normal we would get

$$\mathbf{C}^8 \oplus M_2(\mathbf{C}) \oplus M_2(\mathbf{C}) \oplus M_2(\mathbf{C}) \oplus M_2(\mathbf{C}) \cong \mathcal{A} \cong \ell^\infty(\mathfrak{S}_3) \rtimes H.$$

However, this is impossible due to Lemma 14.28 and Lemma 14.29, and so $G = [\alpha_H] \cdot [\beta_{\mathfrak{S}_3}]$ is not a semi-direct product. Thus, Lemma 14.5, Lemma 14.6, and Lemma

14.7 imply that there exits a unique depth 2 intermediate subfactor $\mathcal{P} \supseteq \mathcal{L} \supseteq \mathcal{R}$ and that $\mathcal{L}$ is the crossed product of $\mathcal{Q}$ by an action of $N$ where $N$ is either $D_{12}$ or $\mathbf{Z}_3 \rtimes \mathbf{Z}_4$. We claim that $\mathcal{L}$ is uniquely determined by the condition that $\mathcal{L}$ is the crossed product of $\mathcal{Q}$ by an action of either $D_{12}$ or $\mathbf{Z}_3 \rtimes \mathbf{Z}_4$. Indeed, if $\mathcal{L}_1$ also satisfies the condition with actions $\alpha^2$ and $\beta^2$ of $\mathbf{Z}_2$ and $N$ satisfying $\mathcal{P} = \mathcal{L} \rtimes_{\alpha^2} \mathbf{Z}_2$ and $\mathcal{Q} = \mathcal{L}^{\beta^2}$. Then, thanks to Lemma 14.17, Lemma 14.18, Lemma 14.19, Lemma 14.22, Lemma 14.23, and Lemma 14.24, our near action $(\alpha^2, \beta^2)$ corresponds to one of (iv), (v), and (vi) of Lemma 14.18. Therefore, we can find an element $n \in N$ such that $[\beta^2_n]$ commutes with $[\alpha^2_{\mathbf{Z}_2}]$ and $H/\{1,n\} \cong \mathfrak{S}_3$. Therefore, we get $\mathcal{R} = \mathcal{L}_1^{\beta^2_n}$. Since $\mathcal{L}_1$ is an intermediate subfactor of $\mathcal{P} \supseteq \mathcal{R}$ such that $\mathcal{L}_1 \supseteq \mathcal{Q}$ is of depth 2, we get $\mathcal{L}_1 = \mathcal{L}$. This finishes the proof. □

PROPOSITION 14.39. *Assume that the underlying algebra structure of both $\mathcal{A}$ and $\hat{\mathcal{A}}$ is*

$$\mathbf{C}^4 \oplus M_2(\mathbf{C}) \oplus M_2(\mathbf{C}) \oplus M_2(\mathbf{C}) \oplus M_2(\mathbf{C}) \oplus M_2(\mathbf{C}).$$

*Let $\mathcal{P} \supseteq \mathcal{Q}$ be a depth 2 inclusion of factors corresponding to the dual Kac algebra $\hat{\mathcal{A}}$. Then, one of the following holds:*
*(i) There exist an intermediate subfactor $\mathcal{P} \supseteq \mathcal{R} \supseteq \mathcal{Q}$, an action $\alpha$ of $\mathfrak{S}_3$, and an action $\beta$ of a group $K$ of order 4 on $\mathcal{R}$ such that $\mathcal{P} = \mathcal{R} \rtimes_\alpha \mathfrak{S}_3$ and $\mathcal{Q} = \mathcal{R}^\beta$.*
*(ii) There exist an intermediate subfactor $\mathcal{P} \supseteq \mathcal{R} \supseteq \mathcal{Q}$, an action $\alpha$ of a group $N$, and an action $\beta$ of $\mathbf{Z}_2$ on $\mathcal{R}$ such that $\mathcal{P} = \mathcal{R} \rtimes_\alpha N$ and $\mathcal{Q} = \mathcal{R}^\beta$. Here, $N$ is either $D_{12}$ or $\mathbf{Z}_3 \rtimes \mathbf{Z}_4$.*
*(iii) There exist an intermediate subfactor $\mathcal{P} \supseteq \mathcal{R} \supseteq \mathcal{Q}$, an action $\alpha$ of $\mathbf{Z}_2$, and an action $\beta$ of $D_{12}$ on $\mathcal{R}$ such that $\mathcal{P} = \mathcal{R} \rtimes_\alpha \mathbf{Z}_2$ and $\mathcal{Q} = \mathcal{R}^\beta$.*

PROOF. Let $\rho : \mathcal{Q} \hookrightarrow \mathcal{P}$ be the inclusion map. Then, the irreducible decomposition of $\rho\bar{\rho}$ is as follows:

$$[\rho\bar{\rho}] = \oplus_{g \in G(\hat{\mathcal{A}})}[\tau_g] \oplus \left(\oplus_{i=1}^5 2[\pi_i]\right)$$

with $d(\pi_i) = 2$. The argument in the proof of Lemma 14.37 shows that there exists an intermediate subfactor $\mathcal{P} \supseteq \mathcal{R} \supseteq \mathcal{Q}$ with $[\mathcal{P} : \mathcal{R}] = 3$, and so the principal graph of $\mathcal{P} \supseteq \mathcal{R}$ is $A_5$. Thus, there exist an intermediate subfactor $\mathcal{R} \supseteq \mathcal{L} \supseteq \mathcal{Q}$ and an action $\alpha$ of $\mathfrak{S}_3$ on $\mathcal{L}$ such that $\mathcal{P} = \mathcal{L} \rtimes_\alpha \mathfrak{S}_3$.

When $\mathcal{Q}$ is the fixed-point algebra of a group action on $\mathcal{L}$, we are in the situation described in (i). Thus, we can assume that $\mathcal{Q}$ is not the fixed-point algebra in the rest. Then, the principal graph of $\mathcal{L} \supseteq \mathcal{Q}$ is $D_k^{(1)}$. (Note that sectors appearing in the above irreducible decomposition are of dimension either 1 or 2.) Let $\rho_1 : \mathcal{L} \hookrightarrow \mathcal{P}$ and $\rho_2 : \mathcal{Q} \hookrightarrow \mathcal{L}$ be the inclusion maps. Then, we get $[\rho_2\bar{\rho}_2] = [id] \oplus [\theta] \oplus [\mu]$ with $d(\theta) = 1$ and $d(\mu) = 2$. Thus, we have

$$[\rho\bar{\rho}] = [\rho_1\bar{\rho}_1] \oplus [\rho_1\theta\bar{\rho}_1] \oplus [\rho_1\mu\bar{\rho}_1].$$

We claim that $[\theta]$ normalizes $\{[\alpha_g]\}_{g \in \mathfrak{S}_3}$. In fact, since

$$\dim(\rho_1\theta\bar{\rho}_1, \rho_1\mu\bar{\rho}_1) = \dim(\bar{\rho}_1\rho_1\theta, \mu\bar{\rho}_1\rho_1) = \dim(\oplus_{g \in \mathfrak{S}_3}\alpha_g\theta, \oplus_{h \in \mathfrak{S}_3}\mu\alpha_h) = 0,$$

$\rho_1\theta\bar{\rho}_1$ contains a 2-dimensional sector with multiplicity 2. Thus, we get

$$6 = \dim(\rho_1\theta\bar{\rho}_1, \rho_1\theta\bar{\rho}_1) = \dim(\theta^{-1}\bar{\rho}_1\rho_1\theta, \bar{\rho}_1\rho_1) = \dim(\oplus_{g \in \mathfrak{S}_3}\theta^{-1}\alpha_g\theta, \oplus_{h \in \mathfrak{S}_3}\alpha_h),$$

proving the claim.

We set $\mathcal{M} = \mathcal{L}^\theta$, which is an intermediate subfactor between $\mathcal{L}$ and $\mathcal{Q}$. If $[\theta]$ commutes with $[\alpha_g]$ for all $g \in \mathfrak{S}_3$, we get (ii). Hence, we assume it does not commute. Then, $\mathcal{P} \supseteq \mathcal{M}$ is a non-trivial depth 2 inclusion of index 12. Therefore, classification of Kac algebras of dimension 12 shows that there exists an intermediate subfactor $\mathcal{P} \supseteq \mathcal{S} \supseteq \mathcal{M}$ such that $\mathcal{M}$ is the fixed-point subalgebra of $\mathcal{S}$ under an $\mathfrak{S}_3$ action. Let $\rho_3 : \mathcal{M} \hookrightarrow \mathcal{P}$ be the inclusion map. Then, the irreducible decomposition of $\overline{\rho}_3\rho_3$ is as follows:

$$[\overline{\rho}_3\rho_3] = [id] \oplus [\theta_1] \oplus [\theta_2] \oplus [\theta_3] \oplus 2[\mu_1] \oplus 2[\mu_2].$$

We may assume that $\{id, \theta_1, \mu_1\}$ comes from $\mathcal{S} \supseteq \mathcal{M}$. Thus, we get $[\theta_2][\mu_1] = [\mu_1][\theta_2] = [\mu_2]$ and

$$[\mu_1^2] = [\mu_2^2] = [id] \oplus [\theta_1] \oplus [\mu_1].$$

Let $\varphi$ be a period two automorphism of $M$ such that $\mathcal{Q} = \mathcal{M}^{\varphi}$. Then, since

$$[\varphi]\{[id], [\theta_1], [\theta_2], [\theta_3], [\mu_1], [\mu_2]\}[\varphi] = \{[id], [\theta_1], [\theta_2], [\theta_3], [\mu_1], [\mu_2]\},$$

the above fusion rule shows that we have

$$[\varphi][\theta_1][\varphi] = [\theta_1], \quad [\varphi][\mu_1][\varphi] = [\mu].$$

This implies that $\mathcal{M} \supseteq \mathcal{Q}$ is a depth 2 inclusion. Since $H^2(\hat{\mathfrak{S}}_3, \mathbf{T})$ is trivial, Proposition 13.5 shows that $\mathcal{Q}$ is the fixed-point algebra of $\mathcal{M}$ under a $D_{12}$ action and we are in the situation described in (iii). □

## 4. Classification

The strategy mentioned at the beginning of §3 enables us to obtain the classification of Kac algebras of dimension 24.

THEOREM 14.40. *A possible underlying algebra of a non-trivial Kac algebra of dimension* 24 *is one of the seven types* (**I**,**II**, $\cdots$,**VII**) *described below, and in each case classification of Kac algebras is given as follows*:

**I.** *Assume*

$$\mathcal{A} \cong \mathbf{C}^{12} \oplus M_2(\mathbf{C}) \oplus M_2(\mathbf{C}) \oplus M_2(\mathbf{C}).$$

*Then, $\mathcal{A}$ is classified as follows*:

(*i*) *There is one Kac algebra with*

$$\hat{\mathcal{A}} \cong \mathbf{C}^{12} \oplus M_2(\mathbf{C}) \oplus M_2(\mathbf{C}) \oplus M_2(\mathbf{C}),$$
$$G(\mathcal{A}) \cong G(\hat{\mathcal{A}}) \cong \mathbf{Z}_2 \times \mathbf{Z}_2 \times \mathbf{Z}_3.$$

*This Kac algebra $\mathcal{A}$ is the tensor product of Kac-Paljutkin's* 8*-dimensional Kac algebra and* $\mathbf{C}(\mathbf{Z}_3)$.

(*ii*) *There is one Kac algebra with*

$$\hat{\mathcal{A}} \cong \mathbf{C}^{8} \oplus M_2(\mathbf{C}) \oplus M_2(\mathbf{C}) \oplus M_2(\mathbf{C}) \oplus M_2(\mathbf{C}),$$
$$G(\mathcal{A}) \cong D_8, \quad G(\hat{\mathcal{A}}) \cong \mathbf{Z}_2 \times \mathbf{Z}_2 \times \mathbf{Z}_3.$$

(*iii*) *There is one Kac algebras with*

$$\hat{\mathcal{A}} \cong \mathbf{C}^{8} \oplus M_2(\mathbf{C}) \oplus M_2(\mathbf{C}) \oplus M_2(\mathbf{C}) \oplus M_2(\mathbf{C}),$$
$$G(\mathcal{A}) \cong Q_8, \quad G(\hat{\mathcal{A}}) \cong \mathbf{Z}_2 \times \mathbf{Z}_2 \times \mathbf{Z}_3.$$

(*iv*) *There is one Kac algebra with*

$$\hat{\mathcal{A}} \cong \mathbf{C}^{6} \oplus M_3(\mathbf{C}) \oplus M_3(\mathbf{C}),$$
$$G(\mathcal{A}) \cong \mathbf{Z}_6, \quad G(\hat{\mathcal{A}}) \cong \mathfrak{A}_4.$$

*(v) There is one Kac algebra with*

$$\hat{\mathcal{A}} \cong \mathbf{C}^4 \oplus M_2(\mathbf{C}) \oplus M_2(\mathbf{C}) \oplus M_2(\mathbf{C}) \oplus M_2(\mathbf{C}) \oplus M_2(\mathbf{C}),$$
$$G(\mathcal{A}) \cong \mathbf{Z}_2 \times \mathbf{Z}_2, \quad G(\hat{\mathcal{A}}) \cong \mathbf{Z}_2 \times \mathbf{Z}_2 \times \mathbf{Z}_3.$$

*(vi) There is one Kac algebra with*

$$\hat{\mathcal{A}} \cong \mathbf{C}^2 \oplus M_2(\mathbf{C}) \oplus M_3(\mathbf{C}) \oplus M_3(\mathbf{C}),$$
$$G(\mathcal{A}) \cong \mathbf{Z}_2, \quad G(\hat{\mathcal{A}}) \cong \mathfrak{A}_4.$$

**II.** *Assume*

$$\mathcal{A} \cong \mathbf{C}^8 \oplus M_4(\mathbf{C}).$$

*There is only one non-trivial Kac algebra of this type. This $\mathcal{A}$ comes from the bicrossed product associated with the product group $\mathfrak{S}_4 = \mathfrak{S}_3 \cdot \mathbf{Z}_4$, and the dual Kac algebras and the intrinsic groups are*

$$\hat{\mathcal{A}} \cong \mathbf{C}^2 \oplus M_2(\mathbf{C}) \oplus M_3(\mathbf{C}) \oplus M_3(\mathbf{C}),$$
$$G(\mathcal{A}) \cong \mathbf{Z}_2, \quad G(\hat{\mathcal{A}}) \cong D_8.$$

**III.** *Assume*

$$\mathcal{A} \cong \mathbf{C}^8 \oplus M_2(\mathbf{C}) \oplus M_2(\mathbf{C}) \oplus M_2(\mathbf{C}) \oplus M_2(\mathbf{C}).$$

*Then, $\mathcal{A}$ is classified as follows:*
*(i) There is one Kac algebra with*

$$\hat{\mathcal{A}} \cong \mathbf{C}^{12} \oplus M_2(\mathbf{C}) \oplus M_2(\mathbf{C}) \oplus M_2(\mathbf{C}),$$
$$G(\mathcal{A}) \cong \mathbf{Z}_2 \times \mathbf{Z}_2 \times \mathbf{Z}_3, \quad G(\hat{\mathcal{A}}) \cong D_8.$$

*(ii) There is one Kac algebra with*

$$\hat{\mathcal{A}} \cong \mathbf{C}^{12} \oplus M_2(\mathbf{C}) \oplus M_2(\mathbf{C}) \oplus M_2(\mathbf{C}),$$
$$G(\mathcal{A}) \cong \mathbf{Z}_2 \times \mathbf{Z}_2 \times \mathbf{Z}_3, \quad G(\hat{\mathcal{A}}) \cong Q_8.$$

*(iii) There is one self-dual Kac algebra with*

$$\hat{\mathcal{A}} \cong \mathbf{C}^8 \oplus M_2(\mathbf{C}) \oplus M_2(\mathbf{C}) \oplus M_2(\mathbf{C}) \oplus M_2(\mathbf{C}),$$
$$G(\mathcal{A}) \cong G(\hat{\mathcal{A}}) \cong \mathbf{Z}_2 \times \mathbf{Z}_2 \times \mathbf{Z}_2.$$

*Furthermore, this $\mathcal{A}$ is isomorphic to $\mathcal{A}_0 \otimes \mathbf{C}(\mathbf{Z}_2)$ with the unique Kac algebra $\mathcal{A}_0$ of dimension 12 (see §5 in Chapter 7) satisfying $\mathcal{A}_0 \cong \hat{\mathcal{A}}_0 \cong \mathbf{C}^4 \oplus M_2(\mathbf{C}) \oplus M_2(\mathbf{C})$ and $G(\mathcal{A}_0) \cong G(\hat{\mathcal{A}}_0) \cong \mathbf{Z}_2 \times \mathbf{Z}_2$.*
*(iv) There are three Kac algebras with*

$$\hat{\mathcal{A}} \cong \mathbf{C}^8 \oplus M_2(\mathbf{C}) \oplus M_2(\mathbf{C}) \oplus M_2(\mathbf{C}) \oplus M_2(\mathbf{C}),$$
$$G(\mathcal{A}) \cong G(\hat{\mathcal{A}}) \cong \mathbf{Z}_2 \times \mathbf{Z}_4.$$

*One of the three Kac algebras here is isomorphic to $\mathcal{A}_1 \otimes \mathbf{C}(\mathbf{Z}_2)$ with the unique Kac algebra $\mathcal{A}_1$ of dimension 12 (see §5 in Chapter 7) satisfying $\mathcal{A}_1 \cong \hat{\mathcal{A}}_1 \cong \mathbf{C}^4 \oplus M_2(\mathbf{C}) \oplus M_2(\mathbf{C})$ and $G(\mathcal{A}_1) \cong G(\hat{\mathcal{A}}_1) \cong \mathbf{Z}_4$.*
*(v) There is one self-dual Kac algebra with*

$$\hat{\mathcal{A}} \cong \mathbf{C}^8 \oplus M_2(\mathbf{C}) \oplus M_2(\mathbf{C}) \oplus M_2(\mathbf{C}) \oplus M_2(\mathbf{C}),$$
$$G(\mathcal{A}) \cong G(\hat{\mathcal{A}}) \cong \mathbf{Z}_8.$$

*(vi) There is one Kac algebra with*

$$\hat{\mathcal{A}} \cong \mathbf{C}^6 \oplus M_3(\mathbf{C}) \oplus M_3(\mathbf{C}),$$
$$G(\mathcal{A}) \cong \mathbf{Z}_6, \quad G(\hat{\mathcal{A}}) \cong \mathbf{Z}_2 \times \mathbf{Z}_2 \times \mathbf{Z}_2.$$

*(vii) There are two Kac algebras with*

$$\hat{\mathcal{A}} \cong \mathbf{C}^4 \oplus M_2(\mathbf{C}) \oplus M_2(\mathbf{C}) \oplus M_2(\mathbf{C}) \oplus M_2(\mathbf{C}) \oplus M_2(\mathbf{C}),$$
$$G(\mathcal{A}) \cong \mathbf{Z}_2 \times \mathbf{Z}_2, \quad G(\hat{\mathcal{A}}) \cong D_8.$$

*(viii) There is one Kac algebra with*

$$\hat{\mathcal{A}} \cong \mathbf{C}^4 \oplus M_2(\mathbf{C}) \oplus M_2(\mathbf{C}) \oplus M_2(\mathbf{C}) \oplus M_2(\mathbf{C}) \oplus M_2(\mathbf{C}),$$
$$G(\mathcal{A}) \cong \mathbf{Z}_2 \times \mathbf{Z}_2, \quad G(\hat{\mathcal{A}}) \cong Q_8.$$

**IV.** *Assume*

$$\mathcal{A} \cong \mathbf{C}^6 \oplus M_3(\mathbf{C}) \oplus M_3(\mathbf{C}).$$

*Then, $\mathcal{A}$ is classified as follows:*
*(i) There is one Kac algebra with*

$$\hat{\mathcal{A}} \cong \mathbf{C}^{12} \oplus M_2(\mathbf{C}) \oplus M_2(\mathbf{C}) \oplus M_2(\mathbf{C}),$$
$$G(\mathcal{A}) \cong \mathfrak{A}_4, \quad G(\hat{\mathcal{A}}) \cong \mathbf{Z}_6.$$

*(ii) There is one Kac algebra with*

$$\hat{\mathcal{A}} = \mathbf{C}^8 \oplus M_2(\mathbf{C}) \oplus M_2(\mathbf{C}) \oplus M_2(\mathbf{C}) \oplus M_2(\mathbf{C}),$$
$$G(\mathcal{A}) \cong \mathbf{Z}_2 \times \mathbf{Z}_2 \times \mathbf{Z}_2, \quad G(\hat{\mathcal{A}}) \cong \mathbf{Z}_6.$$

*(iii) There is one Kac algebra with*

$$\hat{\mathcal{A}} = \mathbf{C}^4 \oplus M_2(\mathbf{C}) \oplus M_2(\mathbf{C}) \oplus M_2(\mathbf{C}) \oplus M_2(\mathbf{C}) \oplus M_2(\mathbf{C}),$$
$$G(\mathcal{A}) \cong \mathbf{Z}_2 \times \mathbf{Z}_2, \quad G(\hat{\mathcal{A}}) \cong \mathfrak{S}_3.$$

**V.** *Assume*

$$\mathcal{A} \cong \mathbf{C}^4 \oplus M_2(\mathbf{C}) \oplus M_2(\mathbf{C}) \oplus M_2(\mathbf{C}) \oplus M_2(\mathbf{C}) \oplus M_2(\mathbf{C}).$$

*Then, $\mathcal{A}$ is classified as follows:*
*(i) There is one Kac algebra with*

$$\hat{\mathcal{A}} \cong \mathbf{C}^{12} \oplus M_2(\mathbf{C}) \oplus M_2(\mathbf{C}) \oplus M_2(\mathbf{C}),$$
$$G(\mathcal{A}) \cong \mathbf{Z}_2 \times \mathbf{Z}_2 \times \mathbf{Z}_3, \quad G(\hat{\mathcal{A}}) \cong \mathbf{Z}_2 \times \mathbf{Z}_2.$$

*(ii) There are two Kac algebras with*

$$\hat{\mathcal{A}} \cong \mathbf{C}^8 \oplus M_2(\mathbf{C}) \oplus M_2(\mathbf{C}) \oplus M_2(\mathbf{C}) \oplus M_2(\mathbf{C}),$$
$$G(\mathcal{A}) \cong D_8, \quad G(\hat{\mathcal{A}}) \cong \mathbf{Z}_2 \times \mathbf{Z}_2.$$

*(iii) There is one Kac algebra with*

$$\hat{\mathcal{A}} \cong \mathbf{C}^8 \oplus M_2(\mathbf{C}) \oplus M_2(\mathbf{C}) \oplus M_2(\mathbf{C}) \oplus M_2(\mathbf{C}),$$
$$G(\mathcal{A}) \cong Q_8, \quad G(\hat{\mathcal{A}}) \cong \mathbf{Z}_2 \times \mathbf{Z}_2.$$

*(iv) There is one Kac algebra with*

$$\hat{\mathcal{A}} \cong \mathbf{C}^6 \oplus M_3(\mathbf{C}) \oplus M_3(\mathbf{C}),$$
$$G(\mathcal{A}) \cong \mathfrak{S}_3, \quad G(\hat{\mathcal{A}}) \cong \mathbf{Z}_2 \times \mathbf{Z}_2.$$

*(v) There is one self-dual Kac algebra with*

$$G(\mathcal{A}) \cong G(\hat{\mathcal{A}}) \cong \mathbf{Z}_2 \times \mathbf{Z}_2.$$

*Furthermore, this $\mathcal{A}$ is the Kac algebra described in §6 of Chapter 7 (the non-trivial cocycle case with $n = 6$).*
*(vi) There is one Kac algebra with*

$$\hat{\mathcal{A}} \cong \mathbf{C}^3 \oplus M_2(\mathbf{C}) \oplus M_2(\mathbf{C}) \oplus M_2(\mathbf{C}) \oplus M_3(\mathbf{C}),$$
$$G(\mathcal{A}) \cong \mathbf{Z}_3, \quad G(\hat{\mathcal{A}}) \cong \mathbf{Z}_2 \times \mathbf{Z}_2.$$

**VI.** *Assume*

$$\mathcal{A} \cong \mathbf{C}^3 \oplus M_2(\mathbf{C}) \oplus M_2(\mathbf{C}) \oplus M_2(\mathbf{C}) \oplus M_3(\mathbf{C}).$$

*There is one Kac algebra with*

$$\hat{\mathcal{A}} \cong \mathbf{C}^4 \oplus M_2(\mathbf{C}) \oplus M_2(\mathbf{C}) \oplus M_2(\mathbf{C}) \oplus M_2(\mathbf{C}) \oplus M_2(\mathbf{C}),$$
$$G(\mathcal{A}) \cong \mathbf{Z}_2 \times \mathbf{Z}_2, \quad G(\hat{\mathcal{A}}) \cong \mathbf{Z}_3.$$

**VII.** *Assume*

$$\mathcal{A} \cong \mathbf{C}^2 \oplus M_2(\mathbf{C}) \oplus M_3(\mathbf{C}) \oplus M_3(\mathbf{C}).$$

*Then, $\mathcal{A}$ is classified as follows:*
*(i) There is one Kac algebra with*

$$\hat{\mathcal{A}} \cong \mathbf{C}^{12} \oplus M_2(\mathbf{C}) \oplus M_2(\mathbf{C}) \oplus M_2(\mathbf{C}),$$
$$G(\mathcal{A}) \cong \mathfrak{A}_4, \quad G(\hat{\mathcal{A}}) \cong \mathbf{Z}_2.$$

*(ii) There is one Kac algebra with*

$$\hat{\mathcal{A}} \cong \mathbf{C}^8 \oplus M_4(\mathbf{C}),$$
$$G(\mathcal{A}) \cong D_8, \quad G(\hat{\mathcal{A}}) \cong \mathbf{Z}_2.$$

PROOF. We recall what was mentioned at the beginning of §3.

Case **II.** This follows from Theorem 14.33.

Case **IV.** This follows from Proposition 14.34 together with
Lemma 14.26
Lemma 14.27
Lemma 14.14,(i) (with Lemma 14.13).
These lemmas give us $1 + 1 + 1$ Kac algebras, corresponding to (ii),(i),(iii) (in **IV**) respectively.

Case **VI.** This follows from Lemma 14.35 together with Lemma 14.14,(ii).

Case **VII.** This follows from proposition 14.36 together with Lemma 14.30. (Note that **VII**,(ii) is the dual case of **II**.)

Case **I.** Lemma 14.32 says that there are two cases

$$\text{either } G(\hat{A}) = \mathfrak{A}_4 \text{ or } G(\hat{A}) = \mathbf{Z}_2 \times \mathbf{Z}_2 \times \mathbf{Z}_3.$$

In the second case, classification comes from $(\mathbf{Z}_2 \times \mathbf{Z}_2 \times \mathbf{Z}_3) \rtimes \mathbf{Z}_2$. Namely,
Lemma 14.12,(i)
Lemma 14.12,(ii)
Lemma 14.12,(iii),(1)
Lemma 14.12,(iii),(2)

give us $1+1+1+1$ Kac algebras, and they correspond to (i),(v),(ii),(iii) (in **I**) respectively. On the other hand, we have $\hat{\mathcal{A}} = \mathbf{C}(\mathfrak{A}_4) \rtimes \mathbf{Z}_2$ in the first case. Theorem 13.22 says that we have two such Kac algebras in this case with the following structure:

$$\hat{\mathcal{A}} \cong \mathbf{C}^6 \oplus M_3(\mathbf{C}) \oplus M_3(\mathbf{C}), \quad G(\mathcal{A}) \cong \mathbf{Z}_6,$$
$$\hat{\mathcal{A}} \cong \mathbf{C}^2 \oplus M_2(\mathbf{C}) \oplus M_3(\mathbf{C}) \oplus M_3(\mathbf{C}), \quad G(\mathcal{A}) \cong \mathbf{Z}_2.$$

Classification in cases **IV** and **VII** was already done, and these two Kac algebras must be the dual of **IV**,(i) and that of **VII**,(i). In this way we get the two Kac algebras described in **I**,(iv) and (vi).

Case **III.** At first we assume that 3 does no divide $\dim G(\mathcal{A})$ so that Proposition 14.38 can be used. From (i) in Proposition 14.38 together with

Lemma 14.29,(i)
Lemma 14.28,(i), Lemma 14.29,(ii),(iii)
Lemma 14.28,(ii)

we get $1+3+1$ Kac algebras, and they correspond to (iii),(iv),(v) (in **III**) respectively. On the other hand, from (ii) in Proposition 14.38 together with

Lemma 14.18,(iv),(v)
Lemma 14.18,(vi)

we get $2+1$ Kac algebras, and they correspond to (vii),(viii) (in **III**) respectively. (Note that none of Lemma 14.17, Lemma 14.19, Lemma 14.22, Lemma 14.23 and Lemma 14.24 gives us situation mentioned in Proposition 14.38,(ii).)

We now assume that $\dim G(\mathcal{A})$ is a multiple of 3. This means the the dual Kac algebra $\hat{\mathcal{A}}$ is of the form **I**, **IV** or **VI**. Classification of these three cases was already done, and **I**,(ii), **I**,(iii), **IV**,(ii) are the only cases (in **I**, **IV**, **VI**) where the dual Kac algebra is of the form **III**. Therefore, by passing to the respective dual Kac algebras we get the classification (i),(ii),(vi) (in **III**).

Case **V.** At first we assume that the dual Kac algebra $\hat{\mathcal{A}}$ is also of the form **V**. Then, Proposition 14.39 together with

Lemma 14.28 ($\mathfrak{S}_3 \rtimes \mathbf{Z}_4$ case)
Lemma 14.29 ($\mathfrak{S}_3 \rtimes (\mathbf{Z}_2 \times \mathbf{Z}_2)$ case)
Lemma 14.17, Lemma 14.18, Lemma 14.19 ($D_{12} \times \mathbf{Z}_2$ case)
Lemma 14.22, Lemma 14. 23, Lemma 14. 24 ($(\mathbf{Z}_3 \rtimes \mathbf{Z}_4) \rtimes \mathbf{Z}_2$ case)

imply that the our $\mathcal{A}$ must be the non-trivial cocycle case in Lemma 14.17, i.e., the one with the non-trivial cocycle (with $n = 6$) in §6 of Chapter 7. Moreover, Corollary 14.20 says that this Kac algebra is self-dual, and we get the uniqueness of a self-dual Kac algebra mentioned in **V**,(v).

We next assume that $\hat{\mathcal{A}}$ is not of the form **V**. But, note that all the other forms have already been classified. Thus, $\hat{\mathcal{A}}$ (not of the form **V**) can be classified, and (i),(ii),(iii),(iv),(vi) (in **V**) are obtained as the dual cases of

**I**,(v), **III**,(vii), **III**,(viii), **IV**,(iii), **VI**

respectively. □

**Remark.** Our proof (see Lemma 14.10) shows: Case **II** (whose dual case is **VII**,(ii)) coming from the product group $\mathfrak{S}_4 = \mathfrak{S}_3 \cdot \mathbf{Z}_4$ (as a bicrossed product) is the only non-trivial Kac algebra of dimension 24 not arising from a suitable semi-direct product group.

# Bibliography


[1] S. Baaj and G. Skandalis, *Unitaires multiplicatifs et dualité pour les produits croisés de $C^*$ algèbres*, Ann. Sci. École Norm. Sup., **26** (1993), 425-488.

[2] D. Bisch, *A note on intermediate subfactors*, Pacific J. Math., **163** (1994), 201-216.

[3] D. Bisch and U. Haagerup, *Composition of subfactors: new examples of infinite depth subfactors*, Ann. Sci. École Norm. Sup., **29** (1996), 329-383.

[4] A. Connes, *Periodic automorphisms of the hyperfinite factor of type $II_1$*, Acta Sci. Math., **39** (1977), 39-66.

[5] A. Connes, *Noncommutative Geometry*, Academic Press, 1994.

[6] A. Connes and M. Takesaki, *Flow of weights on factors of type $III$*, Tohoku J. Math., **29** (1977), 473-575.

[7] H. S. M. Coxeter and W. O. J. Moser, *Generators and Relations for Discrete Groups*, Springer-Verlag, 1957.

[8] J. Cuntz, *Regular actions of Hopf algebras on the $C^*$-algebra generated by a Hilbert space*, in "Operator Algebras, Mathematical Physics, and Low Dimensional Topology", Research Notes in Mathematics Vol. 5, A K Peters, Wellesley Massachusetts, 1993 (p 87-100).

[9] M. C. David, *Paragroupe d'Adrian Ocneanu et algèbre de Kac*, Pacific J. Math., **172** (1996), 331-363.

[10] M. C. David, *Coupe assorti de systèmes de Kac et inclusions de facteurs de type $II_1$*, J. Funct. Anal., **159** (1998), 1-42.

[11] J. De Cannière, *Produit croisé d'une algèbre de Kac par un groupe localement compact*, Bull. Soc. Math. France, **107** (1979), 337-372.

[12] J. De Cannière, *On the intrinsic group of a Kac algebra*, Proc. London Math. Soc., (3) **40** (1980), 1-20.

[13] M. Enock and J. M. Schwartz, *Kac Algebras and Duality of Locally Compact Groups*, Springer-Verlag, 1992.

[14] M. Enock and L. Vainerman, *Deformation of a Kac algebra by an abelian subgroup*, Commun. Math. Phys., **178** (1996), 571-596.

[15] P. Etigof and S. Gelaki, *Semisimple Hopf algebras of dimension pq are trivial*, J. Algebra, **210** (1998), 664-669.

[16] D. Evans and Y. Kawahigashi, *Quantum Symmetry on Operator Algebras*, Oxford Univ. Press, 1998.

[17] N. Fukuda, *Semisimple Hopf algebras of dimension* 12, Tsukuba J. Math., **21** (1997), 43-54.

[18] I. Hochstetter, *Erweiterungen von Hopf-Algebren und ihre kohomologische Beschreibung*, Dissertation, Universität München, 1990.

[19] J. H. Hong, *Subfactors with principal graph $E_6^{(1)}$*, Acta Appl. Math., **40** (1995), 255-264.

[20] J. H. Hong and H. Kosaki, *The group of one-dimensional bimodules arising from composition of subfactors*, J. Math. Soc. Japan, **52** (2000), 293-333.

[21] J. H. Hong and W. Szymański, *Composition of subfactors and twisted bicrossed product*, J. Operator Theory, **37** (1997), 281-302.

[22] J. H. Hong and W. Szymański, *On cohomological deformations of bicrossed product Hopf algebras*, Bull. Austral. Math. Soc., **60** (1999), 365-375.

[23] R. B. Howlett and I. M. Isacs, *On groups of central type*, Math. Z., **179** (1982), 555-569.

[24] N. Iwahori and H. Matsumoto, *Several remarks on projective representations of finite groups*, J. Fac. Sci. Univ. Tokyo Sect. IA Math., **43** (1964), 129-146.

[25] M. Izumi, *Application of fusion rules to classification of subfactors*, Publ. RIMS, Kyoto Univ., **27** (1991), 953-994.

[26] M. Izumi, *Goldman's type theorem for index* 3, Publ. RIMS, Kyoto Univ., **28** (1992), 833-843.

[27] M. Izumi, *Subalgebras of infinite $C^*$-algebras with finite Watatani indices I: Cuntz algebras*, Commun. Math. Phys., **155** (1993), 157-182.

[28] M. Izumi, *On type II and type III graphs for subfactors,* Math. Scand., **73** (1994), 307-319.

[29] M. Izumi, *Subalgebras of infinite $C^*$-algebras with finite Watatani indices II: Cuntz-Krieger algebras*, Duke Math. J., **91** (1998), 409-461.

[30] M. Izumi, *Goldman's type theorems in index theory*, in "Operator Algebras and Quantum Field Theory", International Press, 1997 (p 249-269).

[31] M. Izumi and Y. Kawahigashi, *Classification of subfactors with the principal graph $D_n^{(1)}$*, J. Funct. Anal., **112** (1993), 257-286.

[32] M. Izumi and H. Kosaki, *Finite-dimensional Kac algebras arising from certain group actions on a factor*, IMRN, **8** (1996), 357-370.

[33] M. Izumi and H. Kosaki, in preparation.

[34] M. Izumi, R. Longo, and S. Popa, *A Galois correspondence for compact groups of automorphisms of von Neumann algebras with a generalization to Kac algebras*, J. Funct. Anal., **155** (1998), 25-63.

[35] V. F. R. Jones, *Actions of finite groups on the hyperfinite $II_1$ factor*, Memoirs Amer. Math. Soc., No. 237, 1980.

[36] V. F. R. Jones, *Index for subfactors*, Invent. Math., **72** (1983), 1-25.

[37] G. I. Kac, *Extensions of groups to group rings*, Math. USSR Sbornik, **5** (1968), 451-474.

[38] G. I. Kac and V. G. Paljutkin, *Finite group rings*, Trans. Moscow Math. Soc., 1966, 251-294.

[39] T. Kobayashi and A. Masuoka, *A result extended from groups to Hopf algebras*, Tsukuba J. Math., **21** (1997), 55-58.

[40] H. Kosaki, *Sector theory and automorphisms for factor-subfactor pairs*, J. Math. Soc. Japan., **48** (1996), 428-454.

[41] H. Kosaki, *Automorphisms arising from composition of subfactors*, in "Operator Algebras and Quantum Field Theory", International Press, 1997 (p 236-248).

[42] H. Kosaki, A. Munemasa, and S. Yamagami, *On fusion algebras associated to finite group actions*, Pacific J. Math., **177** (1997), 527-554.

[43] M. Landstad, *Quantization arising from abelian subgroups*, Internat. J. Math., **5** (1994), 897-936.

[44] M. Landstad and I. Raeburn, *Twisted dual-group algebras: Equivariant deformations of $C_0(G)$*, J. Funct. Anal., **132** (1995), 43-85.

[45] R. Longo, *Simple injective factors*, Adv. in Math., **63** (1987), 152-172.

[46] R. Longo, *Index of subfactors and statistics of quantum fields I*, Commun. Math. Phys., **126** (1989), 217-247.

[47] R. Longo, *Index of subfactors and statistics of quantum fields II*, Commun. Math. Phys., **130** (1990), 285-309.

[48] R. Longo, *A duality for Hopf algebras and subfactors I*, Commun. Math. Phys., **159** (1994), 133-155.

[49] S. Majid, *Hopf-von Neumann algebra bicrossed products, Kac algebra bicrossed products, and the classical Yang-Baxter equation*, J. Funct. Anal., **95** (1991), 291-319.

[50] A. Masuoka, *Freeness of Hopf algebras over coideal subalgebras*, Commun. Algebra, **20** (1992), 1353-1373.

[51] A. Masuoka, *Semisimple Hopf algebras of dimension $2p$*, Commun. Algebra, **23** (1995), 361-373.

[52] A. Masuoka, *Some further classification results on semi-simple Hopf algebras*, Commun. Algebra, **24** (1996), 307-329.

[53] A. Masuoka, *Self-dual Hopf algebras of dimension $p^3$ obtained by extension*, J. Algebra, **178** (1996), 791-806.

[54] A. Masuoka, *The $p^n$ theorem for semi-simple Hopf algebras*, Proc. Amer. Math. Soc., **124** (1996), 735-737.

[55] A. Masuoka, *Calculations of some groups of Hopf algebra extensions*, J. Algebra, **191** (1997), 568-588.

[56] S. Montgomery, *Classification of finite-dimensional semisimple Hopf algebras*, Contem. Math., **229** (1998), 265-279.

[57] Y. Nakagami and M. Takesaki, *Duality for crossed products of von Neumann algebras*, LNM Vol. 731, Springer-Verlag, 1979.

[58] S. Natale, *On semisimple Hopf algebras of dimension $pq^2$*, J. Algebra, **221** (1999), 242-278.

[59] D. Robinson, *A Course in the Theory of Groups*, G.T.M. Vol. 80, Springer-Verlag, 1993.
[60] T. Sano, *Commuting and co-commuting squares and finite dimensional Kac algebras*, Pacific J. Math., **172** (1996), 243-253.
[61] T. Sano, *On commuting canonical endomorphisms of subfactors*, to appear in J. Math. Soc. Japan.
[62] N. Sato, *Fourier transform for irreducible inclusions of type $II_1$ factors with finite index and its application to depth two case*, Publ. RIMS, Kyoto Univ., **33** (1997), 189-222.
[63] Y. Sekine, *An example of finite dimensional Kac algebras of Kac-Paljutkin type*, Proc. Amer. Math. Soc., **124** (1996), 1139-1147.
[64] S. Shahriari, *On central factor groups*, Pacific J. Math., **151** (1991), 151-178.
[65] W. M. Singer, *Extension theory for connected Hopf algebras*, J. Algebra, **21** (1972), 1-16.
[66] Y. Sommerhäuser, *Yetter-Drinfel'd Hopf algebras over groups of prime order*, preprint.
[67] S. Strătilă, *Modular Theory in Operator Algebras*, Abacus Press, 1981.
[68] C. Sutherland, *Cohomology and extensions of von Neumann algebras II*, Publ. RIMS, Kyoto Univ., **16** (1980), 135-174.
[69] M. Suzuki, *Group Theory I*, Springer-Verlag, 1982.
[70] W. Szymański, Finite index subfactors and Hopf algebra crossed products, Proc. Amer. Math. Soc., **120** (1994), 519-528.
[71] M. Takeuchi, *Matched pairs of groups and bismashed product of Hopf algebras*, Commun. Algebra, **9** (1981), 841-882.
[72] L. Vainerman, *2-cocycles and twisting of Kac algebras*, Commun. Math. Phys., **191** (1998), 697-721.
[73] A. Wassermann, *Ergodic actions of compact groups on operator algebras II: Classification of full multiplicity ergodic actions*, Canad. J. Math. **40** (1988), 1482-1527.
[74] S. Yamagami, *Group symmetry in tensor categories*, preprint
[75] T. Yamanouchi, *The intrinsic group of Majid's bicrossproduct Kac algebra*, Pacific J. Math., **159** (1993), 185-199.
[76] G. Zeller-Meier, *Produits croisés d'une $C^*$-algèbre par un group d'automorphismes*, J. Math. Pures Appl., **47** (1968), 101-239.
[77] Y. Zhu, *Hopf algebras of prime dimension*, IMRN, **1** (1994), 53-59.

# Index

## Editorial Information

To be published in the *Memoirs*, a paper must be correct, new, nontrivial, and significant. Further, it must be well written and of interest to a substantial number of mathematicians. Piecemeal results, such as an inconclusive step toward an unproved major theorem or a minor variation on a known result, are in general not acceptable for publication. Papers appearing in *Memoirs* are generally longer than those appearing in *Transactions*, which shares the same editorial committee.

As of February 28, 2002, the backlog for this journal was approximately 4 volumes. This estimate is the result of dividing the number of manuscripts for this journal in the Providence office that have not yet gone to the printer on the above date by the average number of monographs per volume over the previous twelve months, reduced by the number of volumes published in four months (the time necessary for preparing a volume for the printer). (There are 6 volumes per year, each containing at least 4 numbers.)

A Consent to Publish and Copyright Agreement is required before a paper will be published in the *Memoirs*. After a paper is accepted for publication, the Providence office will send a Consent to Publish and Copyright Agreement to all authors of the paper. By submitting a paper to the *Memoirs*, authors certify that the results have not been submitted to nor are they under consideration for publication by another journal, conference proceedings, or similar publication.

## Information for Authors

*Memoirs* are printed from camera copy fully prepared by the author. This means that the finished book will look exactly like the copy submitted.

The paper must contain a *descriptive title* and an *abstract* that summarizes the article in language suitable for workers in the general field (algebra, analysis, etc.). The *descriptive title* should be short, but informative; useless or vague phrases such as "some remarks about" or "concerning" should be avoided. The *abstract* should be at least one complete sentence, and at most 300 words. Included with the footnotes to the paper should be the 2000 *Mathematics Subject Classification* representing the primary and secondary subjects of the article. The classifications are accessible from `www.ams.org/msc/`. The list of classifications is also available in print starting with the 1999 annual index of *Mathematical Reviews*. The Mathematics Subject Classification footnote may be followed by a list of *key words and phrases* describing the subject matter of the article and taken from it. Journal abbreviations used in bibliographies are listed in the latest *Mathematical Reviews* annual index. The series abbreviations are also accessible from `www.ams.org/publications/`. To help in preparing and verifying references, the AMS offers MR Lookup, a Reference Tool for Linking, at `www.ams.org/mrlookup/`. When the manuscript is submitted, authors should supply the editor with electronic addresses if available. These will be printed after the postal address at the end of the article.

**Electronically prepared manuscripts.** The AMS encourages electronically prepared manuscripts, with a strong preference for $\mathcal{A}_{\mathcal{M}}\mathcal{S}$-LaTeX. To this end, the Society has prepared $\mathcal{A}_{\mathcal{M}}\mathcal{S}$-LaTeX author packages for each AMS publication. Author packages include instructions for preparing electronic manuscripts, the *AMS Author Handbook*, samples, and a style file that generates the particular design specifications of that publication series. Though $\mathcal{A}_{\mathcal{M}}\mathcal{S}$-LaTeX is the highly preferred format of TeX, author packages are also available in $\mathcal{A}_{\mathcal{M}}\mathcal{S}$-TeX.

Authors may retrieve an author package from e-MATH starting from www.ams.org/tex/ or via FTP to ftp.ams.org (login as anonymous, enter username as password, and type cd pub/author-info). The *AMS Author Handbook* and the *Instruction Manual* are available in PDF format following the author packages link from www.ams.org/tex/. The author package can be obtained free of charge by sending email to pub@ams.org (Internet) or from the Publication Division, American Mathematical Society, P.O. Box 6248, Providence, RI 02940-6248. When requesting an author package, please specify $\mathcal{A}_\mathcal{M}\mathcal{S}$-LaTeX or $\mathcal{A}_\mathcal{M}\mathcal{S}$-TeX, Macintosh or IBM (3.5) format, and the publication in which your paper will appear. Please be sure to include your complete mailing address.

**Sending electronic files.** After acceptance, the source file(s) should be sent to the Providence office (this includes any TeX source file, any graphics files, and the DVI or PostScript file).

Before sending the source file, be sure you have proofread your paper carefully. The files you send must be the EXACT files used to generate the proof copy that was accepted for publication. For all publications, authors are required to send a printed copy of their paper, which exactly matches the copy approved for publication, along with any graphics that will appear in the paper.

TeX files may be submitted by email, FTP, or on diskette. The DVI file(s) and PostScript files should be submitted only by FTP or on diskette unless they are encoded properly to submit through email. (DVI files are binary and PostScript files tend to be very large.)

Electronically prepared manuscripts can be sent via email to pub-submit@ams.org (Internet). The subject line of the message should include the publication code to identify it as a Memoir. TeX source files, DVI files, and PostScript files can be transferred over the Internet by FTP to the Internet node e-math.ams.org (130.44.1.100).

**Electronic graphics.** Comprehensive instructions on preparing graphics are available at www.ams.org/jourhtml/graphics.html. A few of the major requirements are given here.

Submit files for graphics as EPS (Encapsulated PostScript) files. This includes graphics originated via a graphics application as well as scanned photographs or other computer-generated images. If this is not possible, TIFF files are acceptable as long as they can be opened in Adobe Photoshop or Illustrator. No matter what method was used to produce the graphic, it is necessary to provide a paper copy to the AMS.

Authors using graphics packages for the creation of electronic art should also avoid the use of any lines thinner than 0.5 points in width. Many graphics packages allow the user to specify a "hairline" for a very thin line. Hairlines often look acceptable when proofed on a typical laser printer. However, when produced on a high-resolution laser imagesetter, hairlines become nearly invisible and will be lost entirely in the final printing process.

Screens should be set to values between 15% and 85%. Screens which fall outside of this range are too light or too dark to print correctly. Variations of screens within a graphic should be no less than 10%.

**Inquiries.** Any inquiries concerning a paper that has been accepted for publication should be sent directly to the Electronic Prepress Department, American Mathematical Society, P. O. Box 6248, Providence, RI 02940-6248.

# Editors

This journal is designed particularly for long research papers, normally at least 80 pages in length, and groups of cognate papers in pure and applied mathematics. Papers intended for publication in the *Memoirs* should be addressed to one of the following editors. In principle the Memoirs welcomes electronic submissions, and some of the editors, those whose names appear below with an asterisk (*), have indicated that they prefer them. However, editors reserve the right to request hard copies after papers have been submitted electronically. Authors are advised to make preliminary email inquiries to editors about whether they are likely to be able to handle submissions in a particular electronic form.

**Algebra** to KAREN E. SMITH, Department of Mathematics, University of Michigan, 525 University, Suite 2832, Ann Arbor, MI 48109-1109; email: `kesmith@lsa.umich.edu`

**Algebraic geometry and commutative algebra** to LAWRENCE EIN, Department of Mathematics, University of Illinois, 851 S. Morgan (M/C 249), Chicago, IL 60607-7045; email: `ein@uic.edu`

**Algebraic topology and cohomology of groups** to STEWART PRIDDY, Department of Mathematics, Northwestern University, 2033 Sheridan Road, Evanston, IL 60208-2730; email: `priddy@math.nwu.edu`

**Combinatorics and Lie theory** to SERGEY FOMIN, Department of Mathematics, University of Michigan, Ann Arbor, Michigan 48109-1109; email: `fomin@math.lsa.umich.edu`

**Complex analysis and complex geometry** to DUONG H. PHONG, Department of Mathematics, Columbia University, 2990 Broadway, New York, NY 10027-0029; email: `phong@math.columbia.edu`

***Differential geometry and global analysis** to LISA C. JEFFREY, Department of Mathematics, University of Toronto, 100 St. George St., Toronto, ON Canada M5S 3G3; email: `jeffrey@math.toronto.edu`

**Dynamical systems and ergodic theory** to ROBERT F. WILLIAMS, Department of Mathematics, University of Texas, Austin, Texas 78712-1082; email: `bob@math.utexas.edu`

**Functional analysis and operator algebras** to DAN VOICULESCU, Department of Mathematics, University of California, Berkeley, 970 Evans Hall, Floor 9, Berkeley, CA 94720-0001; email: `dvv@math.berkeley.edu`

**Geometric topology, knot theory and hyperbolic geometry** to ABIGAIL A. THOMPSON, Department of Mathematics, University of California, Davis, Davis, CA 95616-5224; email: `thompson@math.ucdavis.edu`

**Harmonic analysis, representation theory, and Lie theory** to ROBERT J. STANTON, Department of Mathematics, The Ohio State University, 231 West 18th Avenue, Columbus, OH 43210-1174; email: `stanton@math.ohio-state.edu`

***Logic** to THEODORE SLAMAN, Department of Mathematics, University of California, Berkeley, CA 94720-3840; email: `slaman@math.berkeley.edu`

**Number theory** to HAROLD G. DIAMOND, Department of Mathematics, University of Illinois, 1409 W. Green St., Urbana, IL 61801-2917; email: `diamond@math.uiuc.edu`

***Ordinary differential equations, partial differential equations, and applied mathematics** to PETER W. BATES, Department of Mathematics, Michigan State University, East Lansing, MI 48824-1027; email: `bates@math.msu.edu`

***Probability and statistics** to KRZYSZTOF BURDZY, Department of Mathematics, University of Washington, Box 354350, Seattle, Washington 98195-4350; email: `burdzy@math.washington.edu`

***Real and harmonic analysis and geometric partial differential equations** to WILLIAM BECKNER, Department of Mathematics, University of Texas, Austin, TX 78712-1082; email: `beckner@math.utexas.edu`

**All other communications to the editors** should be addressed to the Managing Editor, WILLIAM BECKNER, Department of Mathematics, University of Texas, Austin, TX 78712-1082; email: `beckner@math.utexas.edu`.

# Selected Titles in This Series

*(Continued from the front of this publication)*

720 **Palle E. T. Jorgensen,** Ruelle operators: Functions which are harmonic with respect to a transfer operator, 2001

719 **Steve Hofmann and John L. Lewis,** The Dirichlet problem for parabolic operators with singular drift terms, 2001

718 **Bernhard Lani-Wayda,** Wandering solutions of delay equations with sine-like feedback, 2001

717 **Ron Brown,** Frobenius groups and classical maximal orders, 2001

716 **John H. Palmieri,** Stable homotopy over the Steenrod algebra, 2001

715 **W. N. Everitt and L. Markus,** Multi-interval linear ordinary boundary value problems and complex symplectic algebra, 2001

714 **Earl Berkson, Jean Bourgain, and Aleksander Pełczynski,** Canonical Sobolev projections of weak type $(1, 1)$, 2001

713 **Dorina Mitrea, Marius Mitrea, and Michael Taylor,** Layer potentials, the Hodge Laplacian, and global boundary problems in nonsmooth Riemannian manifolds, 2001

712 **Raúl E. Curto and Woo Young Lee,** Joint hyponormality of Toeplitz pairs, 2001

711 **V. G. Kac, C. Martinez, and E. Zelmanov,** Graded simple Jordan superalgebras of growth one, 2001

710 **Brian Marcus and Selim Tuncel,** Resolving Markov chains onto Bernoulli shifts via positive polynomials, 2001

709 **B. V. Rajarama Bhat,** Cocylces of CCR flows, 2001

708 **William M. Kantor and Ákos Seress,** Black box classical groups, 2001

707 **Henning Krause,** The spectrum of a module category, 2001

706 **Jonathan Brundan, Richard Dipper, and Alexander Kleshchev,** Quantum Linear groups and representations of $GL_n(\mathbb{F}_q)$, 2001

705 **I. Moerdijk and J. J. C. Vermeulen,** Proper maps of toposes, 2000

704 **Jeff Hooper, Victor Snaith, and Min van Tran,** The second Chinburg conjecture for quaternion fields, 2000

703 **Erik Guentner, Nigel Higson, and Jody Trout,** Equivariant $E$-theory for $C^*$-algebras, 2000

702 **Ilijas Farah,** Analytic guotients: Theory of liftings for quotients over analytic ideals on the integers, 2000

701 **Paul Selick and Jie Wu,** On natural coalgebra decompositions of tensor algebras and loop suspensions, 2000

700 **Vicente Cortés,** A new construction of homogeneous quaternionic manifolds and related geometric structures, 2000

699 **Alexander Fel′shtyn,** Dynamical zeta functions, Nielsen theory and Reidemeister torsion, 2000

698 **Andrew R. Kustin,** Complexes associated to two vectors and a rectangular matrix, 2000

697 **Deguang Han and David R. Larson,** Frames, bases and group representations, 2000

696 **Donald J. Estep, Mats G. Larson, and Roy D. Williams,** Estimating the error of numerical solutions of systems of reaction-diffusion equations, 2000

695 **Vitaly Bergelson and Randall McCutcheon,** An ergodic IP polynomial Szemerédi theorem, 2000

694 **Alberto Bressan, Graziano Crasta, and Benedetto Piccoli,** Well-posedness of the Cauchy problem for $n \times n$ systems of conservation laws, 2000

For a complete list of titles in this series, visit the AMS Bookstore at **www.ams.org/bookstore/**.